Biogeography of Microscopic Organisms

Is Everything Small Everywhere?

Bringing together the viewpoints of leading experts in taxonomy, ecology and biogeography of different taxa, this book synthesises discussion surrounding the so-called 'Everything is everywhere' hypothesis. It addresses the processes that generate spatial patterns of diversity and biogeography in organisms that can potentially be cosmopolitan.

The contributors discuss questions such as: are microorganisms (e.g. prokaryotes, protists, algae, yeast and microscopic fungi, plants and animals) really cosmopolitan in their distribution? What are the biological properties that allow such potential distribution? Are there processes that would limit their distribution? Are microorganisms intrinsically different from macroscopic ones? What can microorganisms tell us about the generalities of biogeography? Can they be used for experimental biogeography?

Written for graduate students and academic researchers, the book promotes a more complete understanding of the spatial patterns and the general processes in biogeography.

DIEGO FONTANETO is a NERC Advanced Research Fellow at the Division of Biology, Imperial College London, Ascot, UK. His research focuses on spatial patterns and processes in microscopic animals, with a particular interest in rotifers.

The Systematics Association Special Volume Series

SERIES EDITOR

DAVID J. GOWER

Department of Zoology, The Natural History Museum, London, UK

The Systematics Association promotes all aspects of systematic biology by organising conferences and workshops on key themes in systematics, running annual lecture series, publishing books and a newsletter, and awarding grants in support of systematics research. Membership of the Association is open globally to professionals and amateurs with an interest in any branch of biology, including palaeobiology. Members are entitled to attend conferences at discounted rates, to apply for grants and to receive the newsletter and mailed information; they also receive a generous discount on the purchase of all volumes produced by the Association.

The first of the Systematics Association's publications *The New Systematics* (1940) was a classic work edited by its then-president Sir Julian Huxley. Since then, more than 70 volumes have been published, often in rapidly expanding areas of science where a modern synthesis is required.

The Association encourages researchers to organise symposia that result in multi-authored volumes. In 1997 the Association organised the first of its international Biennial Conferences. This and subsequent Biennial Conferences, which are designed to provide for systematists of all kinds, included themed symposia that resulted in further publications. The Association also publishes volumes that are not specifically linked to meetings, and encourages new publications (including textbooks) in a broad range of systematics topics.

More information about the Systematics Association and its publications can be found at our website: www.systass.org.

Previous Systematics Association publications are listed after the index for this volume.

Systematics Association Special Volumes published by Cambridge University Press:
78. *Climate Change, Ecology and Systematics*
Trevor R. Hodkinson, Michael B. Jones, Stephen Waldren and John A. N. Parnell

The Systematics Association Special
Volume 79

Biogeography of Microscopic Organisms

Is Everything Small Everywhere?

EDITED BY

Diego Fontaneto

Imperial College London

CAMBRIDGE UNIVERSITY PRESS
Cambridge, New York, Melbourne, Madrid, Cape Town,
Singapore, São Paulo, Delhi, Tokyo, Mexico City

Cambridge University Press
The Edinburgh Building, Cambridge CB2 8RU, UK

Published in the United States of America by Cambridge University Press, New York

www.cambridge.org
Information on this title: www.cambridge.org/9780521766708

First published 2011

Printed in the United Kingdom at the University Press, Cambridge

A catalogue record for this publication is available from the British Library

ISBN 978-0-521-76670-8 Hardback

Contents

Contributors

TOM ARTOIS Centre for Environmental Sciences, Hasselt University, Diepenbeek, Belgium

IBRAHIM M. BANAT School of Biomedical Sciences, University of Ulster, Coleraine, Northern Ireland, UK

DAVID BASS Zoology Department, The Natural History Museum, London, UK

JENS BOENIGK General Botany, University Duisburg-Essen, Essen, Germany

JULIET BRODIE Department of Botany, The Natural History Museum, London, UK

LUC DE MEESTER Laboratory of Aquatic Ecology and Evolutionary Biology, Katholieke Universiteit Leuven, Leuven, Belgium

ISABEL DRAPER Departamento de Biología (Botánica), Facultad de Ciencias, Universidad Autónoma de Madrid, Madrid, Spain

WILHELM FOISSNER FB Organismische Biologie, Universität Salzburg, Salzburg, Austria

DIEGO FONTANETO Department of Invertebrate Zoology, Swedish Museum of Natural History, Stockholm, Sweden, and Division of Biology, Imperial College London, Ascot, UK

RIMA B. FRANKLIN Department of Biology, Virginia Commonwealth University, Richmond, VA, USA

ANDREA FRANZETTI Department of Environmental Sciences, University of Milano-Bicocca, Milano, Italy

PHILIP F. GANTER Department of Biological Sciences, Tennessee State University, Nashville, TN, USA

JÓZSEF GEML National Herbarium of the Netherlands, Netherlands Centre for Biodiversity Naturalis, Leiden University, Leiden, the Netherlands

NOEMI GUIL Department of Biodiversity and Evolutionary Biology, National Museum of Natural History (CSIC), Madrid, Spain

THIERRY J. HEGER WSL, Swiss Federal Institute for Forest, Snow and Landscape Research, Ecosystem Boundaries Research Unit, Wetlands Research Group, Lausanne, Switzerland; Laboratory of Ecological Systems, École Polytechnique Fédérale de Lausanne (EPFL), Lausanne, Switzerland; Department of Zoology and Animal Biology, University of Geneva, Geneva, Switzerland; and Biodiversity Research Center, University of British Columbia, Vancouver, Canada

JOAQUÍN HORTAL Departamento de Biodiversidad y Biología Evolutiva, Museo Nacional de Ciencias Naturales (CSIC), Madrid, Spain, and Azorean Biodiversity Group - CITA A, Department of Agricultural Sciences, University of the Azores, Angra do Heroísmo, Terceira, Açores, Portugal

WILLIAM D. HUMMON Department of Biological Sciences, Ohio University, Athens, OH, USA

DAVID G. JENKINS Department of Biology, University of Central Florida, Orlando, FL, USA

DONNABELLA C. LACAP School of Biological Sciences, The University of Hong Kong, Hong Kong SAR, China

ENRIQUE LARA Institute of Biology, Laboratory of Soil Biology, University of Neuchâtel, Neuchâtel, Switzerland

FRANCISCO LARA Departamento de Biología (Botánica), Facultad de Ciencias, Universidad Autónoma de Madrid, Madrid, Spain

MAGGIE C.Y. LAU School of Biological Sciences, The University of Hong Kong, Hong Kong SAR, China

ROGER MARCHANT School of Biomedical Sciences, University of Ulster, Coleraine, Northern Ireland, UK

SANDRA J. MCINNES British Antarctic Survey, Cambridge, UK

NAGORE G. MEDINA Departamento de Biología (Botánica), Facultad de Ciencias, Universidad Autónoma de Madrid, Madrid, Spain

KIM A. MEDLEY Department of Biology, University of Central Florida, Orlando, FL, USA

EDWARD A.D. MITCHELL Institute of Biology, Laboratory of Soil Biology, University of Neuchâtel, Neuchâtel, Switzerland

STEPHEN B. POINTING School of Biological Sciences, The University of Hong Kong, Hong Kong SAR, China

HANNO SCHAEFER Ecology and Evolutionary Biology, Imperial College London, Silwood Park Campus, Ascot, UK, and Organismic and Evolutionary Biology, Harvard University, Cambridge, MA, USA

MARTIN V. SØRENSEN Natural History Museum of Denmark, Zoological Museum, Copenhagen, Denmark

M. ANTONIO TODARO Dipartimento di Biologia, Università di Modena and Reggio Emilia, Modena, Italy

SILKE WERTH Biodiversity and Conservation Biology, WSL Swiss Federal Research Institute, Birmensdorf, Switzerland

DAVID M. WILLIAMS Department of Botany, The Natural History Museum, London, UK

ALDO ZULLINI Dipartimento di Biotecnologie & Bioscienze, Università di Milano-Bicocca, Milan, Italy

Preface

This volume is derived from a symposium on 'The importance of being small: does size matter in biogeography?' organised during the first BioSyst meeting, which was held in Leiden in August 2009. The idea for the symposium arose during an informal discussion at the Natural History Museum in London. Biogeography is now a well-established science with its own methods and tools, and a strong theoretical framework. Many journals and books are dedicated to biogeography, and specific meetings are organised by the International Biogeography Society. Nevertheless, most of the ideas in biogeography come from empirical evidence from macroscopic organisms, whereas the spatial patterns of microscopic organisms have mostly been neglected.

The aim of this book is to establish the importance of microorganisms in biogeography. In doing so, this book follows the stimulating discussion on the so-called 'Everything is everywhere' hypothesis of the last decades. Currently, enough empirical evidence is available on the biogeography and phylogeography of many microscopic organisms and on larger organisms with microscopic dispersing stages; thus, this book brings together for the first time all this information in a unifying framework, and discusses patterns, processes and consequences.

The coverage of the taxa is broad, spanning from prokaryotes to plants, fungi, and animals; the approaches are rather different in the different chapters, and I hope that readers will enjoy this book and find many inspirations for their own research.

I am very grateful to the Systematics Association for the opportunity to organise the symposium in Leiden and especially to Juliet Brodie, Peter Olson, Dave Roberts and Alan Warren for their support at an early stage of the organisation of the meeting. Other people were very helpful during the meeting, and I am very grateful to Peter Hovenkamp for his help in Leiden. As for any meeting, its success was due to the high quality of the speakers, and I thank them all.

During the preparation of the book, I received strong support both from the Systematics Association, William Baker, Richard Bateman and David Gower, and from Cambridge University Press, Katrina Halliday, Dominic Lewis and

Megan Waddington. The contributors made a great job in providing interesting, stimulating and easily readable chapters on schedule, and I am very grateful to them. They also provided valuable assistance during the reviewing process. Other people read part of the book and provided comments and suggestions, and for that I sincerely thank Tim Barraclough, Lars Hedenäs, Seraina Klopfstein, Petra Korall, Marc-André Lachance, Ulrike Obertegger, Ibai Olariaga Ibarguren, Maureen O'Malley, Albert Phillimore, Brett Riddle, Cuong Tang, Franco Verni, Martin Westberg, Chris Wilson and others who prefer to remain anonymous.

Diego Fontaneto
Stockholm
Sweden

Part I

Theoretical framework

1

Why biogeography of microorganisms?

DIEGO FONTANETO[1] AND JULIET BRODIE[2]

[1] *Department of Invertebrate Zoology, Swedish Museum of Natural History, Stockholm, Sweden; and Division of Biology, Imperial College London, Ascot, UK*
[2] *Department of Botany, The Natural History Museum, London, UK*

1.1 The problem

The aim of this book is to discuss the idea that for microorganisms 'Everything is everywhere' (EiE hypothesis), that is, large organisms have biogeographies, whereas microscopic ones do not have any large-scale spatial pattern of distribution: microorganisms have no biogeographies.

Size is known to be among the supreme regulators of all biological entities. Shape of any organism is mostly a consequence of its size; how the organism interacts with the environment and all its biological functions are a matter of its size; its life history is influenced by its size (Calder, 1996; Bonner, 2006). The strongest correlate of body size is that body volume and body surface are not linearly related. Thus, larger organisms have different needs from small ones regarding temperature, osmosis, physiology, metabolism and many (if not all) other processes (Schmidt-Nielsen, 1984; Peters, 1986). Body size also determines what an organism may be able to do, even its extinction risk, and the communities and the ecosystems the organisms live and interact in (Colinvaux, 1978; Cardillo and Bromhan, 2001; Hildrew et al., 2007).

Biogeography of Microscopic Organisms: Is Everything Small Everywhere?, ed. Diego Fontaneto. Published by Cambridge University Press.

Most of these theories, hypotheses and models attempt to describe some continuous function relating biological patterns and processes with body size, and universal scaling laws in biology have been formulated and tested (Brown and West, 2000). A notable exception to this is the biogeographic implication of body size. Instead of a gradient in the patterns of diversity from large to small organisms, an abrupt distinction has been hypothesised between large and small organisms, with 2 mm being the empirical threshold value discriminating macroorganisms with biogeography and microorganisms without biogeography (Finlay, 2002; Fenchel and Finlay, 2004).

The underlying biological assumption for this abrupt threshold is that microorganisms are really different from macroscopic ones. The threshold dividing these two groups of organisms should fall between 1 and 10 mm, with 2 mm being the most probable size (Fig 1.1). The peculiar features of microorganisms allow them to attain cosmopolitan distribution, an uncommon characteristic in most large organisms: microorganisms are so small that they can be easily passively dispersed everywhere, they produce resting stages that allow them to survive adverse conditions and to persist in any habitat (see Chapter 5 for a detailed report on the effect of resting stages), and can use asexual or parthenogenetic reproduction to quickly increase in number (Shön et al., 2009; see Chapter 15 for a detailed account of the influence of body size, dispersal rate and abundance). However, the hypothesis that for microorganisms everything is everywhere ('ubiquity hypothesis') is

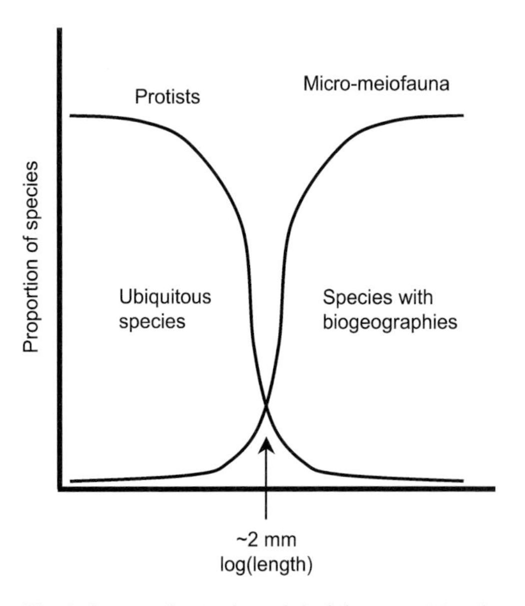

Fig 1.1 Hypothetical model of the transition between larger organisms with biogeography and microscopic organisms without biogeography. Modified from Finlay (2002).

controversial in the current scientific discussion, and in opposition to the 'moderate endemicity model' by Foissner (1999, 2006), which suggests that many microorganisms have restricted distributions (Chapter 5).

Why is the ubiquity hypothesis for microorganisms controversial? Why are scientists still arguing for or against it? Is there evidence supporting or contrasting it? There are many issues in the EiE hypothesis that will continue to be unresolved for a while. They include, for example, our current inability to reliably describe diversity in microorganisms, to quantify the ecological niche and the geographic range of most species, and to define their rates of dispersal. Moreover, knowing if and how microorganisms attain global distributions is a far-reaching topic. For example, is speciation happening also in the absence of geographic barriers isolating populations? What about environmental conservation? If microorganisms are not linked to any geographic area, how can they be conserved? Here we introduce some of the problems of the biogeography of microorganisms, which will be discussed later in more depth in different chapters.

1.2 Taxonomic units

The current formulations of EiE use morphology to define the taxonomic unit under consideration (Fenchel and Finlay, 2006). Nevertheless, finding reliable morphological taxonomic characters for most microorganisms is notoriously difficult; we need to define unambiguous units of diversity in order to map their distribution. Thus, many species considered cosmopolitan may represent species complexes with restricted distributions. Almost all the chapters in the book discuss this topic, reviewing the current knowledge of species reality and identity in each group, and how new tools can help in providing reliable estimates of diversity. On the one hand, molecular taxonomy helps morphological taxonomy to reveal species complexes and potentially identify the correct units of diversity (e.g. Pons et al., 2006; Burns et al., 2008); on the other hand, environmental sequencing by cloning PCR products but mostly by ultrasequencing, is providing distributional data of many taxa otherwise unrecorded or undistinguished (e.g. Robeson et al., 2009; Creer et al., 2010).

1.3 Niche definition

The complete statement of the ubiquity hypothesis for microorganisms is that 'everything is everywhere, but the environment selects' (Baas Becking, 1934; de Wit and Bouvier, 2006). Thus, knowledge of the ecological niche for these organisms ('the environment selects' part of the hypothesis) is needed in order to define their distribution, and niche-based models are used for predicting biogeographic distributions (see Chapter 15). But this is far from reality, because the identification of ecological needs for microorganisms may not be as straightforward as for larger organisms. This is partly due to technical difficulties in measuring environmental parameters at the microscopic scale, but also to the fact that the presence of microorganisms in an area does not mean that the environment is suitable. A detailed example is given in Chapter 4, reporting on thermophilic bacteria in cool soils. This aspect will not be improved by the recent advances in environmental sequencing with ultrasequencing (e.g. Creer et al., 2010), as the results of such analyses will show species even in habitats where they do not live but where they are only present as resting stages.

1.4 Spatial patterns

Reports on findings of species and species lists are the basis for any biogeographic analysis. For microorganisms we already know that there are many problems in identifying 'species' as units of diversity. In this case, new advances in molecular tools will provide more reliable data on the geographic distribution of microorganisms.

Nevertheless, following from the problem of defining ecological niches for these organisms, it will also be difficult to disentangle the contribution of historical vs. ecological biogeography. The difference between historical and ecological biogeography and the implications for the study of the distribution of microorganisms are explained in detail for protists in Chapter 6 (section 6.2): large organisms usually have very small discrepancies between their potential, historically derived biogeographic range and their realised, ecologically derived ranges, whereas microorganisms, due to their resting stages, may be present in areas that are not ecologically optimal for them (Chapter 4). The difference between historical and ecological biogeography highlights one of the major differences in dispersal between micro- and macroorganisms, and one of the difficulties in supporting or falsifying the EiE hypothesis.

To further complicate the scenario, the various chapters on the biogeography of all the microorganisms report both evidence of large distributions and of endemic, restricted distributions. Phylogeographic analyses, using DNA sequences to investigate the spatial patterns of distribution are conveying similar results, providing evidence of both long-distance gene flow and restricted gene flow.

Moreover, the spatial scale of the analysis, the measurement used (distance–decay relationships, taxa–area relationships, local:global taxa richness), and other potential confounding effects may become important when trying to infer the processes driving the spatial patterns (Chapters 15–17).

1.5 The challenge

To take up the challenge of the biogeography of microorganisms, this book brings together empirical observations of the patterns in different groups of microscopic organisms or with microscopic dispersing stages. Different approaches are used in the description of diversity in order to support or not the hypothesis of cosmopolitanism for microorganisms. Moreover, the book goes a step further and discusses the processes and the generalities obtained from these examples. The main result is that in all the groups, evidence of restricted distribution is more common than generally accepted. Thus, the differences between micro- and macroorganisms may not be so strong.

The book is divided into five main parts. The first part (Chapters 1 and 2) provides a brief introduction and an historical and philosophical overview. The following three parts deal with the empirical evidence gathered in different microorganisms, from prokaryotes (Part II, Chapters 3 and 4), to unicellular eukaryotes (Part III, Chapters 5–8), and pluricellular eukaryotes (Part IV, Chapters 9–14). Finally, the last part focuses on the processes and the generalities in the biological properties of microorganisms (Chapters 15–17). The subdivision in different parts serves more as a structure for the book than as a logical continuum. Each chapter can be read alone, as it contains an overview of the problem, empirical evidence and a discussion of the processes. The focus is different in different parts: chapters in Parts II–IV deal with taxon-based evidence and mechanisms, whereas the chapters in Part V are not taxon related, but more mechanism related. So, every chapter presents its own view on the biogeography of microorganisms, providing a theoretical framework, explaining problems and suggesting ways to solve them.

Chapter 2 deals with an historical introduction of the origin of the hypothesis that everything is everywhere, from the first studies at the beginning of the twentieth century; it provides a philosophical background for the hypothesis; it discusses the role of classification of taxa and areas, and then uses the diatoms as an example.

Part II deals with the biogeography of the smallest of all living organisms, prokaryotes. Chapter 3 provides a general review of the current knowledge of prokaryotic biogeography, discussing also problematic topics such as species concepts and identification in prokaryotes, and estimates of diversity. The following chapter (Chapter 4) describes a specific example of thermophilic bacteria in cool temperate areas. It provides also tests on the potential activity of these organisms in a potentially unsuitable habitat and a discussion on mechanisms of transport and the potential sources.

Part III deals with unicellular eukaryotes (protists and yeast); two of the chapters provide general reviews using data from morphological taxonomy

(Chapter 5) and from phylogenetic and phylogeographic analyses (Chapter 6); one reports on the example of testate amoebae (Chapter 7); and one on the example of cactophilic yeast (Chapter 8). Chapter 5 starts by assuming that the moderate endemicity model is a better explanation for the evidence, and then pays attention to the reasons why certain species are cosmopolitan and others are not. The most important factors described in Chapter 5 are resting cysts, geological history and human introductions. The chapter provides a detailed report of the biogeography of many species to support its initial assumptions. Chapter 6 adopts a different approach dealing with the same organisms, free-living protists. It reassesses the fundamental principles behind the EiE concept in the light of recent findings and insights from twenty-first century molecular biology and microbial ecology. The chapter also discusses the implications of species concepts and identification, estimates of diversity, evidence for biogeography and the differences between historical and ecological processes in biogeography. Chapter 7 describes the example of Arcellinida testate amoebae, providing evidence of restricted distributions and endemicity in a group of flagship organisms: species so charismatic and easy to recognise, whose lack of records in certain areas cannot be attributed to lack of research, but to their actual absence. Chapter 8 deals with yeast, a group of fungi that adopted unicellular growth: these organisms are thus included in the section on unicellular eukaryotes. The yeast example in the chapter is a special one, as it lives in symbiosis with cacti. Thus, it provides a different approach to the problem, discussing the generalities and the peculiarities of the spatial patterns of the system.

Part IV contains six chapters dealing with the biogeography of microscopic pluricellular organisms or of larger organisms with microscopic dispersing stages, such as fungi (Chapter 9), lichens (Chapter 10), mosses (Chapter 11), ferns (Chapter 12) and animals (Chapters 13 and 14). Chapter 9 deals with genetic evidence of patterns of intercontinental gene flow in arctic-alpine and boreal-temperate macroscopic fungi, which have single-celled microscopic dispersing spores. Chapter 10 deals with the biogeography of lichens, providing evidence of wide, disjunct and endemic distributions, focusing mostly on the fungus symbiont; moreover, it reports the little evidence available at present for the biogeography of the algal symbiont. Chapter 11 discusses the biogeography of mosses. These plants are macroscopic, but they all have microscopic dispersing spores. As in the previous chapter, there are records of wide, disjunct and endemic distributions, and of both spatially restricted and not-restricted gene flow. Chapter 12 takes a different approach, dealing with ferns. These plants also have microscopic dispersing spores. The chapter deals with chorological and phylogeographic analyses, trying to disentangle whether dispersal limitation or habitat quality shapes the distribution ranges of ferns. Chapter 13 focuses on microscopic animals, which have dormant stages that can act as dispersing propagules. It gathers empirical evidence on a wide range of groups from a morphological perspective in species identification, showing that the biological peculiarities of the taxa are more important than size alone. Chapter 14 deals with the same groups of microscopic animals, but using recently published evidence from molecular taxonomy and phylogeography. The results are qualitatively similar, with evidence of both widespread and restricted distributions.

Part V has three chapters focusing on the processes and on the biological properties of microorganisms. Chapter 15 uses microorganisms to test the generalities of biogeographic principles. It deals with the effect of abundance, body size and niche constraints on the spatial patterns of distribution. Chapter 16 provides a metacommunity perspective, dealing with species sorting, mass effects, patch dynamics and the neutral model. Chapter 17 analyses geographic gradients, suggesting that microorganisms may not be so different from macroscopic ones, and provides a rationale for the use of microorganisms in what can be called 'experimental biogeography'.

1.6 Conclusions

As a summary of the ideas expressed in the chapters of this book some main conclusions arise: several distribution patterns are described and different processes are hypothesised to explain them. Long-distance dispersal is evidently possible, but it is not the rule, as other distribution patterns can be explained by other mechanisms, such as continental drift, stepping-stone migration and anthropogenic introduction.

The EiE hypothesis focuses on one single explanatory factor, dividing organisms into two main groups, larger organisms with biogeography and smaller ones without biogeography. Given the complexity of the spatial patterns in microorganisms, it seems that their biogeography is more likely to depend on a complex set of interacting phenomena, in which size is of course important, but it is not the only driver. The differences between micro- and macroorganisms can thus be included in a gradient, disregarding the hypothesised abrupt threshold assumed by the EiE hypothesis.

References

Baas Becking, L.G.M. (1934). *Geobiologie of inleiding tot de milieukunde*. The Hague: Van Stockum and Zoon.

Bonner, J.T. (2006). *Why Size Matters: From Bacteria to Blue Whales*. Princeton, NJ: Princeton University Press.

Brown, J.H., West, G.B. (2000). *Scaling in Biology*. New York, NY: Oxford University Press.

Burns, J.M., Janzen, D.H., Hajibabaei, M., Hallwachs, W., Hebert, P.D.N. (2008). DNA barcodes and cryptic species of skipper butterflies in the genus

Perichares in Area de Conservacion Guanacaste, Costa Rica. *Proceedings of the National Academy of Sciences USA* **105**, 6350–6355.

Calder, W.A. III (1996). *Size, Function and Life History*. Mineola, NY: Dover Publications.

Cardillo, M., Bromhan, L. (2001). Body size and risk of extinction in Australian mammals. *Conservation Biology* **15**, 1435–1440.

Colinvaux, P. (1978). *Why Big Fierce Animals are Rare*. Princeton, NJ: Princeton University Press.

Creer, S., Fonseca, V.G., Porazinska, D.L. et al. (2010). Ultrasequencing of the meiofaunal biosphere: practice, pitfalls and promises. *Molecular Ecology* **19**, 4–20.

De Wit, R., Bouvier, T. (2006). "Everything is everywhere, but, the environment selects"; what did Baas-Becking and Beijerinck really say? *Environmental Microbiology* **8**, 755–758.

Fenchel, T., Finlay, B.J. (2004). The ubiquity of small species: patterns of local and global diversity. *BioScience* **54**, 777–784.

Fenchel, T., Finlay, B.J. (2006). The diversity of microbes: resurgence of the phenotype. *Philosophical Transactions of the Royal Society B – Biological Sciences* **361**, 1965–1973.

Finlay, B.J. (2002). Global dispersal of free-living microbial eukaryote species. *Science* **296**, 1061–1063.

Foissner, W. (1999). Protist diversity: estimates of the near-imponderable. *Protist* **150**, 363–368.

Foissner, W. (2006). Biogeography and dispersal of micro-organisms: a review emphasizing protists. *Acta Protozoologica* **45**, 111–136.

Hildrew, A.G., Raffaelli, D.G., Edmonds-Brown, R. (2007). *Body Size: The Structure and Function of Aquatic Ecosystems*. Cambridge: Cambridge University Press.

Peters, R.H. (1986). *The Ecological Implications of Body Size*. Cambridge Studies in Ecology. Cambridge: Cambridge University Press.

Pons, J., Barraclough, T.G., Gomez-Zurita, J. et al. (2006). Sequence-based species delimitation for the DNA taxonomy of undescribed insects. *Systematic Biology* **55**, 595–609.

Robeson, M.S. II, Costello, E.K., Freeman, K.R. et al. (2009). Environmental DNA sequencing primers for eutardigrades and bdelloid rotifers. *BMC Ecology* **9**, 25.

Schmidt-Nielsen, K. (1984). *Scaling: Why is Animal Size so Important?* Cambridge: Cambridge University Press.

Schön, I., Martens, K., Dijk, P. (Eds.) (2009). *Lost Sex. The Evolutionary Biology of Parthenogenesis*. Berlin: Springer.

2

Historical biogeography, microbial endemism and the role of classification: everything is endemic

DAVID M. WILLIAMS

Department of Botany, The Natural History Museum, London, UK

2.1 Introduction

Microbial biogeography, the study of the distribution of 'small' organisms, has been said to have gained renewed vigour because of the recently resurrected 'Everything is everywhere' hypothesis (EiE) (Finlay, 2002; Fenchel and Finlay 2003; Finlay and Esteban, 2007). That hypothesis was concisely summarised by the organisers of the conference on the biogeography of microorganisms in Leiden, August 2009, in the promotional material:

> This symposium is based around the hypothesis of everything-is-everywhere (EiE) amongst small organisms. This hypothesis was proposed at the beginning of the twentieth century for microbial diversity and, about ten years ago, extended to describe spatial patterns of diversity for any organism smaller than two mm, under the simple observation that microscopic organisms such as protists seem to be cosmopolitan, at least in habitats that support their growth. Since its recent resurgence, this topic became hotly debated, with evidence apparently supporting and denying the hypothesis.

Biogeography of Microscopic Organisms: Is Everything Small Everywhere?, ed. Diego Fontaneto. Published by Cambridge University Press.

The slogan 'Everything is everywhere, [but] the environment selects' has been attributed to the microbiologist Lourens G.M. Baas Becking (Baas Becking, 1934). The word 'but' is in brackets as sometimes the phrase appears as 'Everything is everywhere, the environment selects' (e.g. Wilkinson, 2001), at other times as 'Everything is everywhere, *and* the environment selects' (e.g. Kuehne et al., 2007, my italics). According to de Wit and Bouvier (2006), the original Dutch – 'Alles is overal: *maar* het milieu selecteert' (Baas Becking, 1934) – translates as 'Everything is everywhere, *but* the environment selects' (e.g. Heino et al., 2010). Either way it does not seem to matter too much as the core idea(s) remain the same (it might be thought that the word 'but' in the original adds a little more precision connecting the two ideas together and that subsequent versions that either omit the word or substitute the word 'and' are merely mistaken).

Credit for the origin of this hypothesis – or pair of hypotheses – goes to Martinus Willem Beijerinck (1851–1931), a Dutch biologist (for a brief biography, see Chung and Ferris, 1996), as well as Baas Becking (1895–1963), with Baas Becking elaborating on an original comment from Beijerinck that was understood to be a general hypothesis explaining the geographic distribution, or lack thereof, in microorganisms with the notion that because microorganisms can get just about anywhere on the planet then they should indeed be everywhere, and the reason they are not is that they are susceptible to different environmental pressures (for biographical details on Baas Becking, see Ferguson Wood, 1963 and Quispel, 1998; for a bibliography, see Westenberg, 1977). The hypothesis might be summarised as meaning that ecological processes are the primary drivers of microbial diversity, there is no 'true' endemicity (in the sense of any taxon being confined to just one region of the globe) and historical factors are either negligible or irrelevant – 'Beijerinck ... declared that 'everything is everywhere; the environment selects', and microbiologists have assumed since then that there is no such discipline as microbial biogeography' (Fitter, 2005, p. R187).

Rather than examine the propositions within the hypothesis for possible empirical content, I offer a short digression on the nature of the hypothesis, as formulated in the phrase: 'Everything is everywhere, but the environment selects'.

2.2 Distributional hypotheses

As noted above, the hypothesis appears to be composed of two parts. The first is 'Everything is everywhere', that microorganisms are ubiquitous: 'According to advocates of the ubiquity hypothesis ... the vast population sizes of microorganisms drive ubiquitous dispersal ... and make local extinction virtually impossible .... Geographic isolation is therefore absent and as a result, allopatric speciation should be rare or nonexistent, which would explain the perceived low global morphospecies diversity of microbial eukaryotes ...' (Vanormelingen et al.,

2008, p. 394). The notion that there are 'vast population sizes of microorganisms ...' is for the most part derived from another notion, that endemicity in microorganisms is scarce, rather than an empirical reality. Still, regardless of detail or explanation, the phrase 'Everything is everywhere' is clearly false, and in any case, one would need to know in what sense the words 'everything' and 'everywhere' are intended. As soon as something is not found everywhere then everything (obviously) is not everywhere. This much could be easily agreed upon.

The second part, 'the environment selects', is designed to protect the first from failing because as everything is clearly *not* everywhere ('Tom Fenchel and Bland Finlay seem determined to perpetuate the myth of ubiquity despite all evidence to the contrary', Lachance, 2004), it is because 'the environment selects'. The structure of the slogan is such that the first part can never be found to be conclusively false as the second part protects that from ever happening, should any particular organism be discovered not to be everywhere. In its original formulation, one would have been expected to confirm the fact that 'Everything is everywhere' by experimental manipulation to discover which particular 'environment selects'. Thus, one might encounter an organism in one place, culture it and then demonstrate that it can, indeed, exist anywhere. But that is no different from observing, for example, a giraffe in London Zoo and as it survives in what would be rather awful conditions relative to its natural environment means giraffes should really be everywhere. This is the key: natural environments and migration. In any case, EiE is neither a theory nor a hypothesis. What is it then?

Maureen O'Malley (2008), in her paper on the history of microbial biogeography, has called the phrase 'Everything is everywhere, but the environment selects' 'a tidy axiom':

> This tidy axiom, often, cited by microbiologists as attributable to Beijerinck due to Baas Becking's self effacing presentation of it ... (O'Malley, 2008, p. 320, but see O'Malley, 2007, p. 651).

An axiom is a self-evident truth, something that can be taken for granted. The word I used above was slogan, defined in this instance as:

> Something that serves perhaps more as a social expression of unified purpose, rather than a projection for an intended audience (http://en.wikipedia.org/wiki/Slogan)

So, if the central, self-evident truth is obscure, wrong or misleading, then what is it that is being investigated?

2.3 Geobiology

Baas Becking's ideas – 'the law of Beijerinck' (Jacobs, 1984, p. 205); 'Beijerinck's laws' (Baas Becking, 1959, p. 48) – were given serious consideration in the context of the developing evolutionary theories in Europe in the 1930s, prior to the establishment of the modern synthesis (Mayr and Provine, 1980). Beijerinck's laws were discussed within a rich constellation of ideas relating to biogeography, most of which have sadly faded into obscurity (for example, see Lam, 1938, p. 84). In 1934 Baas Becking published a book entitled *Geobiologie of inleiding tot de milieukunde* (Baas Becking, 1934; I have not been able to examine a copy). A section has been translated in Anton Quispel's biographical study of Baas Becking:

> These lectures are an effort to describe the relation between the organisms and the earth. The name 'Geobiology' merely expresses this relation. This new word does not want to describe a new discipline. It tries to unite under one point of view, as far as possible, phenomena which already were known in different areas of biology (Quispel, 1998, p. 70).

Baas Becking, then, was interested in the prospect of unifying 'the relation between the organisms and the earth'. Since then geobiology has arisen in a variety of other forms. Koch suggested this version:

> Following the idea of Good (1947), who termed the combination of phytogeography and plant ecology as geobotany, a similar combination of biogeography and ecology can logically be known as 'geobiology' (Koch, 1957, p. 145).

Sylvester-Bradley suggested this version:

> Does it [geobiology] differ at all, except in name, from the venerable science of palaeontology, a science far older than geology? (Sylvester-Bradley, 1972, p. 110).

Geobiology has now become a discipline: 'Geobiology is that unifying discipline that seeks to span and link the geological and biological sciences' (Liebermann, 2005, p. 23). There is a journal called *Geobiology*, which began in 2003 (seven volumes published so far; the editorial for the journal's launch referred to Baas Becking's 1934 book; *Geobiology* **1**, p. 1); and there is a book series entitled *Topics in Geobiology* (with over 30 items published so far).

Even so, what emerges from these varied efforts to define a new discipline is the study of geography and its relevance to species distributions (biology, ecology), past and present (palaeontology and neontology), and their explanation – a subject usually referred to as biogeography.

2.4 Geography and species

> The history of civilisation is a history of wandering, sword in hand, in search of food. In the misty younger world we catch glimpses of phantom races, rising, slaying, finding food, building rude civilisations, decaying, falling under the swords of stronger hands, and passing utterly away. Man, like any other animal,

> has roved over the earth seeking what he might devour; and not romance and adventure, but the hunger-need, has urged him on his vast adventures (Jack London, 1917).

The concept of evolution, the idea that some species might give rise to other, new species, was on the minds of men some time before Charles Darwin proposed his own particular theory as to their origin (Bowler, 2009). Darwin began his *On the Origin of Species* with these words:

> When on board H.M.S. Beagle, as naturalist, I was much struck with certain facts in the distribution of the inhabitants of South America, and in the geological relations of the present to the past inhabitants of that continent. These facts seemed to me to throw some light on the origin of species — that mystery of mysteries, as it has been called by one of our greatest philosophers (Darwin, 1859).

Darwin's inspiration came from two sources: biogeography – 'the distribution of the inhabitants of South America …' – and history – '… the geological relations of the present to the past inhabitants of that continent'.

By a remarkable coincidence, Alfred Russel Wallace – a professional plant collector, working in the tropics of South East Asia – hit upon more-or-less the same theory of species origin as Darwin (Raby, 2002). He communicated his ideas directly to Darwin in a letter from Ternate, Eastern Indonesia. Wallace's essay was published in 1858 as *On the Tendency of Varieties to Depart Indefinitely from the Original Type* (Darwin and Wallace, 1858; Darwin contributed a few items to the presentation but had no input into Wallace's title essay). Wallace's communication has great significance, as it helped inspire Darwin to complete what he called the 'abstract' of his theory, the book that became *On the Origin of Species*.

As Darwin and Wallace both understood, it was not really possible to grasp the concept of the evolution of organisms without some understanding, appreciation and explanation of their geographic distribution (McCarthy, 2009). Viewed from this perspective – and with the benefit of hindsight – it may not be quite so surprising that Darwin and Wallace quickly came to view species as mutable entities (Wallace before Darwin), as both were struck by what they saw in the regions they visited (Desmond and Moore, 1991; Shermer, 2002). In a general sense, the marriage of geography and history established evolution as both a viable concept and tractable research programme.

Further commentary from Darwin is relevant:

> We are thus brought to the question which has been largely discussed by naturalists, namely, whether species have been created at one or more points of the earth's surface … undoubtedly there are many cases of extreme difficulty in understanding how the same species could possibly have migrated from some one point to the several distant and isolated points, where now found. Nevertheless the simplicity of the view that each species was first produced

> within a single region captivates the mind. He who rejects it, rejects the *vera causa* of ordinary generation with subsequent migration, and calls in the agency of a miracle (Darwin, 1859).

If organisms have but one point of origin, what is essential to Darwin's argument is the need for species to migrate from one place to the next, constantly travelling from their point of origin, to achieve any kind of wider distribution: 'ordinary generation with subsequent migration'. Migration of organisms to other places was, of course, well known at the time but Darwin saw it necessary to extend the reasoning to *all* organisms that were considered widespread. That migration was later graphically captured by Ernst Haeckel in his map of wandering humans, which first appeared in the second edition of his *Natürliche Schöpfungsgeschichte* (1870, translated into English as *The History of Creation*, Haeckel, 1876) and represented in American literature by Jack London (see epigraph above), a convinced Darwinian ('As a boy, the first heroes that I put into my Pantheon were Napoleon and Alexander the Great. Later on I destroyed this Pantheon and built a new Pantheon in which I began inscribing names such as David Starr Jordan, as Herbert Spencer, as Huxley, as Darwin, as Tyndall', letter dated 7 September 1915). Nearly 150 years after Haeckel, similar images still appear, depicting almost the same routes that humans supposedly took to get to where they are today (Kohn, 2006; Shreeve, 2006; but on humans and their biogeography see Grehan and Schwartz, 2009 for a plausible alternative; for migration, see Haywood, 2008: 'Migration is one of the defining features of the human race'). Of course, if humans travelled, well then, so the argument goes, everything else must have. Diatoms and other algae obviously could not all travel under their own steam, so how? Many suggestions, for algae at least, have been made: ducks (Atkinson, 1970), birds (Atkinson, 1972), swan faeces (Luther, 1963), 'small aquatic organisms' (Maguire, 1963), aquatic beetles (Milliger and Schlichting, 1968), waterbirds (Proctor, 1959), waterfowl (Schlichting, 1960), the Gizzard Shad, *Dorosoma cepedianum* (Velasques, 1940), and so on – the list is exhausted only by the extent of one's imagination.

Darwin noted 'that each species was first produced within a single region captivates the mind. He who rejects it, rejects the *vera causa* of ordinary generation with subsequent migration, and calls in the agency of a miracle'. Of course, quite unknown to Darwin at that time was the possibility that – as if by miracle – the earth moved. The discovery of plate tectonics and the ever-shifting nature of the earth's crust did indeed confirm that even though organisms move – so does the earth (McCarthy, 2009). Thus migration might indeed captivate the mind but it need not be the sole cause of organism distribution.

The notion of wandering organisms, migrating to a better place, seems to have first arisen from the biblical story of Noah's Ark (Nelson, 1978) but retains a certain significance for dispersalist biogeography, the view that the primary agency in species distribution is their ability to disperse, to travel around the globe. As Moore wrote, 'Evidently organisms possessed adaptive flexibility. They could fly, swim, walk, crawl, or hitchhike to pastures new and still survive. The story of Noah's descent from Ararat contained this truth at least: migration occurred' (Moore, 2005, p. 126), and in the context of the development of the science of biogeography, O'Malley added:

> The victory of the dispersalist school of macrobiogeography after Darwin contributed to the acceptance of Beijerinck's special rule about microbial distribution. A common ground of dispersal placed exceptional ubiquity within a broader framework that joined macrobial and microbial biogeography into a continuum of distribution (O'Malley, 2008, p. 318).

O'Malley also noted that 'environmentally determined ubiquity became a law of microbiology in part because of how well it fit the dispersalism of macrobiogeography, and in part because it fit the standardisation of laboratory practice and the prevailing ideas of 'good science'' (O'Malley, 2008, p. 319); good science, as then understood, was the experimental, such as the manipulation of environments in a laboratory.

If the EiE hypothesis is simply a slogan or an axiom found to be false, and, like much biogeography, has its basis in the biblical dispersalist approach, then what future is there for microbial biogeography? If we discard the slogan, the axiom and the dispersalist view, then there are organisms and continents (areas), without the assumption of either wandering organisms or continents but mindful of the fact that both do so. Relevant is historical biogeography, or more accurately comparative biogeography, which subsumes any notion of geobiology (Parenti and Ebach, 2009). Comparative biogeography addresses problems of classification, the classification of organisms and the classification of areas, a view captured by Parenti and Ebach's title: *Comparative Biogeography: Discovering and Classifying Biogeographic Patterns of a Dynamic Earth.*

2.5 Classification

Classification concerns the relationships of taxa (organisms) and areas (the places taxa occupy). Classifications are best expressed in hierarchical schemes, familiar to all biologists, where organisms are placed in classes, orders, families and genera, each less general than its preceding category.

Biogeographic classification is less definitive. There have been many different competing schemes, such as the biogeographic regions created by Wallace (1876), which have themselves been subdivided. Thus, to a certain extent, biogeographic classification is also understood to be hierarchical. But what groups to recognise, what groups have reality?

Progress in taxon classification identifies three possible groups for organisms: monophyletic, paraphyletic and polyphyletic groups. Such groups relate to both characters and ancestry. The only groups with any reality are monophyletic groups. Monophyletic groups are discovered by shared derived characters, synapomorphies, homologies; monophyletic groups can be explained by the fact that they relate most closely among themselves. Neither para- nor polyphyletic groups have unique ancestors or unique characters. The same notion can be applied to geographic areas, with the presence of shared taxa acting as synapomorphies.

2.5.1 Taxa

Consider the recent paper of Caron et al. (2009) entitled 'Protists are microbes too: a perspective' (see also Caron, 2009). What does the title mean? For a definition of microbes the authors offer the following:

> Strictly speaking, microorganisms are defined by their size; that is, organisms that are smaller than can be resolved with the naked eye .... If we define microbes by cell size, then most protists qualify as microbes. A few single cells and numerous colonial forms exist that are visible to the unaided eye, but the vast majority are microscopic. Similarly, most bacteria and archaea are indeed microscopic, but there are exceptions here as well. In fact, a very large number of protistan taxa are much smaller than the largest bacteria (Caron et al., 2009, p. 6).

Is size a character, in the sense that every organism so measured and of the same or similar size could have that factor considered homologous, the attributes required for natural groups of organisms? No. And in any case, groups of organisms do not gain reality by being defined, as is obvious from Caron et al.'s own text. Definition isn't discovery – it is imposition. It is clear Caron et al. appreciate this difference but their purpose is to create an argument to support future study of protists (a different argument is offered by O'Malley and Dupré, 2007a, 2007b, but for roughly the same cause). Regardless of however noble the purpose, microbes do not exist in any evolutionary (or even biological) sense – they did not come into being, they will not go extinct: they can't be anywhere, or do anything. Interestingly, and given the thrust of the authors' intent (to show that many protists are 'microbes', that is: small), the same reasoning applies to protists, inasmuch as it too is an artificial (non-existent) group (Schlegel and Hülsmann, 2007). Inspection of any tree of relationships for eukaryote organisms reveals protists to be defined only by creating the group after excluding all plants, animals and fungi. Protists are those organisms left behind after the excision of these three monophyletic groups, a classic case of a paraphyletic group, a group named by convention (definition) rather than discovered by evidence. Thus, to study any taxon in a biological (phylogenetic) context, there is a need for it to be demonstrably monophyletic (evidence derived from characters), such as diatoms, the group I study, their size being completely irrelevant to that discovery or their distribution. For the purposes of biogeographic study, it is only their distribution that matters.

2.5.2 Areas

Every study of a taxon's geographic distribution relates to the concept of endemism, regardless of the organism concerned or the extent of its distribution. Endemism depicts the area(s) that any particular organism lives in and lives nowhere else. The absolute area of endemism is the earth itself, notwithstanding the promise of astrobiogeography. Other areas of endemism are smaller, such as islands (Madagascar, the Isle of Man), continents (Australia) or even a river basin (Angara River, Lake Baikal).

There have been a number of schemes describing different kinds of endemism. One is that of Myers and de Grave (2000). I am not advocating Myers and de Grave's scheme here (although it is an interesting approach) but use it to demonstrate that the problem of taxon distribution can and has been looked at from the point of view of varying kinds of endemism. Myers and de Grave name different kinds of distributions. Holoendemic indicates global distribution which they define as having 'unlimited biogeography' – cosmopolitan, in other words; euryendemic represents broadly conjunct distributions, which have more or less (broad) continuous or contiguous distribution; stenoendemic represents conjunct distributions, which are restricted but continuous; and finally, rhoendemic represents disjunct distributions where the same organism occupies different and separate areas of the globe. Thus, the distributional range of every organism can be expressed as a factor of their endemism (Myers and de Grave, 2000).

No two taxa occur in exactly the same areas on the globe but some will coincide enough to suggest that they occupy a natural region of the globe. In this respect, 'An area of endemism is a geographical unit inferred from the combined distributions of endemic taxa' (Ebach et al., 2008). These natural areas are like natural taxa, hypotheses concerning the relationships of particular areas, one to another.

Systems of naming natural regions of the planet have been in place for a long time: categories are provinces, dominions, regions and realms (much like taxon categories: classes, orders, families and genera), with the further possibility of sub-districts, sub-provinces, sub-dominions, sub-regions, and sub-realms (Ebach et al., 2008), each category related in a hierarchical fashion as in taxonomic names (see, for example, Udvardy, 1975; for review and commentary on biogeographic classification, see Nelson and Platnick, 1981, Chapter 6; Nelson, 1983; Williams, 2007; Williams and Ebach, 2007; Parenti and Ebach, 2009).

Various systems have been proposed for classifying areas (e.g. Amorim, 1992) and recently the first *International Code of Area Nomenclature* was published (Ebach et al., 2008, for commentary see Zaragüeta-Bagils et al., 2009; Parenti et al., 2009; and López et al., 2008 for an example of its use).

Thus, correspondence of taxa and areas is the focus for problems in biogeography.

2.6 Biogeography and evolution in diatoms

Darwin made some comments concerning diatoms and their distribution in the first edition of his journal of the voyage of the *HMS Beagle* (1839). In a footnote he remarked on Christian Gottfried Ehrenberg's (1845) study of the face paint used by the local people in Tierra del Fuego:

> This substance, when dry, is tolerably compact, and of little specific gravity: Professor Ehrenberg has examined it: he states ... that it is composed of infusoria [which included diatoms] ... He says that they are all inhabitants of fresh-water; this is a beautiful example of the results obtainable through Professor Ehrenberg's microscopic researches ... (Darwin, 1839, p. 127, footnote).

Darwin then delivers not more facts but two opinions:

> It is, moreover, a striking fact that in the geographical distribution of the infusoria, which are well known to have very wide ranges, that all the species in this substance, although brought from the extreme southern point of Tierra del Fuego, are old, known forms (Darwin, 1839, p. 127, footnote).

The first opinion is that '... the geographical distribution of the infusoria ... are well known to have very wide ranges'; the second is that '... all the species in this substance, although brought from the extreme southern point of Tierra del Fuego, are old, known forms'.

Darwin suggests that diatoms have very wide geographic ranges and, because these are old known forms, diatom distributions are very old. It would be overstating the case to suggest that Darwin had any direct influence on subsequent interpretations of diatom distributions from that point on but his viewpoint was topical (further commentary on Darwin and Ehrenberg is in Jardine, 2009). Here it is worth a digression to note Ehrenberg's ideas on distribution, as it was he who had studied the diatomite used as face paint.

Ehrenberg found a total of 18 species in the Tierra del Fuego face paint, of which 11 were diatoms (Ehrenberg, 1845). When he published the *Mikrogeologie* in 1854, the total number of species was reduced to 17, of which only eight were diatoms (Fig 2.1, from Ehrenberg, 1854; the table includes the 8 diatom species – numbered 2–9). Of these eight species, three are actually confined to parts of South America, according to Ehrenberg (Fig 2.2, from Ehrenberg, 1854, pl. 35, fig V, illustrations of organisms from Tierra del Fuego, diatoms numbered 2–9). Darwin's assumption that they were known forms is correct – but they were known and described by Ehrenberg from restricted parts of the world. In other words, Ehrenberg had seen

V.

BERGMEHL ALS SCHMINKE DER FEUERLÄNDER. AMERIKA, SÜDSPITZE.

Monatsberichte der Berliner Akad. der Wissensch. 1845 S. 68. Text S. 297 Nr. 636.

Polygastern:

Fig. 1. Chaetotyphla saxipara.
" 2. Eunotia tridentula.
" 3. Navicula Silicula.
" 4. Pinnularia inaequalis α.
" 5. " " β.
" 6. " mesogongyla.
Fig. 7. Pinnularia viridis.
" 8. Stauroneïs Phoenicenteron.
" 9. " Baileyi.
" 10. Trachelomonas coronata.
" 11.} " laevis.
" 12.}
" 13. " granulata.

Phytolitharien:

Fig. 14. Lithodontium furcatum.
" 15. Lithostylidium crenulatum.
" 16. " laeve.
" 17. " Securis, nachträglich.

Fig 2.1 Reproduced from Ehrenberg's *Mikrogeologie* (Ehrenberg, 1854), a list of organisms found in Tierra del Fuego (17 species, eight are diatoms, numbered 2–9).

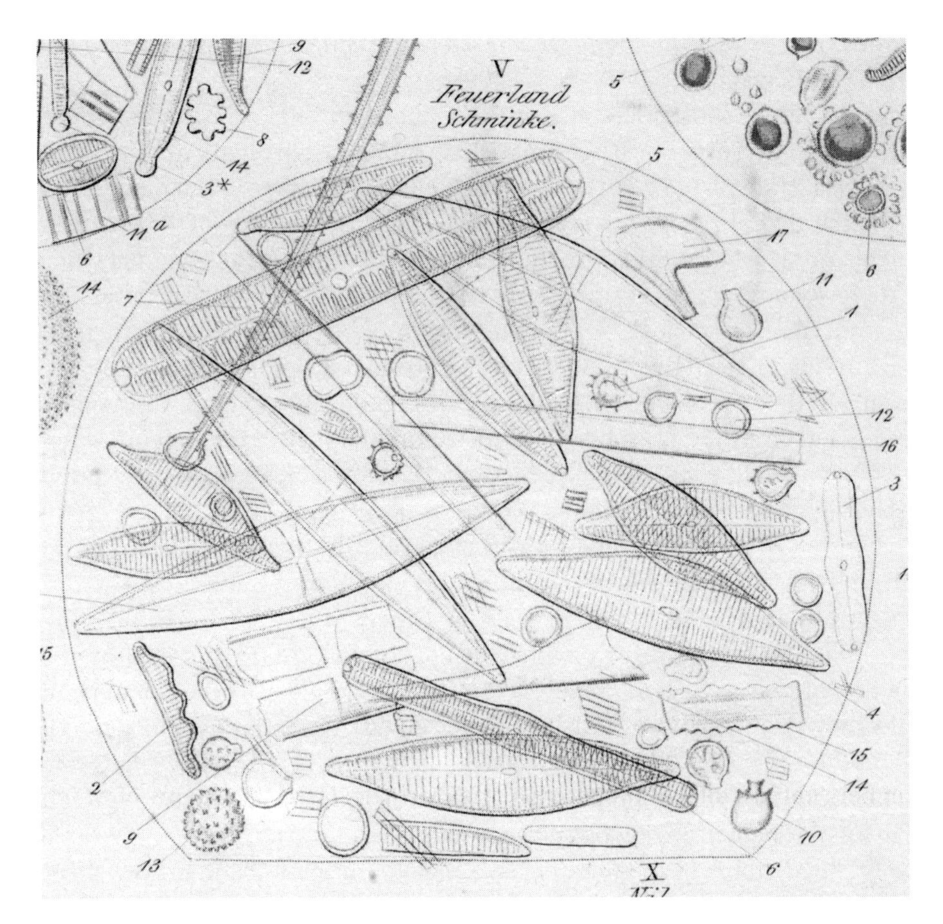

Fig 2.2 Reproduced from Ehrenberg's *Mikrogeologie* (Ehrenberg, 1854, pl. 35, fig. V), illustrations of organisms listed in Fig 2.1 from Tierra del Fuego, diatoms numbered 2–9. Of the eight diatom species, three are actually confined to parts of South America.

the usual mix one encounters in any flora, diatom or otherwise – some species are widespread, some not.

Earlier Ehrenberg made some general comments on distribution, first published in 1849:

> ... the Rocky Mountains are a more powerful barrier between the two sides of America, than the Pacific Ocean between America and China; the infusorial forms of Oregon and California being wholly different from those of the east side of the mountains, while they are partly identical with Siberian species' (Ehrenberg 1850, modified and translated from the original German in Ehrenberg, 1849).

Ehrenberg understood that a relationship of some kind existed between the diatoms from both coasts of the Pacific Ocean, even if he could only explain it by coincidence (O'Malley, 2008).

Ehrenberg offered some potential examples. The freshwater genus *Tetracyclus* is primarily composed of extinct species (Williams, 1996). At present there are some 40–45 species, of which only four or five are still living. Ehrenberg first named and described the genus in 1843 with the name *Biblarium* but because of the principle of nomenclatural priority *Tetracyclus*, a name used by William Smith in 1843 for a living representative, is the correct name (Williams, 1986). Ehrenberg described species from two fossil deposits – one from Oregon, USA, the other from Barguzin, Siberia. Ehrenberg noted that some species occurred only in the Siberian deposit, others only in the Oregon deposit, and a few from both. Even though subsequent studies have revised the taxonomy and distributional limits, Ehrenberg's *interpretation* remains the same today. Extinct species of *Tetracyclus* are mostly found in the continental areas surrounding the Pacific Ocean: in China, Siberia, Japan and, rather strangely – and possibly by a species that still survives – in Java; and on the opposite coast in the USA (Oregon and California), British Columbia, Mexico – and possibly by a species that still survives – in Chile.

The primary interest in this example was not so much to try and explain the distribution of species of *Tetracyclus* – a remarkable number of other plants and animals show the same kind of trans-Pacific divide (Humphries and Parenti, 1999) – but to encourage the search for more diatom taxa that reflect the same or similar kinds of distribution patterns. Lake Baikal, the source of Ehrenberg's Siberian fossils, seemed an obvious place to look.

The benthic diatoms in the flora of Lake Baikal are remarkable for the sheer quantity of endemic taxa present. The number and diversity was recognised early in the twentieth century, by two Russian scientists Boris Skvortzov and Konstantin I. Meyer (Skvortzov and Meyer, 1928; Meyer, 1930; Skvortzov, 1937; see Williams and Reid, 2001). Skvortzov's and Meyer's studies were based on material collected in the early 1900s. Skvortzov noted the total number of diatoms at around 300, with roughly half (148) endemic to the lake.

Recent estimates of Lake Baikal's diversity suggest that there are in excess of 500 taxa, with about 200–250 endemic (Flower, personal communication). It is a major task to document a flora as diverse as that encountered in Lake Baikal. Thus, species that may yield data of interest relative to the origin of the flora, for example, are best tackled first. Such species are those that have varying geographic distributions. While Lake Baikal harbours only two living species of *Tetracyclus* (fossil material is also available, with more species; Williams, 2004; Williams et al., 2006), another species, *Eunotia clevei*, was of greater significance for three reasons: (1) It had previously been considered a widespread, cold-water species, found only in deep lakes; (2) although it was classified as a species of *Eunotia*, it was clearly different from the more typical kind; and (3) Skvortzov had previously published descriptions and illustrations of three conspicuous Lake Baikal endemics, all evidently related to *Eunotia clevei*, as he named two as varieties (Williams and Reid, 2006a): *Eunotia clevei* var. *hispida*, distinguished by prominent valve spines, and *Eunotia clevei* var. *baicalensis*; the third taxon he called *Eunotia lacusbaicalii* was distinguished by the parallel sides of its valves and its disorganised valve surface but evidently also related to *Eunotia clevei*.

Later, another Russian diatomist, Alexander Pavlovich Skabichevsky (Skabichevskaya and Strelnikova, 2003; Skabichevskaya et al., 2004) also published on *Eunotia clevei* var. *baicalensis* (Skabichevsky, 1977). He changed its taxonomic rank from the rather ill-defined category of variety to subspecies. Subspecies were understood to depart from the type in some geographic and morphological dimension, departing from *Eunotia clevei* as it is normally recognised.

Many characters were evident that separated *Eunotia clevei* and its endemic varieties from other species in the genus *Eunotia*, enough to place them all in a separate genus, named *Amphorotia* (Williams and Reid, 2006a). Among the species in *Amphorotia*, five are recognised as endemic to Lake Baikal: *Amphorotia lacusbaikalii*, *A. baicalensis*, *A. hispida*, *A. lineare* and *A. lunata*. *Amphorotia hispida* has large bifurcating spines, said to be characteristic of this species, although a similar type of spine is found on *A. clevei*. The common possession of spines provides evidence for their close relationship, remembering that *hispida* is endemic to Baikal while *clevei* occurs across the Boreal in deep cold-water lakes – Lake Ladoga, Lake Onega and a few other places. However, a number of other specimens were discovered in various museum collections that were either only partly described, buried and forgotten in the literature or had been identified as something completely unrelated. For example, some specimens from the Mekong River delta in Vietnam remained undescribed until they were placed in *Amphorotia* as *A. mekonensis* (Williams and Reid, 2006a); other specimens, first called *Eunotia clevei* var. *asiatica*, which occur in South China, and possibly marine, are now known as *Amphorotia asiatica*; the taxon first named *Eunotia clevei* var. *sinica* by Skvortzov, is now recognised as *Amphorotia sinica* (Williams and

Reid, 2006a). From a geographic point of view, the living species have two relationships, one a Northern cold-water, Boreal range, the other south towards the more tropical parts of Asia. Consideration of the fossils adds another dimension. These occur – from various deposits – in the USA, Japan and China, and are probably interrelated among themselves, but all located around the Pacific Ocean. Thus a third trans-Pacific relationship appears related to the South East Asian living group. Remarkably, the trans-Pacific relationship mirrors that of the fossils from the genus *Tetracyclus* (Williams, 1996) discussed above (Williams and Reid, 2006a).

Further, the genus *Colliculoamphora*, also recently described, has several species occurring either side of the Pacific Ocean as well as either side of the Isthmus of Panama (Williams and Reid, 2006b, 2009); *Colliculoamphora* is closely related to *Amphorotia* (it belongs in the same family) but unlike the remaining members of that group is wholly marine. The genus *Eunophora*, also closely related to

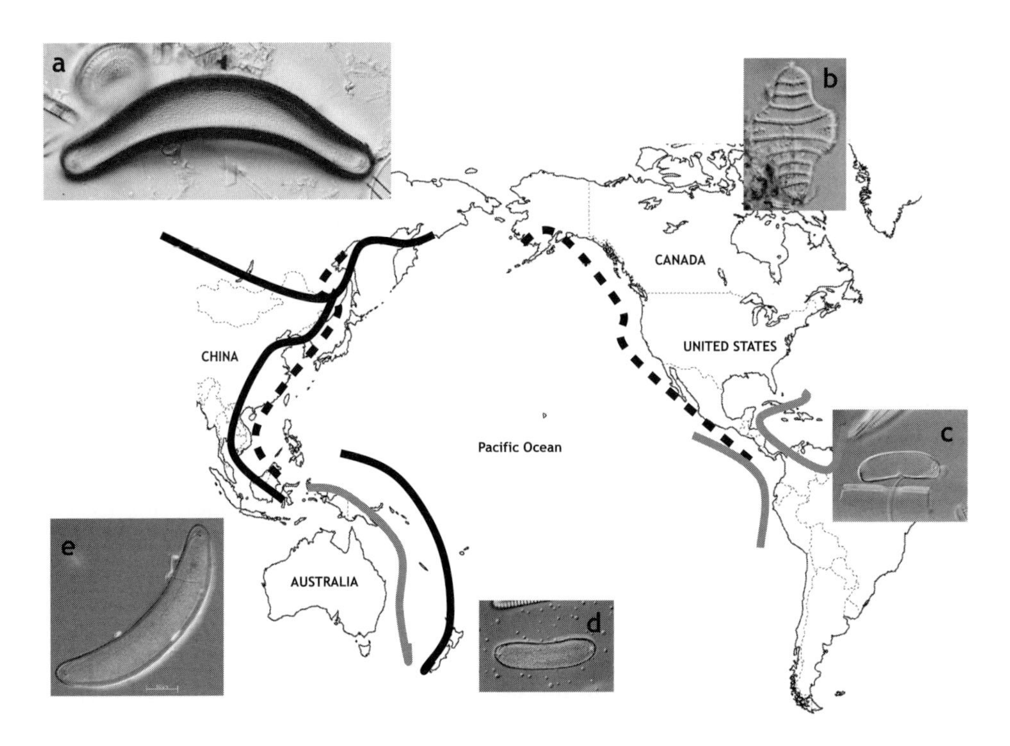

Fig 2.3 Map of the world, orientated at the Pacific Ocean, showing the distributions of species around the Pacific Ocean margin: Solid black line: *Amphorotia* (A, *Amphorotia baikalensis*); black dotted lines: *Tetracyclus* (B, *Tetracyclus tschermissionvae*) and fossil *Amphorotia* (E, *Amphorotia asiatica*); grey lines: *Colliculoamphora* (C, *Colliculoamphora reichardtiana*, D, *Colliculoamphora reedii*); hatched line: *Eunophora* (not illustrated).

Amphorotia and *Colliculoamphora*, occurs in Australia and New Zealand. From the South American perspective, there are many species and genera of significance, *Peronotia*, for example (but see Mezeltin and Lange-Bertalot, 2007, for an extensive, illustrated document of the diversity within *Eunotia*, and Kociolek et al., 2001 for *Actinella*).

In summary, a picture is beginning to emerge with respect to this family of diatoms that centres on the Pacific Ocean and the southern hemisphere (Fig 2.3). It does not, however, 'stress the importance of dispersal and migration in structuring diatom communities at regional to global scales' (Vyverman et al., 2007), even if 'microbes that adhere to Saharan dust can live for centuries and easily survive transport across the Atlantic' (Gorbushina et al., 2007, p. 2911; see also Campbell Smith in Bannermann, 1922; Kellogg and Griffin, 2006), any more than it suggests complex and ancient vicariance processes acting on immobile organisms. What is emerging is a common pattern of disjunction, centred on the Pacific and, while complex, remains a puzzle.

One might speculate that an ancient pattern has been preserved simply because more recent Quaternary changes (from 1.8 million years to 2.6 million years ago) would not affect species already extinct. Nevertheless, it is natural to assume more evidence will elucidate. Yet even at this stage, it seems that geological, rather than ecological, changes have played a major part in establishing diatom patterns of distribution.

2.7 Conclusions

To study geographic distributions, general patterns are required – or at least to search for them rather than assuming they do not exist. So what does endemicity mean in microorganisms? About as much as it does with any other organism: every study of geographic distribution relates to the concept of endemism, and even if areas of endemism are of varying sizes, they may relate to earth history rather than the ability of organisms to wander. Size, shape and dispersal abilities of any organism are irrelevant to the study of its geographic dimension. The slogan (axiom) 'Everything is everywhere' is simply false: 'Everything is endemic' is a more meaningful starting point, a more meaningful axiom, if you wish.

Acknowledgements

I am grateful to Diego Fontaneto for his kind invitation to speak at the EiE symposium in Leiden, Holland, and to Juliet Brodie and Maureen O'Malley for useful comments on an earlier draft of this paper. The presentation in Leiden was dedicated to Chris Humphries (1947–2009), as is this written version. I only hope he would have approved of its content.

References

Amorim, D.d.S. (1992). An empirical system of ranking of biological classifications using biogeographical components. *Revista brasileira de Entomologia* **36**, 281–292.

Atkinson, K.M. (1970). Dispersal of phytoplankton by ducks. *Wildfowl* **21**, 110–111.

Atkinson, K.M. (1972). Birds as transporters of algae. *British Phycological Journal* **7**, 319–321.

Baas Becking, L.G.M. (1934). *Geobiologie of inleiding tot de milieukunde.* The Hague: Van Stockum and Zoon.

Baas Becking, L.G.M. (1959). Geology and microbiology. *Information Series New Zealand, Department for Scientific and Industrial Research*, no. 22 [*New Zealand Oceanographic Institute Memoir no. 3, Contributions to Marine Microbiology*], pp. 48–64.

Bowler, P.J. (2009). *Evolution: The History of an Idea*. 4th edition. Berkeley, CA: University of California Press.

Campbell Smith, W. (1922). Note on a fall of dust, "Blood-rain", at Gran Canaria, 8th to 11th February 1920. In Bannerman, D.A. (ed.), *The Canary Islands: Their History, Natural History and Scenery*, pp. 321–327. London: Gurney and Jackson.

Caron, D.A. (2009). Protistan biogeography: why all the fuss? *Journal of Eukaryote Microbiology* **56**, 105–112.

Caron, D.A., Worden, A.Z., Countway, P.D., Demir, E., Heidelberg, K.B. (2009). Protists are microbes too: a perspective. *ISME* [*International Society for Microbial Ecology*] *Journal* **3**, 4–12.

Chung, K.-T., Ferris, D.H. (1996). Martinus Willem Beijerinck (1851–1931), Pioneer of general microbiology. *ASM* [*American Society for Microbiology*] *News* **62**, 539–543.

Darwin, C.D. (1839). *Journal of Researches into the Natural History and Geology of the Countries visited during the Voyage of H.M.S. Beagle round the World.* London: John Murray.

Darwin, C. (1859). *On the Origin of Species by Means of Natural Selection, or, the Preservation of Favoured Races in the Struggle for Life*. London: John Murray.

Darwin, C.D., Wallace, A.R. (1858). On the tendency of species to form varieties; and on the perpetuation of varieties and species by natural means of selection. *Journal of the Proceedings of the Linnean Society* **3**, 45–62.

De Wit, R., Bouvier, T. (2006). "Everything is everywhere, but, the environment selects"; what did Baas-Becking and Beijerinck really say? *Environmental Microbiology* **8**, 755–758.

Desmond, A.J., Moore, J. (1991). *Darwin.* London: Michael Joseph.

Ebach, M., Morrone, J.J., Parenti, L.R., Vilora, A.L. (2008). International Code of Area Nomenclature. *Journal of Biogeography* **35**, 1153–1157.

Ehrenberg, C.G. (1845). Neue Untersuchungen über das kleinste Leben als geologisches Moment. *Bericht über die zurBekanntmachung geeigneten Verhandlungen der Königlich-Preussischen Akademie der Wissenschaften zu Berlin* **1845**, 53–87.

Ehrenberg, C.G. (1849). Über das mächtigste bis jetzt bekannt gewordene (angeblich 500 Fuß mächtige) Lager von mikroscopischen reinen Kieselalgen Süswasser-Formen am Wasserfall-Flusse im Oregon. *Bericht über die zur Bekanntmachung geeigneten Verhandlungen der Königlich-Preussischen Akademie der Wissenschaften zu Berlin* **1849**, 76–87.

Ehrenberg, C.G. (1850). On infusorial deposits on the River Chutes in Oregon. *American Journal of Science,* **2nd ser. 9**, 140.

Ehrenberg, C.G. (1854). *Mikrogeologie. Das Erden und Felsen schaffende Wirken der unsichtbar kleinen selbständigen Lebens auf der Erde.* Leipzig: Leopold Voss.

Finlay, B.J. (2002). Global dispersal of free-living microbial eukaryote species. *Science* **296**, 1061–1063.

Finlay, B.J., Esteban, G.F. (2007). Body size and biogeography. In Hildrew, A., Raffaelli, D., Edmonds-Brown, R. (eds.), *Body Size: The Structure and Function of Aquatic Ecosystems*, pp. 167–185. Cambridge: Cambridge University Press.

Fitter, A. (2005). Common ground. *Current Biology* **15**, R185–R187.

Fenchel, T., Finlay, B.J. (2003). Is microbial diversity fundamentally different from biodiversity of larger animals and plants? *European Journal of Protistology* **39**, 486–490.

Fenchel, T., Finlay, B.J. (2004a). The ubiquity of small species: patterns of local and global diversity. *BioScience* **54**, 777–784.

Fenchel, T., Finlay, B.J. (2004b). Response from Fenchel and Finlay. *BioScience* **54**, 884–885.

Ferguson Wood, E.J. (1963). Professor L. G. M. Baas Becking. *Limnology and Oceanography* **8**, 312.

Good, R. (1947). *The Geography of the Flowering Plants.* London: Longmans Green.

Gorbushina, A.A., Kort, R., Schulte, A. et al. (2007). Life in Darwin's dust: intercontinental transport and survival of microbes in the nineteenth century. *Environmental Microbiology* **9**, 2911–2922.

Grehan, J.R., Schwartz, J.H. (2009). Evolution of the second orangutan: phylogeny and biogeography of hominid origins. *Journal of Biogeography* **36**, 1823–1844.

Haeckel, E. (1870). *Natürliche Schöpfungsgeschichte. Gemeinverständliche wissenschaftliche Vorträge über die Entwicklungslehre im Allgemeinen und diejenige von Darwin, Goethe und Lamarck im Besonderen über die Anwendung derselben auf den Ursprung des Menschen und andere damit zusammenhängende Grundfragen der Naturwissenschaft.* 2nd edition. Berlin.

Haeckel, E. (1876). *The History of Creation, or, The Development of the Earth and its Inhabitants by the Action of Natural Causes: Doctrine of Evolution in General, and of that of Darwin, Goethe, and Lamarck in Particular* (translation revised by E. Ray Lankester). London: Henry S. King.

Haywood, J. (2008). *The Great Migrations: From the Earliest Humans to the Age of Globalisation.* London: Quercus Books.

Heino, J., Bini, L.M., Karjalainen, S.M. et al. (2010). Geographical patterns of micro-organismal community structure: are diatoms ubiquitously distributed across boreal streams? *Oikos* **119**, 129–137.

Humphries, C.J., Parenti, L.R. (1999). *Cladistic Biogeography. Interpreting Patterns of Plant and Animal Distributions.* 2nd Edition. Oxford: Oxford University Press.

Jacobs, M. (1984). *Herman Johannes Lam (1892–1977): the Life and Work of a Dutch Botanist.* Amsterdam: Rodopi.

Jardine, B. (2009). Between the Beagle and the barnacle: Darwin's microscopy, 1837–1854. *Studies in History and Philosophy of Biological and Biomedical Sciences* **40**, 382–395.

Kellogg, C.A., Griffin, D.W. (2006). Aerobiology and the global transport of desert dust. *Trends in Ecology and Evolution* **21**, 638–644.

Koch, L.F. (1957). Index of biotal dispersity. *Ecology* **38**, 145–148.

Kociolek, J.P., Lyon, D., Spaulding, S.A. (2001). Revision of the South American species of *Actinella*. In Jahn, R., Kociolek, J.P., Witkowski, A., Compére, P. (eds.), *Lange-Bertalot-Festschrift: Studies on Diatoms,* pp. 131–166. Koenigstein: Koeltz Scientific Books.

Kohn, M. (2006). Made in Savannahstan. *New Scientist* **191**(2558), 34–39.

Kuehne, H.A., Murphy, H.A., Francis, C.A., Sniegowski, P.D. (2007). Allopatric divergence, secondary contact, and genetic isolation in wild yeast populations. *Current Biology* **17**, 407–411.

Lachance, M.-A. (2004). Here and there or everywhere? *Bioscience* **54**, 884.

Lam, H.J. (1938). Over de theorie der arealen (chorologie). *Vakblad voor Biologen* **5**, 77–87.

Liebermann, B.S. (2005). Geobiology and paleobiology: tracking the coevolution of the earth and its biota. *Paleogeography, Paleoclimatology, Paleoecology* **2119**, 23–33.

London, J. (1917) *The Human Drift.* New York: The Macmillan Company.

López, H.L., Menni, R.C., Donato, M., Miquelarena, A.M. (2008). Biogeographical revision of Argentina (Andean and Neotropical Regions): an analysis using freshwater fishes. *Journal of Biogeography* **35**, 1564–1579.

Luther, H. (1963). Botanical analysis of mute swan faeces. *Acta Vertebrata* **2**, 266–267.

Maguire, B. (1963). The passive dispersal of small aquatic organisms and their colonization of isolated bodies of water. *Ecological Monographs* **33**, 161–185.

Mayr, E. , Provine, W.B. (eds.) (1980). *The Evolutionary Synthesis.* Cambridge, MA: Harvard University Press.

McCarthy, D. (2009). *Here be dragons. How the Study of Animal and Plant Distributions Revolutionized our Views of Life and Earth.* Oxford: Oxford University Press.

Meyer, K. (1930). Einfuhrung in die Algenflora des Baicalsees. *Byulleten' Moskovskogo obshchestva ispytatelei prirody, odtel biologicheskii* **39**, 179–396.

Mezeltin, D., Lange-Bertalot, H. (2007). Tropical Diatoms, II: Special remarks on biogeographic disjunction. *Annotated Diatom Micrographs.* Vol. **18**.

Milliger, L.E., Schlichting, H.E. (1968). The passive dispersal of viable algae and protozoa by an aquatic beetle. *Transactions of the American Microscopical Society* **87**, 443–448.

Moore, J. (2005). Revolution of the Space Invaders: Darwin and Wallace on the geography of life. In Livingstone, D.N., Withers, C.W.J. (eds.), *Geography and Revolution,* pp. 106–132. Chicago: University of Chicago Press.

Myers, A.A., de Grave, S. (2000). Endemism: Origins and implications. *Vie et Milieu* **50**, 195–204.

Nelson, G. (1978). From Candolle to Croizat: Comments on the history of biogeography. *Journal of the History of Biology* **11**, 269–305.

Nelson, G. (1983). Vicariance and cladistics: Historical perspectives with implications for the future. In Sims, R.W., Price, J.H., Whalley, P.E.S. (eds.), *Evolution, Time and Space: The Emergence of the Biosphere*, pp. 469–492. London: Academic Press.

Nelson, G., Platnick, N.I. (1981). *Systematics and Biogeography: Cladistics and Vicariance*. New York, NY: Columbia University Press.

O'Malley, M.A. (2007). The nineteenth-century roots of 'everything is everywhere'. *Nature Reviews Microbiology* **5**, 647–651.

O'Malley, M.A. (2008). *'Everything is everywhere: but the environment selects'*: Ubiquitous distribution and ecological determinism in microbial biogeography. *Studies in History and Philosophy of Biological and Biomedical Sciences* **39**, 314–325.

O'Malley, M.A., Dupré, J. (2007a). Size doesn't matter: towards a more inclusive philosophy of biology. *Biology and Philosophy* **22**, 155–191.

O'Malley, M.A., Dupré, J. (2007b). Towards a philosophy of microbiology. *Studies in History and Philosophy of Biological and Biomedical Sciences* **38**, 775–779.

Parenti, L.R., Ebach, M.C. (2009). *Comparative Biogeography: Discovering and Classifying Biogeographical Patterns of a Dynamic Earth*. Berkeley, CA: University of California Press.

Parenti, L.R., Viloria, A.L., Ebach, M.C., Morrone, J.J. (2009). On the International Code of Area Nomenclature (ICAN): a reply to Zaragüeta-Bagils et al. *Journal of Biogeography* **36**, 1619–1621.

Proctor, V.W. (1959). Dispersal of freshwater algae by migratory waterbirds. *Science* **130**, 623–624.

Quispel, A. (1998). Lourens G.M. Baas Becking (1895–1963), Inspirator for many (micro)biologists. *International Microbiology* **1**, 69–72.

Raby, P. (2002). *Alfred Russel Wallace: A Life*. Princeton, NJ: Princeton University Press.

Schlegel, M., Hülsmann, N. (2007). Protists – a textbook example for a paraphyletic group. *Organisms, Diversity and Evolution* **7**, 166–172.

Schlichting, H.E. (1960). The role of waterfowl in the dispersal of algae. *Transactions of the American Microscopical Society* **79**, 160–166.

Shermer, M. (2002). *In Darwin's Shadow: The Life and Science of Alfred Russel Wallace: A Biographical Study on the Psychology of History*. Oxford: Oxford University Press.

Shreeve, J. (2006). The greatest journey. *National Geographic* **209**(3), 60–69.

Skabichevsky, A.P. (1977). Vodoroslevye obrastaniya khetomorfy sublitorali vostochnogo berega Bajkala. In Popova, T.G. (ed.), *Prirodnye kompleksy nizshikh rastenji zapadnoj sibiri*, pp. 121–132. Akademiya Nauk SSSR Sibirskoe Otdelenie tsentral'nyj sibirskij botanicheskij sad. Izsatel'stvo 'Nauka' sibirskoe otdelenie Novosibirsk.

Skabichevskaya, N.A., Kuzmina, A.E., Strelnikova, N.I., Potapova, M.G., Williams, D.M. (2004). Alexander Pavlovich Skabichevsky (November 23, 1904 – May 5, 1990): An obituary and bibliography. *Diatom Research* **18**, 385–398.

Skabichevskaya, N.A., Strelnikova, N.I. (2003). [In memoriam: Alexander Pavlovich Skabichevsky (1904 - 1990)]. *Botanische Zhurnal* **88**, 142–147.

Skvortzov, B.W. (1937). Bottom diatoms from Olhon Gate of Baikal Lake, Siberia. *Philippine Journal of Science* **62**, 293–377.

Skvortzov, B.W., Meyer, C.I. (1928). A contribution to the diatoms of Baikal Lake. *Proceedings of the Sungaree River Biological Station* **1**, 1–55.

Sylvester-Bradley, P.C. (1972). Geobiology and the future of palaeontology. *Journal of the Geological Society* **28**, 109–117.

Udvardy, M.D.F. (1975). *A classification of the biogeographical provinces of the world*. IUCN Occasional Paper 18.

Vanormelingen, P., Verleyen, E., Vyverman, W. (2008). The diversity and distribution of diatoms: from cosmopolitanism to narrow endemism. *Biodiversity and Conservation* **17**, 393–405.

Velasques, G.T. (1940). On the viability of algae obtained from the digestive tract of the gizzard shad, *Dorosoma cepedianum*. *American Midland Naturalist* **22**, 376–412.

Vyverman, W., Verleyen, E., Sabbe, K. et al. (2007). Historical processes constrain patterns in global diatom diversity. *Ecology* **88**, 1924–1931.

Wallace, A.R. (1876). *The Geographical Distribution of Animals: With a Study of the Relations of Living and Extinct Faunas as Elucidating the Past Changes of the Earth's Surface*. London: Macmillan.

Westenberg, J. (1977). *A Bibliography of the Publications of Lourens G.M. Baas Becking*. Amsterdam: North Holland Publishing Company.

Wilkinson, D.M. (2001). What is the upper size limit for cosmopolitan distribution in free-living microrganisms? *Journal of Biogeography* **28**, 285–291.

Williams, D.M. (1986). Proposal to conserve the generic name *Tetracyclus* against *Biblarium* (Bacillariophyta). *Taxon* **35**, 730–731.

Williams, D.M. (1996). Fossil species of the diatom genus *Tetracyclus* (Bacillariophyta, '*ellipticus*' species group): Morphology, interrelationships and the relevance of ontogeny. *Philosophical Transactions of the Royal Society, London*, series B **351**, 1759–1782.

Williams, D.M. (2004). On diatom endemism and biogeography: *Tetracyclus* and Lake Baikal Endemic Species. *Proceedings of the 17th International Diatom Symposium*, pp. 433–459. Bristol: BioPress Ltd.

Williams, D.M. (2007). Ernst Haeckel and Louis Agassiz: trees that bite and their geographical dimension. In Ebach, M.C., Tangey, R. (eds.), *Biogeography in a Changing World*, pp. 1–59. Boca Raton, FL: CRC Press.

Williams, D.M., Ebach, M.C. (2007). *The Foundations of Systematics and Biogeography*. New York: Springer.

Williams, D.M., Reid, G. (2001). A bibliography of the botanical work of Boris V. Skvortzov (1896–1980) with commentary on the publications concerning diatoms (Bacillariophyta). *Bulletin of the British Museum (Natural History), Botany* **31**, 89–106.

Williams, D.M., Reid, G. (2006a). *Amphorotia* nov. gen., a new genus in the family Eunotiaceae (Bacillariophyceae), based on *Eunotia clevei* Grunow in Cleve et Grunow. *Diatom Monographs* **6**.

Williams, D.M., Reid, G. (2006b). Fossils and the tropics, the Eunotiaceae (Bacillariophyta) expanded: The Upper Eocene fossil diatom *Eunotia reedi* and the Recent marine diatom *Amphora reichardtiana* from the tropics. *European Journal of Phycology* **41**, 147–154.

Williams, D.M., Reid, G. (2009). New species in the genus *Colliculoamphora* Williams and Reid with commentary on species concepts in diatom taxonomy. *Beihefte zur Nova Hedwigia* (Eugene Stoermer Festschrift) **135**, 185–200.

Williams, D.M., Khursevich, G.K, Fedenya, S.A, Flower, R.J. (2006). The fossil record in Lake Baikal: Comments on the diversity and duration of some benthic species, with special reference to the genus *Tetracyclus*. *Proceedings of the 18th International Diatom Symposium*, pp. 465–478. Bristol: BioPress Ltd.

Zaragüeta-Bagils, R., Bourdon, E., Ung, V., Vignes-Lebbe, R., Malécot, V. (2009). On the International Code of Area Nomenclature (ICAN). *Journal of Biogeography* **36**, 1617–1619.

Part II

Prokaryotes

3

Biogeography of prokaryotes

DONNABELLA C. LACAP, MAGGIE C.Y. LAU
AND STEPHEN B. POINTING

School of Biological Sciences, The University of Hong Kong, Hong Kong SAR, China

3.1 Introduction

Prokaryotic microorganisms are critical to terrestrial and aquatic ecosystem function due to their involvement in key biogeochemical processes and interaction with macroorganisms (Bell et al., 2005a). The Bacteria are assumed to occur ubiquitously as a result of their large population sizes, rapid generation times and high dispersal rates. Increasingly the Archaea are also being recognised as key components of many biomes (Auguet et al., 2010). The long-held tenet in microbiology that 'everything is everywhere, the environment selects' (Baas Becking, 1934) has been employed as a de facto null hypothesis against which to test the existence of spatio-temporal patterns in prokaryotic distribution. Demonstrating the existence of such patterns and their underlying drivers is key to understanding microbial biogeography. This has wide-reaching implications for understanding ecosystem function, conservation value for microorganisms and bioprospecting for strains with biotechnology potential (Prosser et al., 2007).

Biogeography of Microscopic Organisms: Is Everything Small Everywhere?, ed. Diego Fontaneto.
Published by Cambridge University Press.

3.2 The prokaryotic species concept

A major limitation to the elucidation of prokaryotic biogeography lies with the species concept as applied to prokaryotes. It is necessary to be able to recognise the diversity of species in community-level studies, and also the abundance of a given species in population studies. The traditional view of a species as a group of individuals that interbreed and are isolated from other species by barriers to recombination (Mayr, 1957) is generally assumed not to be applicable to the asexual lifestyle of prokaryotes, although it is emerging that recombinant events may be more widespread than earlier assumed for some microorganisms (Fraser et al., 2007). It has been suggested that an ecological species concept be applied to prokaryotes, where a species (ecotype) within a given niche will arise due to selective pressures, and this will vary with different niches (Cohan, 2002). Sequence-based modelling of this concept has resulted in the proposal of 'ecovar' epithets to bacterial species binomials in order to delineate putative ecotypes (Koeppel et al., 2008).

The issue of how to identify a prokaryotic species presents challenges at the genetic level due to the horizontal transfer of genes. These may represent new traits that become incorporated (non-homologous recombination) or replace existing genes (homologous recombination). Analysis of cyanobacterial genomes indicates they comprise a stable core of critical genes that are not subject to horizontal gene flow, and a variable shell of genes that encode traits that may be acquired or lost (Shi and Fakowlski, 2008). If this applies to all prokaryotes then the stable core (if not subject to homologous recombination) might inform a traditional evolutionary species concept whereas the variable shell suggests an ecological species concept could be applied. The vast majority of ecological studies of prokaryotes have employed analysis of rDNA genes, part of the proposed stable core of the genome, thus the phylogenetic species concept has been the basis for most interpretations thus far. However, this approach does not consider the adaptive traits encoded by loci that are responsible for their distribution.

3.3 Estimating prokaryotic diversity

Identification of prokaryotic species from environmental samples based upon morphological or physiological traits has largely been surpassed in ecological studies by the use of molecular genetic tools. If cultivated strains are available then a multi-locus sequence analysis (MLSA) approach can yield relatively robust data on relationships among taxa. Most ecological studies, however, have centred on the use of rDNA as a marker for phylogenetically defined taxa. Such approaches have generally employed community fingerprinting techniques such as denaturing gradient gel electrophoresis (DGGE), amplified ribosomal intergenic spacer analysis (ARISA) and terminal restriction fragment length polymorphism (TRFLP). These approaches may also be used to generate sequence-based data sets but they yield relatively short fragments with limited value in phylogenetic reconstructions.

An alternative and commonly used approach has been to generate clone libraries of near full-length rDNA amplicons and sequences from which to make phylogenetic comparisons. The cloning approach is also commonly used to infer quantitative data on relative abundance within a community, although even when sampling relatively large libraries considerable under-sampling of communities is likely for most environments (Gans et al., 2005), and so this must be considered when interpreting such data. Application of high-throughput techniques such as pyrosequencing of rDNA markers from environmental samples (e.g. Edwards et al., 2006; Sogin et al., 2006) are emerging as a tool to significantly improve estimates of species richness within a community. This approach also has the potential to elucidate functional capabilities through complete metagenome assembly, and this will undoubtedly emerge as a powerful tool in microbial ecology studies.

A major barrier to current sequence-based approaches to community diversity studies remains the specificity of PCR primers. Simulation analysis in our laboratory has revealed that several 'universal' primer sets for bacteria and archaea are unable to amplify certain phyla. There are also no truly universal PCR primers that amplify phyla across all domains and so information on the overall community is lacking for many microbial habitats. One way to address this shortcoming is to construct datasets for each domain (Archaea, Bacteria, Eukarya) and then use quantitative tools such as real-time PCR to estimate relative abundance of domains. We have applied this approach to lithic microbial communities with some success (Pointing et al., 2009).

A further issue is that whilst most studies have used DNA markers, it is likely that, given the ability of prokaryotes to remain dormant for long periods, in some cases findings may not reflect the active fraction of a given community. This could be addressed by using RNA-based approaches, and in combination with metatranscriptomic and metaproteomic techniques will help to reveal functional relationships and define ecotypes.

3.4 Evidence for biogeography among prokaryotes

A key aspect of the 'everything is everywhere' hypothesis for microorganisms has been the assumption that dispersal is ubiquitous and therefore spatial patterns should not arise. Selective pressures in different environmentally defined niches would then determine the subsequent growth of microorganisms. This scenario assumes that the resultant high levels of gene flow will outweigh any variation that arises due to adaptation, genetic drift or mutation, and so a given niche should support similar organisms regardless of geographic scale. This is in contrast to macroorganisms where biogeographic patterns are manifest due to greater restrictions on gene flow, often attributed to size-related (allometric) dispersal limitations (Rosenzweig, 1995). A large number of studies have demonstrated that bacterial and archaeal taxa display non-random environmental distribution (Hughes Martiny et al., 2006). Relatively few have attempted to address patterns with respect to spatial scales, environmental factors and temporal scales. Some excellent reviews on this subject have highlighted major challenges in this area (Green and Bohannan, 2006; Hughes Martiny et al., 2006; Prosser et al., 2007).

A taxa–area relationship has been demonstrated for tree-hole bacteria and salt-marsh bacteria, where species richness increased with the area sampled (Horner Devine et al., 2004; Bell et al., 2005b). These findings suggested that prokaryotic distribution might not be ubiquitous, since under such constraints diversity should display less variability with spatial scale. A major focus of recent research has been to identify whether spatial scaling of microorganisms is primarily determined by geographic distance in a distance–decay relationship (dispersal limitation > environmental selection) or by environmental heterogeneity (environmental selection > dispersal limitation). The existence of a distance–decay pattern for soil microbial assemblages has been demonstrated on small scales (metres) using DNA fingerprinting (Franklin and Mills, 2003). A study of *Pseudomonas* phylotypes from soil using BOX-PCR identified a negative correlation between genetic similarity and geographic distance on local scales (< 80 km) but not between continents (Cho and Tiedje, 2000). A DGGE study at the regional scale suggested that for lake bacterial assemblages distance was more important than environmental heterogeneity in determining species richness (Reche et al., 2005).

Other studies have concluded that the influence of environmental heterogeneity was more important than distance to community composition. Comparison of bacterial rDNA sequences in a salt marsh revealed distance–decay effects at scales up to hundreds of metres, but this was attributed to environmental heterogeneity rather than distance (Horner Devine et al., 2004). A DNA fingerprint-based study of bacterial diversity in lakes concluded that no distance–decay effect existed and that bacterial diversity was explained largely by pH (Fierer et al., 2007). A DGGE-based study of bacterial colonisation of desert rocks revealed a strong influence for moisture availability but no distance–decay effects (Warren-Rhodes et al., 2006). A DGGE-based study of soils from North and South America concluded that variation in bacterial assemblages could largely be explained by soil pH and was independent of geographic distance (Fierer and Jackson, 2006). A study of sequence data from bacterial assemblages in diverse aquatic and soil environments on a global scale indicated that salinity was the variable best able to explain differences in diversity (Lozupone and Knight, 2007).

In order to fully resolve potential spatial scaling for prokaryotes, a high degree of taxonomic resolution is required. Relatively few studies have employed such approaches. The existence of spatial patterns in diversity has been demonstrated using MLSA for archaea and bacteria. A distance–decay pattern was recorded for hyperthermophilic *Sulfolobus* strains isolated from hot springs across three continents, and the effect of variation in temperature, pH or sulphide levels was not significant (Whitaker et al., 2003). Similarly geographic lineages of thermophilic hot-spring cyanobacteria were also identified at an intercontinental scale with no significant influence from a wide range of abiotic variables (Papke et al., 2003). These studies contradict the notion of ubiquitous dispersal, and imply that dispersal limitations contribute to the existence of endemic variants. We have recently conducted a study of the cyanobacterial genus *Chroococcidiopsis* from deserts on every continent on Earth (Bahl et al., 2011), which suggests that whilst distance-decay may be a regional phenomenon, it is not apparent on a global scale for this terrestrial bacterium. Rather, the distribution is determined by contemporary climate and historical legacies.

A further issue is the influence of temporal scales on prokaryotic diversity. A DGGE-based study revealed a taxa–time relationship where diversity decreased due to selective pressure of environmental factors in bioreactors over a 154-day period (van der Gast et al., 2008). A comparison of environmental rDNA sequences for bacteria and archaea in hot springs revealed that pronounced seasonality occurred but that stochastic disturbance events had little long-term effect on diversity (Lacap et al., 2007). This highlights the importance of gathering temporal data sets for both biotic and abiotic variables. The power of phylogenetic data sets has opened opportunities to explore temporal relationships on a geological timescale where appropriate fossil calibration points are available, as with the cyanobacteria. We have recently constructed a multi-locus temporal phylogeny for the cyanobacterial genus *Chroococcidiopsis* recovered from deserts on every continent on Earth (Bahl et al., 2011). This revealed that regionally endemic variants have persisted over timescales of tens of millions of years with no evidence for recent inter-regional gene flow. This concept of 'ancient endemism' is also supported by a study of thermophiles in calderas of Yellowstone National Park, where variation in rDNA sequences could be explained by geological events over the last two million years but not by contemporary environmental heterogeneity or distance (Takacs-Vesbach et al., 2008).

The issue of spatio-temporal scaling is far from resolved but we postulate that from an evolutionary viewpoint: at local geographic scales where dispersal limitations are minimal, environmental heterogeneity is the major driver of diversity over time. Across larger distances the effects of dispersal limitation become more influential and endemic variants may arise for a given niche. Distance–decay effects may occur at certain geographic scales. A revision of Baas Becking's descriptor for influences on prokaryotic distribution can therefore be articulated as:

> 'The environment selects, with dispersal effects.'

3.5 Filling the gaps

Further work is required to fully resolve and confirm the drivers of spatio-temporal scaling in prokaryotic biogeography. These should be guided by the following questions: first, do microorganisms exhibit endemism indicative of biogeographic patterns on a global scale? Second, what is the cause of the geographic signal, dispersal limitations or adaptation to a specific niche? Third, if microbial biogeography is manifest, then what timescales are involved in evolution of endemic variants? This will require improvements in the approach to sampling and analysis for both biotic and abiotic variables. Sample collection requires relevant geographic scales, and importantly a randomised hierarchical sampling approach in order to avoid sampling bias. For estimating evolutionary species (i.e. as defined by stable core genes) sufficient sampling effort must be made and confidence levels expressed. Greater use can be made of phylogenetic data sets for estimating evolutionary divergences. Functional aspects can be incorporated using transcriptomic or proteomic approaches to infer ecotypes. Long-term monitoring of abiotic variables and biotic data will also allow greater understanding of the temporal aspects of biogeography.

References

Auguet, J-C., Barberan, A., Casamayor, E.O. (2010). Global ecological patterns in uncultured archaea. *ISME Journal* **4**, 182–190.

Baas Becking, L.G.M. (1934). *Geobiologie of inleiding tot de milieukunde*. The Hague: Van Stockum and Zoon.

Bahl, J., Lau, M.C.Y., Smith, G.J.D. et al. (2011). Ancient orgins determine global biogeography of hot and cold desert cyanobacteria. *Nature Communications* **2**, 163.

Bell, T., Newman, J.A., Silverman, B.W., Turner, S.L., Lilley, A.K. (2005a).The contribution of species richness and composition to bacterial services. *Nature* **436**, 1157–1160.

Bell, T., Ager, D., Song, J.-I.et al. (2005b). Larger islands house more bacterial taxa. *Science* **308**, 1884.

Cho, J.-C., Tiedje, J.M. (2000). Biogeography and degree of endemicity of fluorescent Pseudomonas strains. *Applied Environmental Microbiology* **66**, 5448–5456.

Cohan, F.M. (2002). What are bacterial species? *Annual Review of Microbiology* **56**, 457–487.

Edwards, R.A., Rodriguez-Brito, B., Wegley, L. et al. (2006). Using pyrosequencing to shed light on deep mine microbial ecology. *BMC Genomics* **7**, 57.

Fierer, N., Jackson, R.B. (2006). The diversity and biogeography of soil bacterial communities. *Proceedings of the National Academy of Sciences USA* **103**, 626–631.

Fierer, N., Morse, J.L., Berthrong, S.T., Bernhardt, E.S. (2007). Environmental controls on the landscape-scale biogeography of stream bacterial communities. *Ecology* **88**, 2162–2173.

Franklin, R.B., Mills, A.L. (2003). Multi-scale variation in spatial heterogeneity for microbial community structure in an eastern Virginia agricultural field. *FEMS Microbiology Ecology* **44**, 335–346.

Fraser, C., Hanage, W.P., Spratt, B.G. (2007). Recombination and the nature of bacterial speciation. *Science* **315**, 476–480.

Gans, J., Wolinsky, M., Dunbar, J. (2005). Computational improvements reveal great bacterial diversity and high toxicity in soil. *Science* **309**, 1387–1390.

Green, J., Bohannan, B.J.M. (2006). Spatial scaling of microbial biodiversity. *Trends in Ecology and Evolution* **21**, 501–507.

Horner Devine, C., Lange, M., Hughes, J.B., Bohannan, B.J.M. (2004). A taxa–area relationship for bacteria. *Nature* **152**, 750–753.

Hughes-Martiny, J.B., Bohannan, B.J.M., Brown, J.H. et al. (2006). Microbial biogeography: putting microorganisms on the map. *Nature Reviews Microbiology* **4**, 102–112.

Koeppel, A., Perry, E.B., Sikorski, J. et al. (2008). Identifying the fundamental units of bacterial diversity: A paradigm shift to incorporate ecology into bacterial systematics. *Proceedings of the National Academy of Sciences USA* **105**, 2504–2509.

Lacap, D.C., Barraquio, W., Pointing, S.B. (2007). Thermophilic microbial mats in a tropical geothermal location display pronounced seasonal changes but appear resilient to stochastic disturbance. *Environmental Microbiology* **9**, 3065–3076.

Lozupone, C.A., Knight, R. (2007). Global patterns in bacterial diversity. *Proceedings of the National Academy of Sciences USA* **104**, 11436–11440.

Mayr, E. (ed.) (1957). *The Species Problem*. Washington, DC: American Association for the Advancement of Science.

Papke, R.T., Ramsing, N.B., Bateson, M.M., Ward, D.M. (2003). Geographical isolation in hot spring cyanobacteria. *Environmental Microbiology* **5**, 650–659.

Pointing, S.B., Chan, Y., Lacap, D.C. et al. (2009). Highly specialized microbial diversity in hyper-arid polar desert. *Proceedings of the National Academy of Sciences USA* **106**, 19964–19969.

Prosser, J.I., Bohannan, B.J.M., Curtis, T.P. et al. (2007). The role of ecological theory in microbial ecology. *Nature Reviews Microbiology* **5**, 384–392.

Reche, I., Pulido-Villena, E., Morales-Bacquero, R., Casamayor, E.O. (2005). Does ecosystem size determine aquatic bacterial richness? *Ecology* **86**, 1715–1722.

Rosenzweig, M.L. (1995). *Species Diversity in Space and Time.* Cambridge: Cambridge University Press.

Shi, T., Fakowlski, P.G. (2008). Genome evolution in cyanobacteria: The stable core and the variable shell. *Proceedings of the National Academy of Sciences USA* **105**, 2510–2515.

Sogin, M.L., Morrison, H.G., Huber, J.A. et al. (2006). Microbial diversity in the deep sea and the unexplored 'rare' biosphere. *Proceedings of the National Academy of Sciences USA* **103**, 12115–12120.

Takacs-Vesbach, C., Mitchell, K., Jackson-Weaver, O., Reysenbach, A.-L. (2008). Volcanic calderas delineate biogeographic provinces among Yellowstone thermophiles. *Environmental Microbiology* **10**, 1681–1689.

van der Gast, C.J., Ager, D.A., Lilley, A.K. (2008). Temporal scaling of bacterial taxa is influenced by both stochastic and deterministic ecological factors. *Environmental Microbiology* **10**, 1411–1418.

Warren-Rhodes, K.A., Rhodes, K.L., Pointing, S.B. et al. (2006). Hypolithic cyanobacteria, dry limit of photosynthesis and microbial ecology in the hyperarid Atacama Desert, Chile. *Microbial Ecology* **52**, 389–398.

Whitaker, R.J., Grogan, D.W., Taylor, J.W. (2003). Geographic barriers isolate endemic populations of hyperthermophilic archaea. *Science* **301**, 976–978.

4

Thermophilic bacteria in cool soils: metabolic activity and mechanisms of dispersal

ROGER MARCHANT[1], IBRAHIM M. BANAT[1]
AND ANDREA FRANZETTI[2]

[1] *School of Biomedical Sciences, University of Ulster, Coleraine, Northern Ireland, UK*
[2] *Department of Environmental Sciences, University of Milano-Bicocca, Milano, Italy*

4.1 Introduction

The biogeographic patterns of plants and animals, i.e. the distribution of biodiversity over space and time has been studied for many years; however, the question whether microorganisms display similar biogeographic patterns remains unanswered (Fenchel et al., 1997). A fundamental assumption that 'everything is everywhere, but the environment selects' was generally promulgated by the Dutch microbiologist Martinus Wilhelm Beijerinck early in the twentieth century and further supported by Baas Becking in 1934. This hypothesis strongly influenced the scientific community throughout the century, leading to widespread acceptance (O'Malley 2008). If

Biogeography of Microscopic Organisms: Is Everything Small Everywhere?, ed. Diego Fontaneto. Published by Cambridge University Press.

the environment is indeed responsible for 'selecting' the organisms in a particular habitat, then we should expect to be able to identify the specific controlling factors for particular organisms. As a consequence of this speculation and as suggested by the literature and experience, the presence of thermophilic bacteria is to be expected in hot environments, from which many of these organisms have been indeed isolated. However, the presence of thermophilic bacteria in cooler environments has been known for many years but few investigations have been carried out to assess their physiology, their ecological roles and to interpret their presence in the framework of biogeographic theory. In 2002 Marchant and colleagues initiated the investigation of the occurrence of highly thermophilic bacteria in cool soil environments, isolating five bacterial strains able to grow aerobically only above 40 °C, a temperature never achieved in these soils. Based on this finding, several questions were raised. What is the frequency of these microorganisms in soils? What kind of microorganisms are they? What is the extent of their diversity? Are they metabolically active? Are there potential transport mechanisms that sustain their presence? This chapter reviews the results of the subsequent research activities carried out at the University of Ulster under the coordination of Professor I.M. Banat and Professor R. Marchant together with some other worldwide collaborating institutions aimed at answering the aforementioned questions.

4.2 Thermophile community in cool temperate soils

The occurrence of thermophilic bacteria in soils from different temperate and cool environments was highlighted by Marchant et al. (2002a, 2002b). The investigation focused on soil samples from a range of sites in Ireland, and from one site in Bolivia. Two sites were in Aghadowey, Northern Ireland under established mixed coniferous and deciduous trees with no ground cover plants (Irish Grid reference C881 216), the third site was from an established mixed wet meadow area (Irish Grid reference C881 215). The soil type at both sites was basalt till and the sites had been undisturbed for at least 15 years. The fourth site was in Coleraine, Northern Ireland, Irish Grid reference C843 349 from a cultivated basalt till area. Records of soil temperature at a depth of 50 mm taken over a 30-year period in Coleraine (at the University of Ulster meteorological station) indicated that the maximum temperature reached during that time was less than 25 °C. The fourth site was close to the Salar de Uyuni, near the village of Colchani in Bolivia at an altitude of 3653 m (66°54′W, 20°22′S). The area was only sparsely covered in vegetation at a distance of 100 m from the salt plain. All soil samples were taken at a depth of 50 mm into the mineral layer of the soil. Enrichments were carried out on rich growth medium at 70 °C and 80 °C and five pure isolates able to grow at 70 °C were selected, namely B70, F70, T70, I80 and T80. Other non-investigated thermophilic strains have been isolated from a wide range of samples collected in different part of the world (Greece, Italy, Turkey, North America and India). The morphological characterisation of these five strains revealed that the cells were narrow rods (0.5–0.7 μm × 1–5 μm), and showed variable Gram-staining, although transmission electron microscopy analyses revealed Gram-positive cell wall architecture; only B70 showed terminal endospores. The metabolic fingerprinting obtained testing several biochemical parameters resulted in three different metabolic groups constituted by B70 and F70 individually and the cluster T70, T80 and I80 characterised by extensive growth on hydrocarbons. The minimum and the maximum growth temperature under aerobic conditions ranged from 40 to 45 °C and 75 to 80 °C, respectively. Significant were the short generation times of these organisms at the optimal temperatures (around 75 °C) that were less than 30 min. Phylogenetic relationships and taxonomic affiliations of the isolates were performed by means of Amplified Ribosomal DNA Restriction Analyses (ARDRA) and subsequent sequencing of near full-length of the 16S rRNA gene. These molecular tools allowed all the isolates to be assigned to the bacterial domain and to the closest phylogenetic neighbours in the EMBL database. T70, F70 and T80 were closely related (> 99%) to *Geobacillus thermoleovorans* strains while 99% of similarity was found for B70 and F70 with *Geobacillus caldoxylolyticus* (EMBL, AF067651) and *Bacillus* sp. SK-1 (EMBL, AF 326278), respectively. Since the isolation of these thermophilic bacterial strains from enrichment culture provided only evidence of their presence in the soil, quantitative data were obtained by plate counting demonstrating that they were present in high numbers in the soil samples collected in Northern Ireland ($1.5–8.8 \times 10^4$ colony forming unit (cfu) per g) and compare to 10% of the total culturable mesophilic bacteria.

The thermophilic bacterial genus *Geobacillus* is a relatively recent creation through the separation of a number of existing species of *Bacillus* and the addition of some new species isolated from deep oil reservoirs (Nazina et al., 2001). The type species for the genus is *Geobacillus* (*Bacillus*) *stearothermophilus*, a long-established and well-studied species. Subsequent to the publication of the new genus a number of other interesting species have been described from a variety of geothermal and ambient temperature environments (e.g. Sung et al., 2002; Banat et al., 2004; Nazina et al., 2004, 2005).

To describe the biodiversity of the thermophilic community using a higher number of isolates, 52 thermophilic bacterial strains were analysed by ARDRA and sequencing of the 16S rRNA gene. The comparison of the profiles obtained using four different restriction endonucleases allowed differentiation of the isolates in 20 different clusters. Eighteen of these clusters were composed of single isolates while 12 isolates exhibited restriction patterns indistinguishable from those of isolate B70, while 19 isolates had identical restriction patterns to T7. The ARDRA analyses and subsequent sequencing of 16S rRNA indicated that

G. thermoleovorans and *G. caldoxylolyticus*, accounted for 50% and 34.6% respectively. The results of these first studies concerning the presence and diversity of thermophilic bacteria in cool soil clearly demonstrated that there is a great biodiversity of these bacteria within the community and that the *Geobacillus* genus is dominant (Rahman et al., 2004).

4.3 Activity of thermophiles in cool environments

Some species have been described as able to grow anaerobically by denitrification and examination of genome sequence information confirms the presence of all the necessary genes for denitrification activity. For this reason the growth tests reported in the first papers did not completely exclude the possibility that growth at low temperatures could take place anaerobically using nitrate as the final electron acceptor. Thus, the effect of temperature on the denitrification process by *G. thermoleovorans* T-80 culture was investigated (Marchant et al., 2008). The cultures were incubated at 35, 40, 45, 50, 60 and 70 °C. This assay was conducted in 160 ml serum bottles which were pre-flushed with He. After transferring the culture media (100 ml) the serum bottles were autoclaved at 121 °C for 30 min. Glucose, yeast extract and nitrate were added resulting in initial concentrations of 1 g/l, 50 mg/l and 100 mg of nitrogen N/l, respectively. This assay was conducted using triplicate serum bottles. One serum bottle was used for gas sampling and the other two serum bottles were used for liquid sampling. In addition to the cultures, an abiotic control was also set up with the denitrifying culture media which was amended with nitrate (100 mg N/l, without biomass and electron donor).

During the 20 days of incubation nitrate reduction was not observed in the cultures incubated at 35 °C (Fig 4.1). In all of the cultures the initial acetate concentration was approximately 25 mg/l which was contributed by the inoculum. For the cultures incubated at 40 °C, after 15 days of lag, complete reduction of nitrate to nitrite was observed in one serum bottle. Nitrate reduction was also observed in the second serum bottle after 15 days of incubation, but the reduction rate was slow, indicating that 40 °C is the borderline between activity and non-activity. For the cultures incubated at 45 °C, nitrate reduction was observed after 3 days of a lag period. Nitrite reduction was complete within 15 days of incubation in both serum bottles (Fig 4.1). At the end of a 20-day incubation period, all nitrate was reduced to N_2O. For the cultures incubated at 50 °C, nitrate reduction started after 1 day of lag period and complete nitrite reduction required approximately 13 days for both serum bottles. At the end of the 20-day incubation period, all nitrate was reduced to N_2O. For the cultures incubated at 60 °C, immediate nitrate and nitrite reduction took place in both cultures. At the end of the 20-day incubation period,

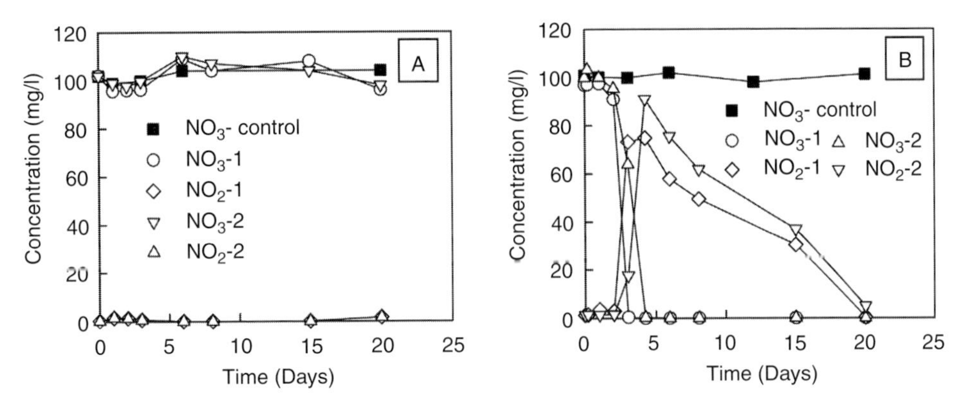

Fig 4.1 Time course of nitrate and nitrite during the denitrification assay with the *G. thermoleovorans* T80 culture at 35 °C (A) and 45 °C (B); 1 and 2 denote replicate 1, 2, respectively. Reproduced with permission from Marchant et al. (2008).

all nitrate was reduced to N_2O. After 15 days of incubation, N_2O reduction to N_2 was observed, but the reduction rate was very low.

The results from the denitrification study gave parallel results to those previously reported for the aerobic activity of *Geobacillus*, i.e. there is no discernable activity at temperatures below about 40 °C and certainly none at temperatures close to normal soil temperatures.

The growth experiments with geobacilli both in aerobic and anaerobic conditions were limited by being carried out in a rich laboratory medium using a pure culture as inoculum, thus not taking into account unculturable geobacilli and the soil environment. To overcome this limitation, the use of microcosm experiments and molecular techniques were necessary (Marchant et al., 2008). The Fluorescence *In Situ* Hybridisation Technique (FISH) was chosen as the tool to quantify the total active geobacilli directly in environmental samples. The FISH technique uses short fluorescence-labelled oligonucleotides that specifically hybridise with the rRNA allowing microscopic discrimination of the different taxa. The fluorescence signal associated with each bacterium reveals both the presence and the activity of the organisms, since a high number of ribosomes are necessary within the cell to provide the signal. Soil collected in Northern Ireland was used in microscosm experiments both with and without the addition of an inoculum of *G. thermoleovorans* T80.

Ten grams of soil collected in Northern Ireland were placed at different temperatures: external ambient; 25 °C; 37 °C; 46 °C and 60 °C. In another set of microcosms 10 g of soil, spiked with 1 ml of washed *G. thermoleovorans* T80 culture in PBS buffer (initial OD 600 nm = 0.3) was placed at the same temperatures. After 15 days, 1 g of soil was sampled at each temperature and FISH analysis was carried out. Separation of microorganisms from the soil matrix was achieved according to a published protocol (Caracciolo et al., 2005). FISH analyses were carried out on cut filter pieces as previously described (Pernthaler et al., 2001). The oligonucleotide probes used had the following sequences: EUB338 (Amann et al., 1990) – GCT GCC TCC CGT AGG AGT Fluorochrome 5′ Fluorescein targeting bacteria and GEOB Tbcil832 (Harmsen et al., 1997) – GGG TGT GAC CCC TCT AAC Fluorochrome 5′ Cy3, targeting *Geobacillus* spp. Images were captured using a Nikon ECLIPSE E 400 epifluorescence microscope. The estimation of microorganisms that bound the probes was determined in at least five randomly selected fields. Using the corrected dilution factor ($2 \times 8.22 \times 10^5$), the results were referred to 1 g of soil. Figure 4.2 reports the results of counting total bacteria (EUB probe) and geobacilli (GEOB probe) at different temperatures.

In the unspiked soil the only temperature at which geobacilli are detectable is 60 °C both after 15 days and 4 months. At this temperature 0.9×10^6 active geobacilli/g of soil were detected after 15 days while 0.7×10^6/g of soil were still present after 4 months representing 6% and 9% of the total active bacterial community, respectively. These data indicated that at this temperature geobacilli in the soil are metabolically active but they do not effectively grow, thus indicating that 10^6 is a good estimate of the presence of total (culturable and unculturable) geobacilli in one gram of soil sample. In the T80 spiked microcosms, after 15 days geobacilli were detected at all tested temperatures; this can be explained by the hypothesis that *G. thermoleovorans* is able to maintain a certain number

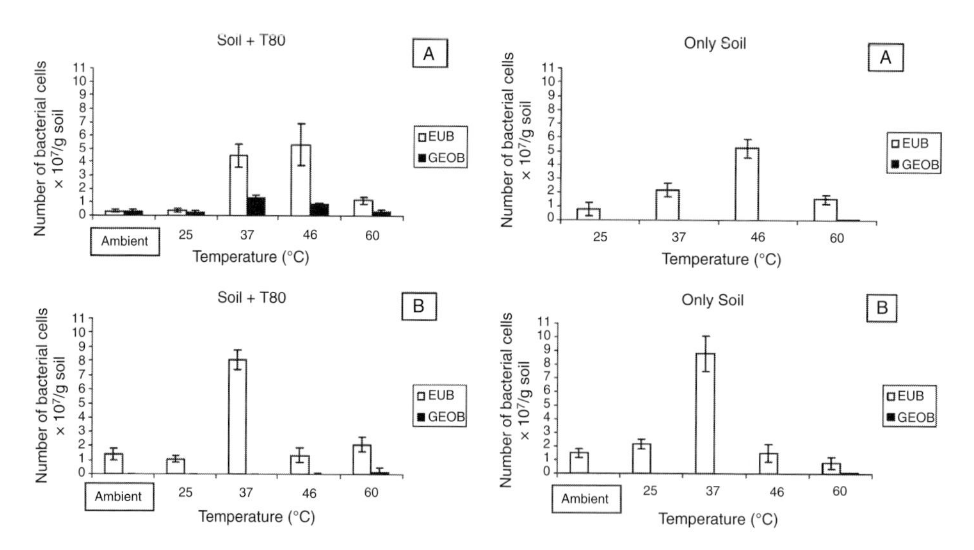

Fig 4.2 Counts of FISH-stained cells using a bacterial probe (EUB) and a specific *Geobacillus* probe (GEOB) in soil microcosms after 15 days (A) and 4 months (B). Graphs on the left represent the microcosms spiked with *G. thermoleovorans* T80, while the graphs on the right the microcosms without addition. Reproduced with permission from Marchant et al. (2008).

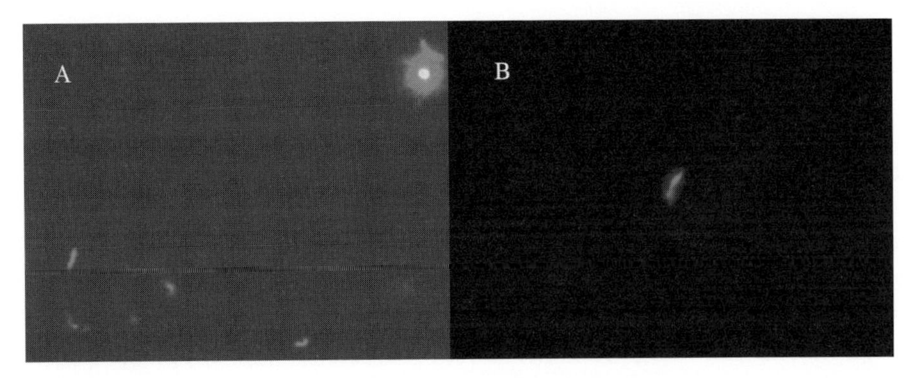

Fig 4.3 Bacterial cell using a *Geobacillus* (GEOB) probe in T80 spiked microcosm at 46 °C (A) and 60 °C (B).

of ribosomes over the first two weeks at a wide range of temperatures. After 4 months geobacilli were countable only at 46 °C and 60 °C reaching a value of 1.3×10^6/g of soil at 60 °C which is comparable with the unspiked microcosms. Figure 4.3 shows *Geobacillus* cells detected by GEOB probe in T80 spiked microcosms at 46 °C and 60 °C.

FISH experiments confirmed the relationship between growth and temperature obtained in liquid growth experiments both in aerobic and anaerobic conditions. In the soil total thermophilic geobacilli seem to be inactive below 40 °C, at least as detected by FISH technique.

Another proof of this behaviour under *in situ* conditions has been obtained by analysing the expression of the alkane mono-oxygenase gene (*alk*B) by thermophilic soil bacteria (Marchant et al., 2006). Many geobacilli and particularly *G. thermoleovorans* are well-known hydrocarbon degraders with high similarity with mesophilic bacteria for the sequences of one of the key enzymes for alkane biodegradation (i.e. *alk*B: alkane-1-mono-oxygenase). In this study, soil microcosms were established in glass universals using glass wool as an interface between 1 ml n-hexadecane and 10 g of soil. The microcosms were then incubated for up to 14 days at room temperature, and 25, 30, 37 and 55 °C. A further soil microcosm was also inserted *in situ* just below the surface soil layer in the natural environment for the same period of time. Control microcosms containing no alkane were also prepared. On soil samples, RT-PCR analyses were performed.

A positive detection of the expression of *alk*B gene by RT-PCR was detected only in soil samples kept for some days at least at 55 °C in the presence of the hydrocarbon. It was also demonstrated that *alk*B production did not originate from mesophilic microorganisms. The same results were obtained using pure cultures of *G. thermoleovorans* T80. These results highlighted the specificity of thermophiles for the degradation of hydrocarbons. It is worth noting that in soil samples taken from a cool region that had never been contaminated with hydrocarbons, raising the temperature is the condition in which the expression of *alk*B gene is induced in the microbial community in the presence of added hydrocarbons. This property was further investigated as an innovative bioremediation alternative (Perfumo et al., 2007). In this study, soil samples were collected in an undisturbed area in Northern Ireland unpolluted with hydrocarbons. Microcosm experiments were prepared in sterile bottles containing 5 g of soil with 2% (v/w) n-hexadecane added. Five different bioremediation techniques were tested: (1) degradation potential by indigenous microorganisms (natural attenuation); (2) supplementation with inorganic nutrients (Biostimulation); (3) supplementation with microbial surfactants (Biosolubilisation); (4) supplementation with selected hydrocarbon-degrading bacteria (Bioaugmentation); and (5) Biostimulation + Biosolubilisation + Bioaugmentation. Sterile soil with no amendments was used as an abiotic control. Microcosms were incubated in the dark at 60 °C and at room temperature (approximately 18 °C), and monitored at 0, 5, 15, 30 and 40 d for the estimation of hydrocarbon content by gas chromatography and the bacterial populations. Results showed that for all the conditions the biodegradation of the hydrocarbon was almost doubled at 60 °C compared with room temperature. It is already well established that one of the limiting factors in bioremediation is the bioavailability of the contaminants. Particularly, for slightly soluble compounds, such as hydrocarbons, the rate of mass transfer to cells limits the overall biodegradation rate (Boopathy, 2000). Increasing temperature leads to a decreased viscosity, higher solubility and faster diffusion of hydrophobic contaminants to the cell thus enhancing the biodegradation rates.

The results reported above concerning growth and activity of thermophiles at different temperatures seem to demonstrate unequivocally that no growth and activity can be postulated for thermophilic bacteria in cool soils. However, the most unexpected and important results came from simple long-term pure culture growth experiments in liquid medium. Sealed replicate universal tubes containing 10 ml of nutrient broth were inoculated with *G. thermoleovorans* and incubated at 4 °C, 25 °C, 37 °C, 46 °C and 60 °C for 9 months. The tubes were observed periodically for visible signs of growth. Tubes incubated at a temperature of 40 °C rapidly showed evidence of growth; even after 9 months, the tubes incubated at 25 °C showed no visible evidence of growth, but once transferred to 60 °C showed visible growth after 24 h indicating that viable thermophile cells remained in the culture. Surprisingly, after 9 months, the tubes maintained at 4 °C showed visible evidence of extensive growth. To avoid misinterpretation of the results due to contamination by psychrophilic bacteria, subcultures were taken from these tubes and incubated at 70 °C, showing that the growth was that of a thermophile. Furthermore microscopic observation and 16S rRNA sequencing confirmed that the organism was identical to the original inoculum.

4.4 Transport mechanisms and potential sources

The establishment of no or low activity of thermophiles in cool soils led to the speculation that they could be transported from warmer places in which these organisms are better able to grow and divide. Two potential sources for thermophiles were hypothesised: a short-range transport by wind from the local environment in which high temperature conditions can transiently occur (heating facilities, composting plants ...) and a long-range transport by clouds and rainwater from warmer geographic regions. Both quantitative and qualitative analyses of the bacterial populations in rainwater and air in Northern Ireland were carried out (Marchant et al., 2008). Air samplings were carried out from 6 until 30 October 2005 both for counting and isolation of thermophilic microorganisms and weather conditions were recorded. Fifty rainwater samples for thermophilic bacteria counting were collected from 1 February until 14 May 2004, while eight samples for isolation were collected from 10 October until 2 November 2005. Wind direction, wind speed and rainfall were further recorded during samplings. Fourteen rainwater and fourteen airborne microorganisms were isolated that were able to grow at 70 °C. Partial 16S rDNA sequences (at least 900 bases) were determined to assess the microbial communities of thermophilic bacteria in air and in rainwater and compare them with the already published characterisation of the thermophilic population in soil (Marchant et al., 2002b). Figure 4.4 shows the phylogenetic tree built with the sequences of all the isolates and the sequences of some type strains that showed more than 97% sequence similarity with the isolates. In the rainwater community, all 14 isolates are assigned to *Geobacillus*; particularly, 12 of them have, as nearest phylogenetic neighbours, *Geobacillus thermodenitrificans* DSM 465T and *Geobacillus subterraneus* T34T while two stand very close to *Geobacillus stearothermophilus* DSM 22TT, *Geobacillus thermocatenulatus* DSM 730T, *Geobacillus vulcani* 3S-1T (Nazina et al., 2004), *Geobacillus kaustophilus* NCIMB 8547T and *Geobacillus thermoleovorans* DSM 5366T.

In the air community, seven isolates are assigned to *Bacillus*, one to *Ureibacillus* and six to *Geobacillus*. Two air isolates (A9.11 and A9.13) showed very high similarity (> 99%) with a thermophilic environmental isolate submitted as *Bacillus aestuarii* (GenBank: AB062696) that was included in the tree. Considering all isolates, four Operational Taxonomic Units (OTUs) have been defined (each OTU comprised sequences that shared > 97% sequence identity) by distance analysis. The two communities showed very different distributions of their isolates along the OTUs. All the isolates of the rainwater community belong to OTU 1, while in the air community, six isolates belong to OTU 1, one to OTU 2, three to OTU 3 and four to OTU 4. Furthermore, only in the air samples were three microorganisms isolated able to grow at 60 °C but not at 70 °C, which were morphologically

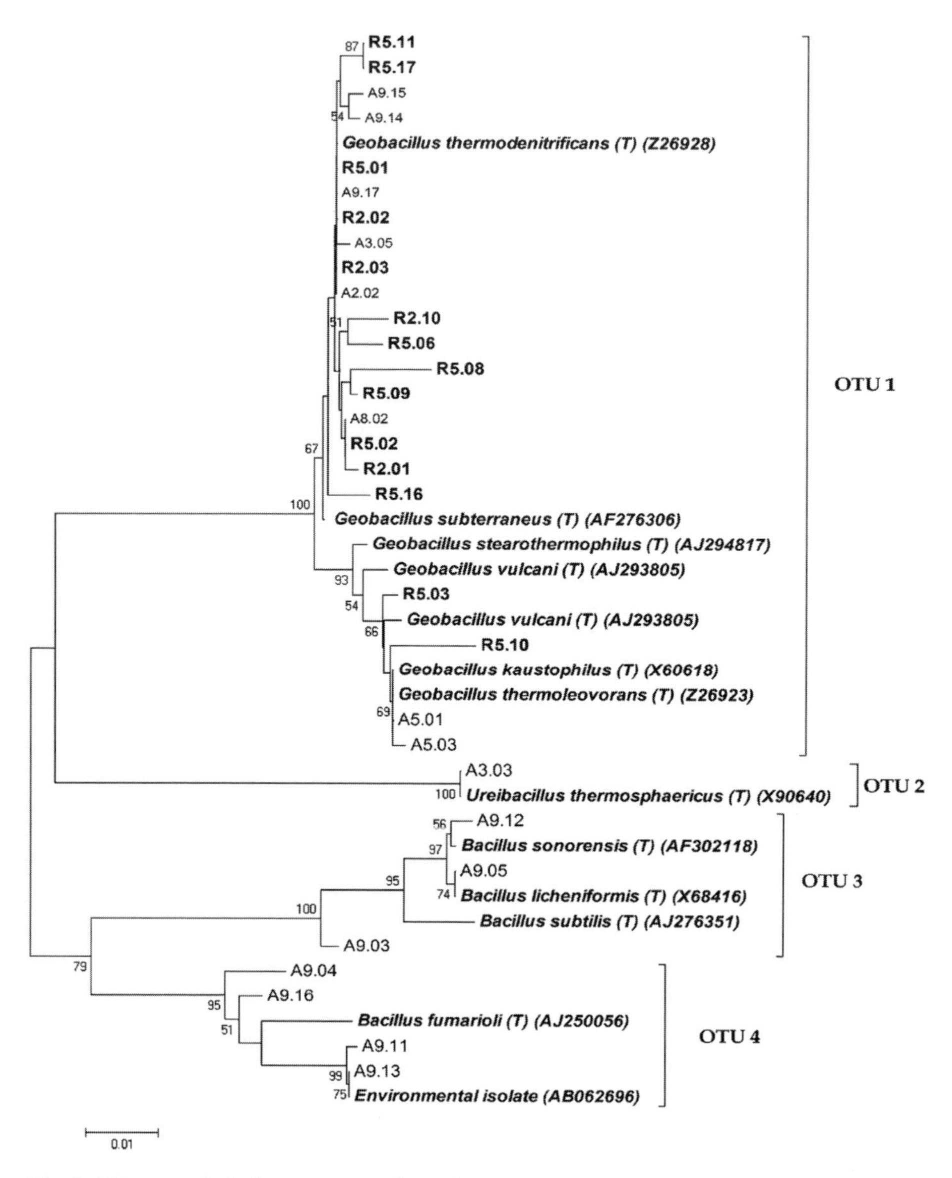

Fig 4.4 Unrooted phylogenetic tree based on 16S rRNA gene comparison showing the position of rainwater (bold) and air isolates and the type strains most closely related (italic). Bootstrap probability values less than 50% were omitted from the figure. The scale bar indicates substitutions per nucleotide position. The GenBank accession numbers of type strains are in parentheses. Reproduced with permission from Marchant et al. (2008).

identified as thermotolerant *Actinomycetales*. This distribution led to the conclusion that the rainwater community of thermophilic bacteria is characterised by a lower biodiversity than the air one. To confirm this lack of biodiversity in rainwater and to avoid any influence due to seasonality, another two rainwater samplings were carried out in January 2006. Ten microorganisms were isolated and 8 showed very high similarity with *Geobacillus thermodenitrificans* DSM 465T while two isolates shared > 99% sequence similarity with *Ureibacillus thermosphericus* P-11T. The structure of the rainwater community seems to be more similar to the structure of the soil community than does the air community suggesting that rainwater deposition could be a mechanism that sustains the thermophilic community in these soils.

For quantitative analyses, thermophilic bacteria were determined by the Most Probable Number (MPN) technique at 70 °C in liquid medium, while airborne thermophiles were quantified by direct sampling and growth at 70 °C on agar-rich medium. For air samplings a mean value of 1.55 cfu/1000 l of air was obtained. For rainwater counting, thermophilic bacteria were found in detectable numbers only in nine samples. The mean value of these nine samples was 8.5 cells/100 ml, while considering the whole amount of sampled water, the mean was 1.1 cells/100 ml. Furthermore, from rainfall data and MPN counting, it is possible to calculate the total number of thermophilic microorganisms that have been deposited onto 1 m^2 of soil during the sampling period; a value of 9.5×10^3, or an average of 140 for each millimetre of rainfall. Considering an average rainfall value in this part of Ireland of 1000 mm per year, the total annual input of thermophilic microorganisms from rainwater to soil could be estimated as $1.4 \times 10^5/m^2$ of soil surface. Marchant et al. (2002a) found values of thermophiles for this area ranging from $1.5–8.8 \times 10^4$ cfu/g soil at 50 mm. Also considering 50 mm as the maximum depth affected by rainwater deposition the number of thermophiles on 1 m^2 of soil ranges from 0.75 to 4.4×10^9 cfu/m^2. This means that only a small fraction ($0.3–2 \times 10^{-5}$) of thermophiles can be replaced yearly by rainwater deposition. Therefore, assuming the rainwater deposition as the only source of thermophiles, it would not be possible to maintain a high and constant viable population of thermophiles without their growth or survival in the soil.

Although the studies of thermophilic geobacilli in rainwater provided an important insight into the origin of these organisms in cool soils they still did not identify the ultimate origin of the microorganisms. Since the advent of satellite imaging it has become apparent that considerable global transport of dust occurs and that one of the major origins for these dust storms is the Sahara/Sahel region of Africa (Griffin et al., 2002). Dust from this region travels northwards over the Mediterranean countries where it is often deposited in a dry form. Further movement then takes the dust over northern Europe and across the Atlantic to the Caribbean and southern USA. Dry dust deposition does not occur in northern Europe but the dust is precipitated in rainfall. To test the hypothesis that the ultimate origin of thermophilic geobacilli in rainfall in northern Europe is Saharan dust storms, samples of dry dust were collected and were examined for the presence of thermophilic geobacilli using culture methods (Perfumo and Marchant, 2010). Samples of dust collected in Turkey and Greece following two distinct desert storm events contained viable thermophilic organisms of the genus *Geobacillus*, namely *G. thermoglucosidasius* and *G. thermodenitrificans*, and the recently reclassified *Aeribacillus pallidus* (formerly *Geobacillus pallidus*). These results provided evidence that African dust storms create an atmospheric bridge between distant geographic regions and that they are also probably the source of thermophilic geobacilli later deposited over northern Europe by rainfall or dust plumes themselves. The same organisms (99% similarity in the 16S rDNA sequence) were found in dust collected in the Mediterranean region and inhabiting cool soils in Northern Ireland (Perfumo and Marchant, 2010).

4.5 Conclusions

In recent years, the Baas Becking hypothesis has provoked intense discussions and some investigations have been carried out to verify it. A recent review has drawn attention to the two components of biogeographic distribution, the province and the habitat (Hughes Martiny et al., 2006). The province represents the legacy of historical events while the habitat represents the existing environment for the organism. This review identified four possible hypotheses to describe microbial biogeography patterns and attempted to differentiate between the possible alternatives with the aim of evaluating whether the biogeographic pattern of macroorganisms can be applied to microorganisms. The first alternative is that microorganisms are randomly distributed over space while in a second hypothesis only environmental conditions affect the geographic distribution of microorganisms. The latter is the Baas Becking hypothesis: 'everything is everywhere, but the environment selects'. In the third hypothesis, the present distribution of microorganisms is influenced only by historical events, while the fourth one allows both legacy of past historical events and present environmental conditions to influence the biogeography of microorganisms. The results of some investigations led the authors to conclude that the environment actually selects and shares with the legacy of historical events the responsibility for the spatial variation of microorganisms. However, as reported by the authors of the review, the idea that all the attributes of the organism that can potentially influence their spatial distribution can be described allometrically does not capture the complexity of the microbial world. Dispersal, colonisation, extinction and diversification are processes that shape the biogeography of the organisms and the wide metabolic diversity among Bacteria, Archaea and microscopic Eukarya makes it difficult to find a simple correlation between size and the rate of these processes. For instance, the dispersal rate is strongly influenced by tolerance to the extreme environmental conditions that microorganisms experience during passive transport. Spore-forming bacteria have an advantage in being transported for long distances. Thus, the spatial dispersal of microorganisms depends not only on their size, but also on their specific attributes.

The results reported in this chapter report the presence and activity of bacteria in an environment that potentially does not allow their growth due to the low temperatures. Since it has been demonstrated that *Geobacilli* are extremely resistant to space vacuum, UV radiation and gamma-ray exposure and are the only microorganisms selected from hot environments under these extreme conditions (Saffary et al., 2002), it can be hypothesised that members of the genus *Geobacillus* are present wordwide due to this resistance to stress. In this sense, the assumptions of the Baas Becking hypothesis hold true for this kind of bacterium. Moreover, for geobacilli the selection of the cold environments seems not to hold since at least a slow growth is supposed to sustain their large population in cold soils. This growth is probably explained by the balance of growth and death rate at low temperatures. In geobacilli, at temperatures above 40 °C, the growth rate exceeds the death rate, and this differential increases up to 60 °C; at temperatures between ambient and 37 °C, the death rate is higher than the growth rate, preventing any increase in biomass in freshly inoculated cultures (Pavlostathis et al., 2006). Thus, it can be supposed that the growth at 4°C over protracted time periods can be explained by a low growth rate but an even lower death rate. These specific features make members of the genus *Geobacillus* cosmopolitan microorganisms.

References

Amann, R.I., Binder, B.J., Olson, R.J. et al. (1990). Combination of 16S rRNA-targeted oligonucleotide probes with flow cytometry for analyzing mixed microbial populations. *Applied Environmental Microbiology* **56**, 1919–1925.

Banat, I.M., Marchant, R., Rahman, T.J. (2004). *Geobacillus debilis* sp. nov., a novel obligately thermophilic bacterium isolated from a cool soil environment, and reassignment of *Bacillus pallidus* to *Geobacillus pallidus* comb. nov. *International Journal of Systematics and Evolutionary Microbiology* **54**, 2197–2201.

Boopathy, R. (2000). Factors limiting bioremediation technologies. *Bioresource Technology* **74**, 63–67.

Caracciolo, A.B., Grenni, P., Cupo, C., Rossetti, S. (2005). *In situ* analysis of native microbial communities in complex samples with high particulate loads. *FEMS Microbiology Letters* **253**, 55–58.

Fenchel, T., Esteban, G.F., Finlay, B.J. (1997). Local versus global diversity of microorganisms: cryptic diversity of ciliated protozoa. *Oikos* **80**, 220–225.

Griffin, D.W., Kellogg, C.A., Garrison, V.H., Shinn, E.A. (2002). The global transport of dust. *American Scientist* **90**, 228–235.

Harmsen, H., Prieur, D., Jeanthon, C. (1997). Group-specific 16S rRNA-targeted oligonucleotide probes to identify thermophilic bacteria in marine hydrothermal vents. *Applied and Environmental Microbiology* **63**, 4061–4068.

Hughes Martiny, J.B., Bohannan, B.J.M., Brown, J.H. et al. (2006). Microbial biogeography: putting microorganisms on the map. *Nature Reviews Microbiology* **4**, 102–112.

Marchant, R., Banat, I.M., Rahman, T.J., Berzano, M. (2002a). The frequency and characteristics of highly thermophilic bacteria in cool soil environments. *Environmental Microbiology* **4**, 595–602.

Marchant, R., Banat, I.M., Rahman, T.J., Berzano, M. (2002b). What are high temperature bacteria doing in cold environments? *Trends in Microbiology* **10**, 120–121.

Marchant, R., Sharkey, F.H., Banat, I.M., Rahman, T.J., Perfumo, A. (2006). The degradation of n-hexadecane in soil by thermophilic geobacilli. *FEMS Microbiology Ecology* **56**, 44–54.

Marchant, R., Franzetti, A., Pavlostathis, S.G. et al. (2008). Thermophilic bacteria in cool temperate soils: are they metabolically active or continually added by global atmospheric transport? *Applied Microbiology and Biotechnology* **78**, 841–852.

Nazina, T.N., Tourova, T.P., Poltaraus, A.B. et al. (2001). Taxonomic study of aerobic thermophilic bacilli: descriptions of *Geobacillus subterraneus* gen. nov., sp. nov. and *Geobacillus uzenensis* sp. nov. from petroleum reservoirs and transfer of *Bacillus stearothermophilus, Bacillus thermocatenulatus, Bacillus thermoleovorans, Bacillus kaustophilus, Bacillus thermoglucosidasius and Bacillus thermodenitrificans* to *Geobacillus* as the new combinations *G. stearothermophilus, G. thermocatenulatus, G. thermoleovorans, G. kaustophilus, G. thermoglucosidasius* and *G. thermodenitrificans. International Journal of Systematics and Evolutionary Microbiology* **51**, 433–446.

Nazina, T.N., Lebedeva, E.V., Poltaraus, A.B. et al. (2004). *Geobacillus gargensis* sp. nov., a novel thermophile from a hot spring, and the reclassification of *Bacillus vulcani* as *Geobacillus vulcani* comb. nov. *International Journal of Systematics and Evolutionary Microbiology* **54**, 2019–2024.

Nazina, T.N., Sokolova, D.S., Grigoryan, A.A. et al. (2005). *Geobacillus jurassicus* sp. nov., a new thermophilic bacterium isolated from a high-temperature petroleum reservoir, and the validation of the *Geobacillus species. Systematic and Applied Microbiology* **28**, 43–53.

O'Malley, M.A. (2008). Everything is everywhere: but the environment selects: ubiquitous distribution and ecological determinism in microbial biogeography. *Studies in History and Philosophy of Biological and Biomedical Sciences* **39**, 314–325.

Pavlostathis, S.G., Marchant, R., Banat, I.M., Ternan, N.G., McMullan, G. (2006). High growth rate and substrate exhaustion results in rapid cell death and lysis in the thermophilic bacterium *Geobacillus thermoleovorans. Biotechnology and Bioengineering* **95**, 84–95.

Perfumo, A., Marchant, R. (2010). Global transport of thermophilic bacteria in atmospheric dust. *Environmental Microbiology Reports* **2**, 333–339.

Perfumo, A., Banat, I.M., Marchant, R., Vezzulli, L. (2007). Thermally enhanced approaches for bioremediation of hydrocarbon-contaminated soils. *Chemosphere* **66**, 179–184.

Pernthaler, J., Glöckner, F-O., Schönhuber, W., Amann, R. (2001). Fluorescence *in situ* hybridization (FISH) with rRNA-targeted oligonucleotide probes. *Methods in Microbiology* **30**, 207–226.

Rahman, T.J., Marchant, R., Banat, I.M. (2004). Distribution and molecular investigation of highly thermophilic bacteria associated with cool soil environments. *Biochemical Society Transactions* **32**, 209–213.

Saffary, R., Nandakumar, R., Spencer, D. et al. (2002). Microbial survival of space vaccum and extreme ultraviolet irradiation: strain isolation and analysis during a rocket flight. *FEMS Microbiology Letters* **215**, 163–168.

Sung, M.-H., Kim, H., Bae, J.-W. et al. (2002). *Geobacillus toebii* sp nov., a novel thermophilic bacterium isolated from hay compost. *International Journal of Systematics and Evolutionary Microbiology* **52**, 2251–2255.

Part III

Unicellular eukaryotes

5

Dispersal of protists: the role of cysts and human introductions

WILHELM FOISSNER

FB Organismische Biologie, Universität Salzburg, Salzburg, Austria

5.1 Introduction

While the distribution of flowering plants and larger animals is easy to determine, this is almost impossible in microorganisms, which are smaller than human beings by a factor of 1.8×10^6, assuming an average size of 100 μm and 180 cm, respectively. Thus, the subject has been searched with varied success and in heated debates (Foissner, 2004; Fenchel and Finlay, 2005), resulting in two hypotheses: the 'cosmopolitan model' (Finlay, 2002; Finlay et al., 2004) and the 'moderate endemicity model', which suggests that one-third of protists has restricted distribution (Foissner, 1999, 2006, 2008). The cosmopolitan model is based on ecological theory, while the moderate endemicity model emphasises flagship species which are so showy, or so novel, that it is unlikely that they would be overlooked if indeed they were widely distributed (Tyler, 1996). The debate has stimulated many investigations whose conclusions frequently read as follows (Bass et al., 2007): 'Our results strongly suggest that geographic dispersal in macroorganisms and microbes is not fundamentally different: some taxa show restricted and/or patchy distributions while others are clearly cosmopolitan.

Biogeography of Microscopic Organisms: Is Everything Small Everywhere?, ed. Diego Fontaneto. Published by Cambridge University Press.

These results are concordant with the 'moderate endemicity model' of microbial biogeography. Rare or continentally endemic microbes may be ecologically significant and potentially of conservational concern. We also demonstrate that strains with identical 18S but different ITS1 rDNA sequences can differ significantly in terms of morphological and important physiological characteristics, providing strong additional support for global protist biodiversity being significantly higher than previously thought.'

Thus, there is hardly any need to enlarge this subject again (for recent reviews, see Dolan, 2005; Martiny et al., 2006; Foissner, 2006, 2008; Caron, 2009). In contrast, little attention has been paid to the reasons why certain species are cosmopolitan and others are not, as evident from the reviews just cited. Wilkinson (2001) and Smith et al. (2008) suggested size and/or air currents as important dispersal factors. However, this has been abandoned by Foissner (2008). He emphasised that microfungi, mushrooms, mosses and ferns are not cosmopolitan although their dispersal means, the spores, are very abundant and usually less than 100 μm in size, corresponding to the trophic and cystic size of most protists (but see Chapters 8–12).

Thus, the reasons for cosmopolitan or restricted distribution must be different. In my opinion, the most important factors are the resting cysts, the geological history and human introductions. The overwhelming structural and chemical diversity of resting cysts becomes meaningful if one considers cysts not only as a simple dormant stage but as dispersal means. This has been widely neglected in the 'everything is everywhere' debate, and thus I shall devote half of this review to the demonstration of cyst diversity, hoping to revive cystology (Gutiérrez and Walker, 1983). The second important factor is the break of Pangaea into Laurasia and Gondwana about 120 million years ago. This has been discussed in several reviews (Foissner, 2006; Smith et al., 2008) and is thus excluded from the present one. The third main factor is human introductions, frequently underweighted by protist biogeographers (Foissner, 2006).

5.2 Dispersal of protists

I recognise four main routes: dispersal in active (non-encysted) state; dispersal by protective resting cysts; dispersal by humans; and dispersal by geological processes, especially the break of Pangaea and continental drift. As mentioned in section 5.1, the palaeobiogeographic route is not treated here.

5.2.1 Dispersal in active state

Usually, live protists are very fragile. Thus, it is reasonable that dispersal occurs mainly in the cystic state (see section 5.2.2). Nonetheless, dispersal in the active state is possibly also rather common, especially in marine environments, where large water currents might disperse species over large areas or even globally. However, benthic and planktonic foraminifera have distinct areals (Darling and Wade, 2008; Pawlowski and Holzman, 2008), just as other marine protists, for instance, the coccolithophores (Winter et al., 1994).

On land, step-by-step dispersal might be of considerable significance, especially in euryoecious species and at local, regional and continental scales (Green and Bohannan, 2006). Many experiments show that new habitats are often colonised within a few weeks (for reviews, see Maguire, 1963, 1971). Unfortunately, species have been rarely identified, for instance, by Wanner and Dunger (2001) and Meisterfeld (1997), who studied testacean communities from reforested opencast mining sites. Colonisation was fast, but only euryoecious species developed, and most humus-specific species disappeared within a year at a site that was amended with humus from a primary forest to stimulate succession. This matches my (unpublished) observations on ciliates. Only nine euryoecious species colonised three small, artificial ponds within a year, in spite of excess food (for details, see legend to Fig 5.1), and few freshwater ciliate species survived when added to soil (Foissner, 1987, table 15).

Some of the most widespread ciliates, e.g. *Glaucoma scintillans*, *Colpidium colpoda* and species of the *Paramecium aurelia* and the *Tetrahymena pyriformis*-complex very likely lack the ability to produce resting cysts, although they have

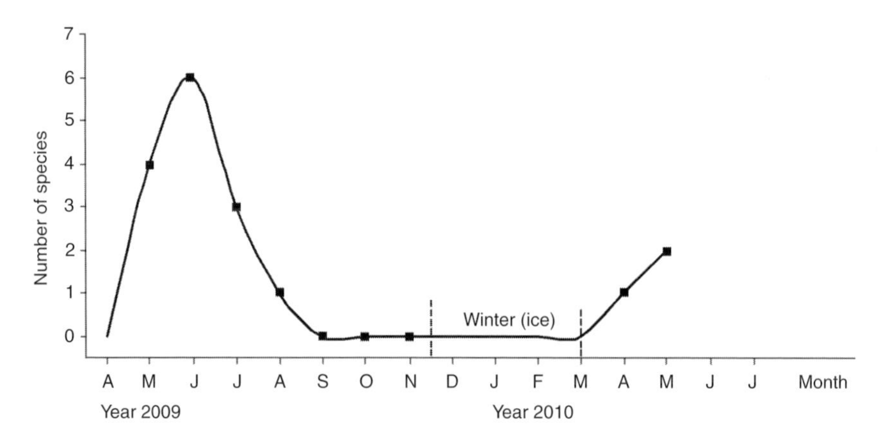

Fig 5.1 Number of ciliate species developing in three artificial ponds containing 1.5 l, 6 l and 12 l tap water and 0.1 g, 0.4 g and 0.8 g porridge oats. The experiment started on 1 April 2009 and is still running. A detailed description will be published later. Altogether, nine species were recognised: *Apocyclidium terricola, Chilodonella uncinata, Colpoda inflata, Epistylis opercularia, Odontochlamys alpestris, Pseudochilodonopsis algivora, Stylonychia pustulata, Tetrahymena rostrata* and *Vorticella infusionum*. With the exception of *E. opercularia*, which is possibly an 'air ciliate', all species are common, euryoecious inhabitants occurring in both limnetic and terrestrial habitats. Note the disappearance of the ciliates during a bloom of various algae and cyanobacteria.

been reported in both *Paramecium* (for a review, see Wichterman, 1986) and *Tetrahymena pyriformis* (Nilsson, 2005). However, the evidence is not convincing and not supported by my data. I never found any *Paramecium* in over 1000 air-dried and then rewetted soil samples from a great variety of habitats globally, including soil from flood plains and the surface of dry, ephemeral puddles (Foissner, 1998; Chao et al., 2006). Likewise, I did not find any *Paramecium* in about 200 samples from tank bromeliads of Central and South America, although it occurred in rivers and streams nearby (Foissner, unpublished). Further, my own experiments with *G. scintillans* and *Colpidium kleinii* failed. Thus, I join that group of scientists who believes that certain *Paramecium* and *Tetrahymena* species cannot make protective resting cysts.

Certainly, cystless species are a challenge to all dispersal models, including my cyst theory. While wide dispersal in the active state could be possible in encased species or in species with a thick cortex, as in *Paramecium*, this appears unlikely for fragile species like *Tetrahymena* and *Glaucoma*. I speculate that some of these species, especially those with a wide ecological range, may have distributed step by step or are older than the break of Pangaea. Further, we cannot exclude that such species were originally able to perform anabiosis (anhydrobiosis), i.e. to dry up without forming a special cyst, and becoming viable again when water becomes available. Although anhydrobiosis is extremely rare in present-day ciliates (I know it from only one *Podophrya*-like suctorian ciliate), it might have been more common in certain developmental stages of the species millions of years ago.

As protists are very small and thus of low weight, it is widely believed that air currents and animal vectors are the main distribution agents (Maguire, 1963; Cowling, 1994; Hamilton and Lenton, 1998; Wilkinson, 2001; Smith et al., 2008). Wilkinson (2001) showed by a detailed analysis of Arctic and Antarctic testacean communities that only large species (> 150 μm) are possibly not cosmopolitan.

All the data and hypotheses reviewed above, and many more not mentioned, are in conflict with a simple fact (Foissner, 2006, 2008; Fig 5.2; see also Chapters 9–12): mushrooms, mosses, ferns, lichens and horsetails have restricted distributions although their distribution means (spores) are produced in masses and in the size of most protists (≤ 100 μm). Further, hundreds of bacterial and fungal pests had regional or continental distribution before they were dispersed by humans. This is why I believe that, for example, air currents and the size of the organisms have little influence on their distribution. This has been supported by a study on microscopic fungi (Taylor et al., 2006). Actually, we do not know the amount of stable populations established by dispersal in the active state. Based on the data discussed above, step-by-step distribution of both, in active and cystic states, may play a significant role in at least the euryoecious species and if many similar habitats occur in a certain region.

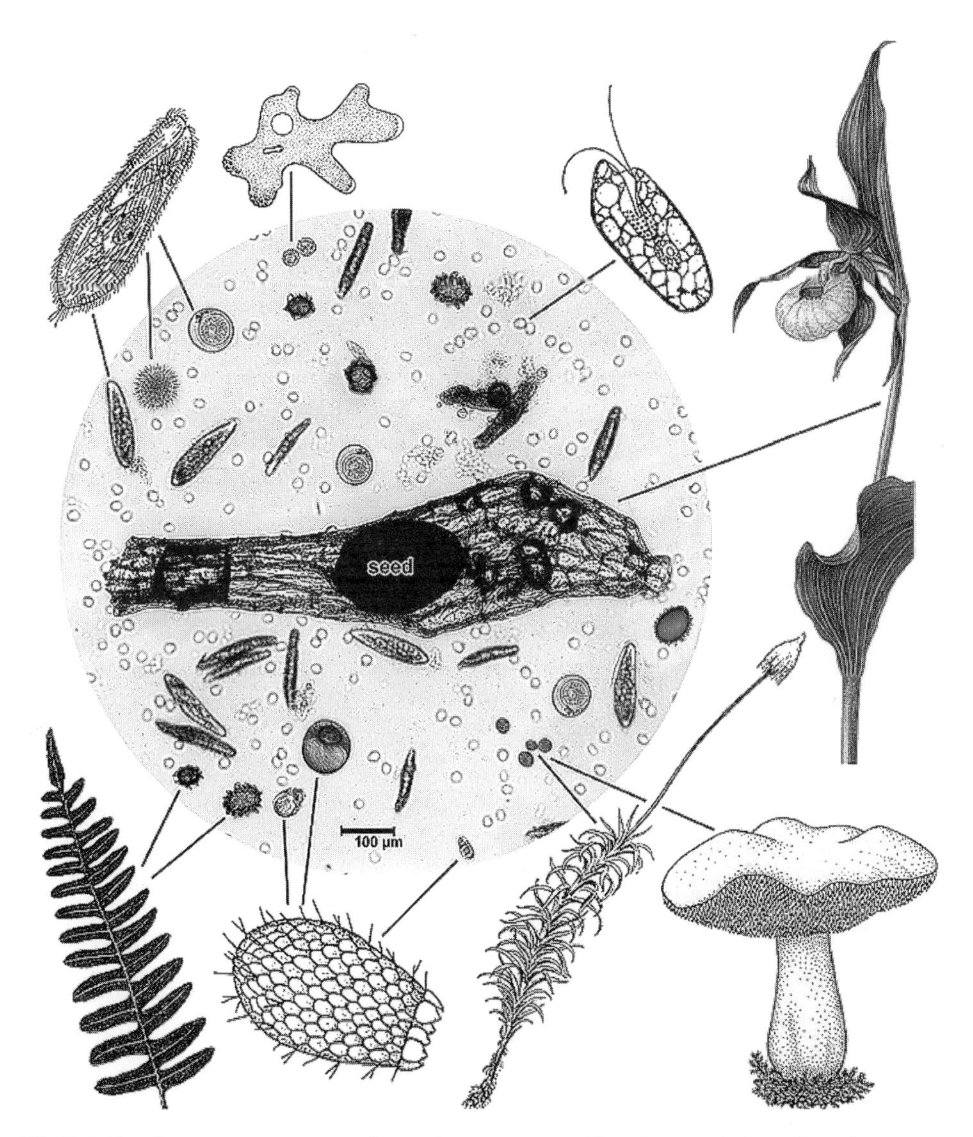

Fig 5.2 This figure compares, at about the same magnification, trophic and cystic protists (ciliates, flagellates, naked and testate amoebae) with spores of macrofungi (mushrooms), mosses, ferns and the minute seed of an orchid (*Vanda caerulescens*). Obviously, all are of minute size and very abundant, for instance, a single *Agaricus campestris* (mushroom) releases 1.6×10^{10} spores within 6 days (Webster, 1983), which exceeds the abundance of ciliates in 1 m^2 of forest soil by several orders of magnitude (Meyer et al., 1989). While nobody denies that mushrooms, mosses and ferns have biogeographies, protists are widely assumed to be cosmopolitan because their small size and high abundance favour air dispersal, an opinion flawed by this figure. Further, protist cysts lack adaptations for air dispersal, while seeds of many flowering plants have such adaptations, for instance, the orchid seed shown which has wings of large-sized, air-filled cells. Reproduced with permission from Foissner (2008).

5.2.2 Dispersal by resting cysts

Many protists can produce a dormant stage, named protective resting cyst, resting cyst (my preferred term), cyst, spore or stomatocyst, depending on the group under investigation. Resting cysts are widely assumed to be the major dispersal agents of unicellular organisms because they are much more stable than live cells (for reviews, see Corliss and Esser, 1974; Foissner, 1987; Gutiérrez and Martin-González, 2002). However, the biogeographic research and discussion ignored almost completely the very different morphological and physiological properties of resting cysts, depending on intrinsic (phylogenetic) and extrinsic (habitats s.l.) factors. Thus, I shall review here some recent studies, showing the overwhelming resting cyst diversity. For instance, the resting cyst of *Maryna umbrellata* is covered with glass granules, representing the first record of biomineralised silicon in ciliates (Foissner et al., 2009). It was just this diversity and some 'simple' observations reported below, which convinced me that cysts are possibly the most important factor for the dispersal of species (cosmopolitan or of restricted distribution) and for their presence/absence in a certain habitat, at a certain time, and under certain environmental conditions.

Unfortunately, our knowledge on the physiology, morphology and macromolecular composition of resting cysts is very limited. Thus, it is not yet possible to ascribe a certain function to the individual cyst layers. However, some general knowledge is available and has been reviewed by Corliss and Esser (1974), Foissner (1987, 2005, 2009), Gutiérrez and Martin-González (2002), Gutiérrez et al. (2003), and Foissner et al. (2005). Very briefly, a 'typical' cyst of a ciliate consists of a pericyst, an ectocyst, a mesocyst, an endocyst and, in certain taxa, a metacyst (Figs 5.11, 5.27). The chemical composition of these layers is known in only a few species (for an example, see Fig 5.11). Generally, acid mucopolysaccharides are frequent in the pericyst, while proteins, glycoproteins, glycogen and chitin are frequent in the mesocyst and endocyst. Unfortunately, the chemical composition of the ectocyst, which is often very thin, is unknown.

5.3 Resting cysts of ciliates from rain forests and hot deserts

Table 5.1 shows cyst survival of soil ciliates from rain forests in Borneo and Malaysia and from various habitats of Namibia, including the Namib Desert (Foissner et al., 2002). In the Namibian samples, there was no loss of species when the air-dried samples were stored for up to seven years, while most species of the rain forest soil could be not activated when the air-dried samples were older than a year, suggesting that the resting cysts died. Obviously, rain-forest ciliates have

Table 5.1 Ciliate species numbers in air-dried and rewetted[a] soil habitats of Namibia[b] and in rain forests of Borneo and Malay.

	Namibia		Rain forests	
Time elapsed since collection	**Number of species ($\bar{x}$)**	**Number of samples**	**Number of species ($\bar{x}$)**	**Number of samples**
≤ 10 h	None	Many	25.0	8
Up to 9 months	30.8	23	30.0	7
Up to 65 months	25.9	10	6.4	5
Up to 82 months	41.5	17	1.8	5

[a] Non-flooded Petri dish method as described in Foissner et al. (2002).

[b] Only 'typical' dry soil habitats were selected from the 73 samples investigated, viz., the samples: 1, 2, 4, 5, 7, 8, 9, 11, 12, 13, 16, 17, 18, 20, 23, 24, 26, 27, 29, 31, 32, 33, 35, 36, 37, 38, 39, 41, 42, 43, 44, 48, 49, 50, 53, 54, 56, 57, 58, 59, 60, 61, 62, 63, 64, 65, 67, 69, 70, 73.

'weak' cysts not adapted to long periods of dryness. Accordingly, they have little chance to disperse via cysts over large areas. This contrasts with the 'strong' cysts from Namibia. The meaning of 'weak' and 'strong' is demonstrated by *Exocolpoda augustini* (Figs 5.3–5.5) collected in Austria and the dry west coast of Namibia (Foissner et al., 2002: site 37). While ordinary cysts with a rather thin (0.5–1 μm) wall are produced under the moderate Austrian climate (Fig 5.5), the wall of the Namibian specimens is about 7 μm thick (Fig 5.4), surpassing the volume of the encysted cell proper three times (3500 μm^3 vs. 14 000 μm^3).

These data match observations from laboratory cultures, where frequently most of the cysts made do not excyst when fresh medium and food are added, and cells sometimes lose the ability to make cysts at all, especially on prolonged cultivation. This makes sense under the constant laboratory conditions, where the populations select for non-encysters or switch off the encystment genes. However, the matter is complex, i.e. both encystment and excystment are influenced by many factors, as shown for example by Meier-Tackmann (1982) and Meier-Tackmann and Wenzel (1988) in a common soil ciliate, *Colpoda cucullus*, and by Müller et al. (2006) in *Meseres corlissi*, a plankton ciliate from ephemeral fresh waters (Fig 5.22). The often highly varying, 'mysterious' encystment and excystment rates give support to the 'scout theory' of Epstein (2009). This theory suggests that microbial populations consist of a mix of active and dormant cells. Faced with an adverse environmental change, more cells are included into dormancy, and survive the challenge. Individual cells would then periodically exit dormancy as a result of infrequent and essentially random events, such as a change in the expression of a master regulatory gene. I call such awakened cells 'scouts'. If the adverse conditions persist, the scout dies. If a scout forms under

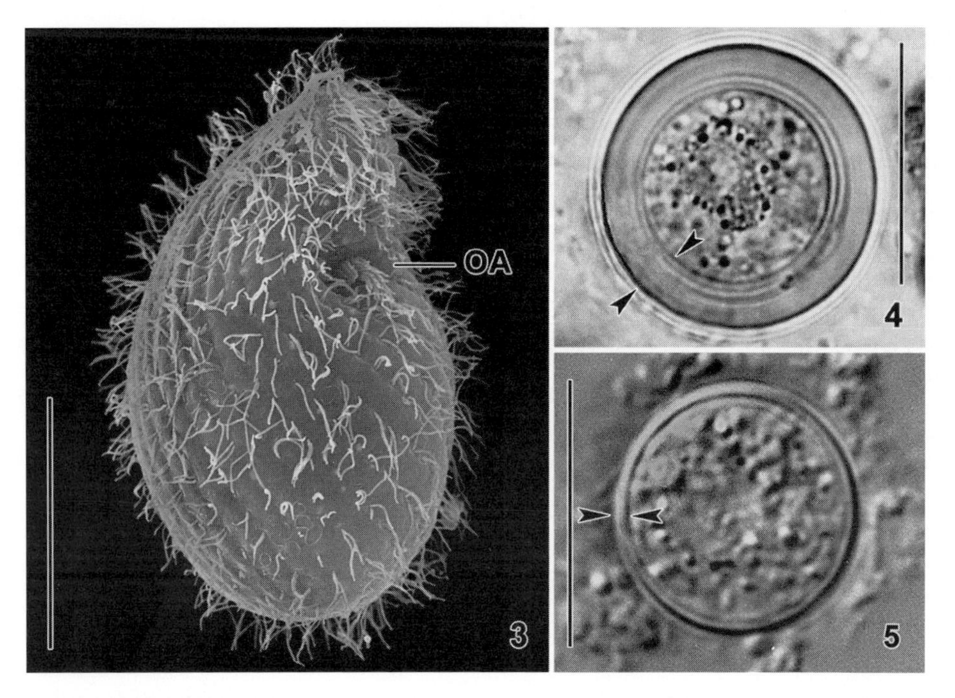

Figs 5.3–5.5 *Exocolpoda augustini*, a small, terrestrial ciliate (**3**), makes extremely thick-walled resting cysts (**4**, opposed arrowheads) in the Namib desert, but makes ordinary-walled cysts in Austria (**5**, opposed arrowheads). OA – entrance to oral apparatus. Scale bars 25 µm. From Foissner et al. (2002) and original (**5**).

growth-permissive conditions, it starts a new population. In some species, scouts might even use growth-inducing signalling compounds to wake up the rest of the dormant population.

5.4 Some remarkable ciliate resting cysts

5.4.1 *Maryna umbrellata* (Colpodea)

This mushroom-shaped ciliate is about 100 µm in size, and is common in ephemeral limnetic habitats, such as rock pools and meadow puddles (Figs 5.6–5.10). The globular resting cyst is conspicuous because it is as large as the trophic cell. *Maryna umbrellata* is restricted to the northern hemisphere. In Africa, Australia, Central America and South America occurs a similar but smaller species having larger (up to 5 µm) silicon granules.

The fine structure and chemical composition of the resting cyst of *M. umbrellata* were studied by Foissner (2009) and Foissner et al. (2009). This showed the following peculiarities: (i) the cyst wall is about 13 µm thick and thus amounts for half of the total cyst volume (Fig 5.8); (ii) the external, about 4 µm-thick layer is made

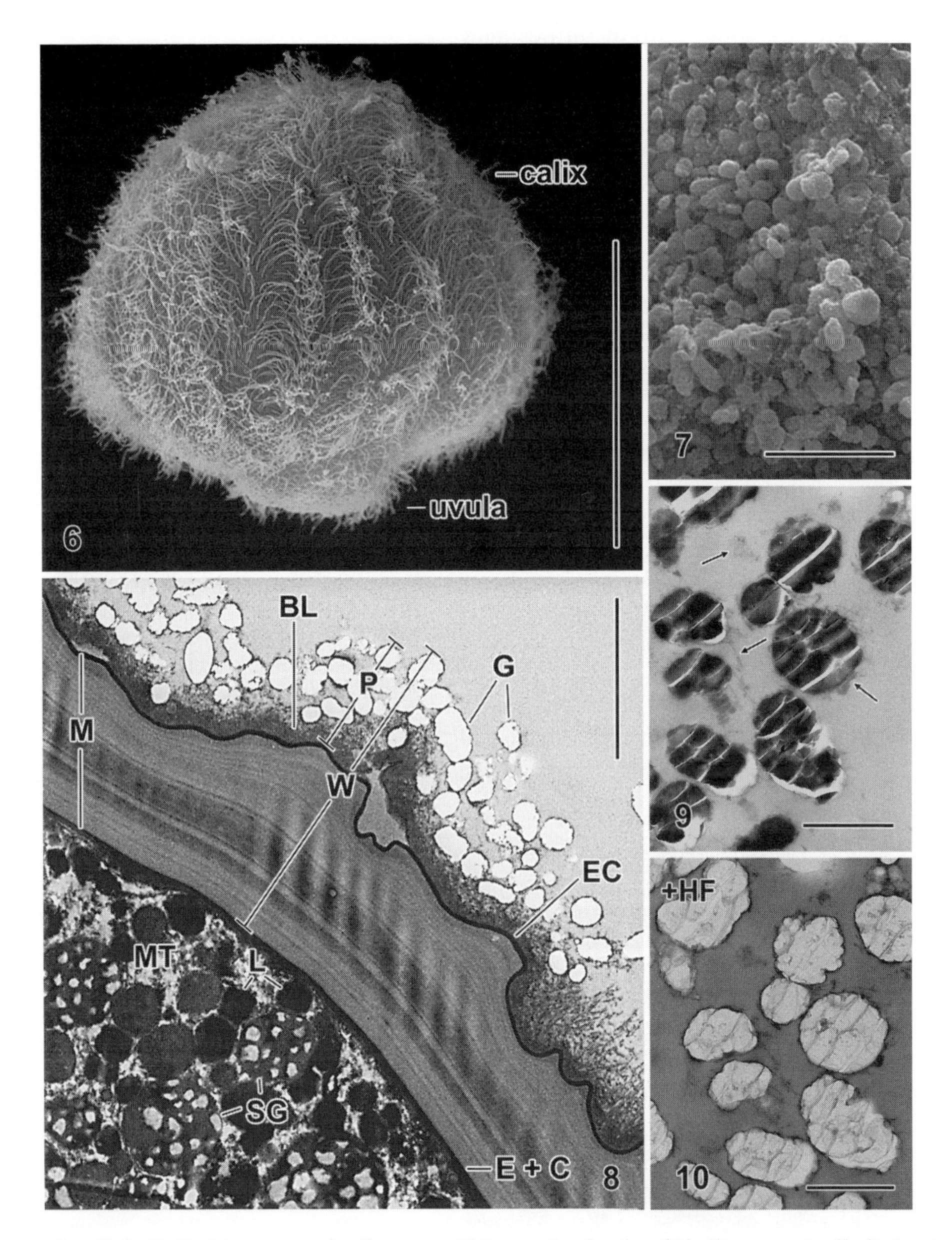

Figs 5.6–5.10 *Maryna umbrellata*, a *c.* 100 µm-sized colpodid ciliate, typically living in ephemeral puddles, makes globular resting cysts covered with a layer of glass (silicon) granules (**7–10**). **6**: Overview of a trophic specimen in the SEM. **7**: The cyst surface is covered by about 1 µm-sized glass granules well recognisable in the SEM. **8**: The resting cyst has a *c.* 13 µm-thick wall composed of many layers, each having a specific macromolecular composition (see Fig 5.11). The glass granules (G) were solved by hydrofluoric acid. **9**, **10**: Glass layer before and after treatment with hydrofluoric acid. Arrows mark mucous material that holds together the silicon granules. BL – basal layer, EC – ectocyst, E + C – endocyst and ciliate cortex, G – glass granules, L – lipid droplets, M – mesocyst, MT – mitochondria, P – pericyst, SG – spongy globules, W – cyst wall. Scale bars 1 µm (**9, 10**), 4 µm (**8**), 5 µm (**7**), and 50 µm (**6**). With permission from Foissner (2009) and Foissner et al. (2009).

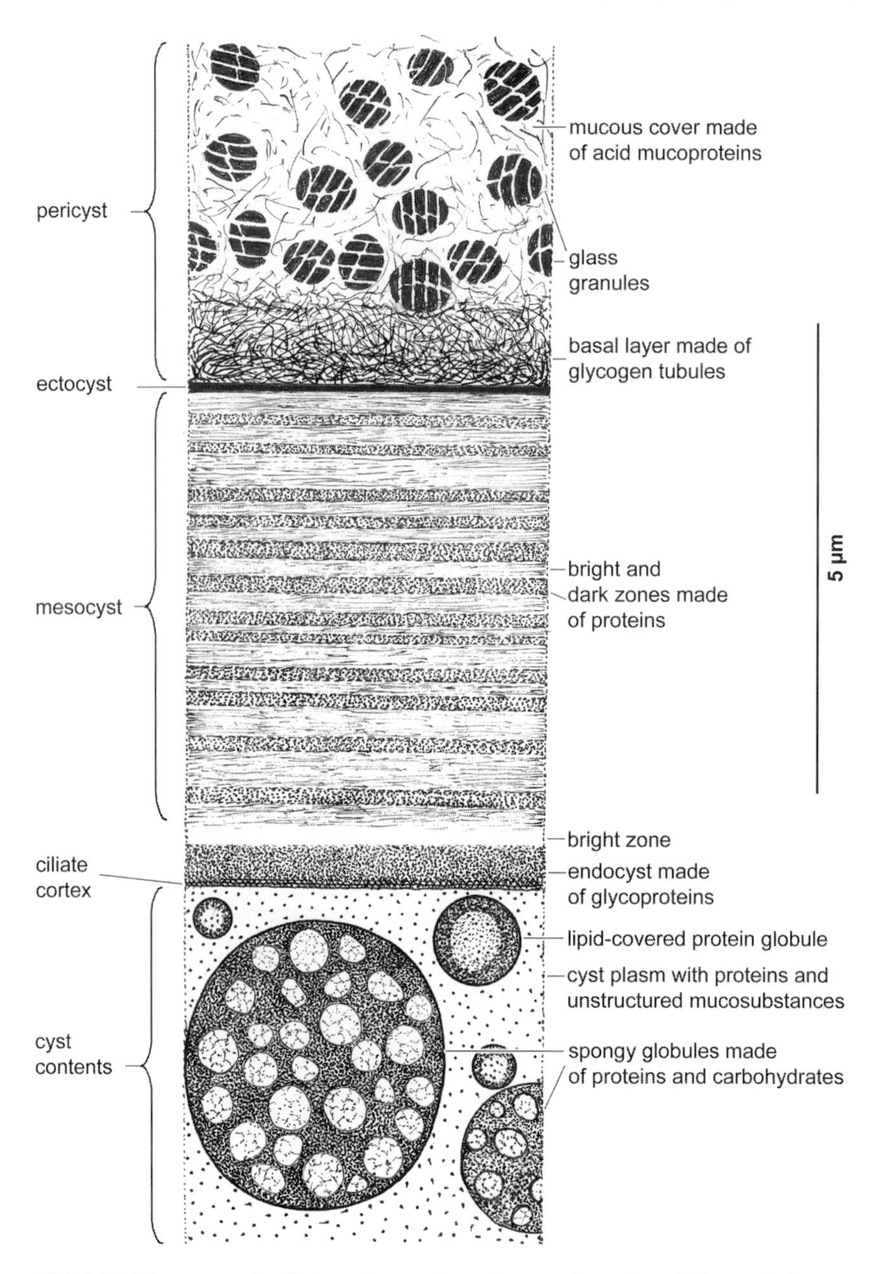

Fig 5.11 *Maryna umbrellata*, scheme of resting cyst based on light-and electron-microscopical observations and cytochemical tests (compare Figs 5.6–5.10). The complex cyst wall very likely determines the dispersal success, i.e. results in a cosmopolitan or restricted distribution. With permission from Foissner (2009).

of minute glass (silicon) granules with a size of about 1 µm (Figs 5.8–5.10); (iii) the mesocyst and endocyst show a high elasticity; (iv) the cytoplasm is studded with about 4 µm-sized globules consisting of a proteinaceous matrix burrowed by electron-lucent strands of glycogen (Figs 5.8, 5.11); (v) the fluid portion of the cyst plasm contains large amounts of acid mucopolysaccharides possibly originating from decomposed mucocysts; and (vi) the ectocyst precursors are released via the parasomal sacs of the kinetids. The most remarkable feature, the glass granules, is produced by the trophic cell and released during the early encystment processes.

Most, possibly all of these peculiarities are related to the ephemeral nature of the habitat, for instance, the thick wall protects the cell from desiccation, while the high elasticity and the glass cover might prevent the cell from mechanical stress (Yang et al., 2009), for instance, when cysts and sand are mixed by a storm. Certainly, all these properties will influence excystment and cyst viability, and thus the dispersal success.

5.4.2 *Pseudomaryna australiensis* and *Sandmanniella terricola* (Colpodea)

These small ciliates (~50 µm), which live in floodplain soils (Figs 5.12–5.19), have been described by Foissner (2003) and Foissner and Stoeck (2009). One of these, *P. australiensis* lives in a mineralic envelope, making cells and cysts appearing like inorganic soil particles, possibly protecting them from predators (Figs 5.12–5.14). Possibly, *P. australiensis* and *S. terricola* are restricted to the Australian and African region, respectively.

Before encysting, most ciliates digest food and expell the remnants, thus becoming rather hyaline when entering the cystic stage. *Pseudomaryna australiensis* and *S. terricola* do the opposite: they feed, but do not digest, in the trophic stage, becoming packed with large, compact food vacuoles (Figs 5.13, 5.15–5.17), which they digest in the resting cyst, using the energy provided for division (Figs 5.18, 5.19). Possibly, this is an extreme adaptation to the ephemeral nature of the habitat, making it possible to use even very short periods of optimal environmental conditions. The *P. australiensis* and *S. terricola* way must not be mixed with the division cysts of, for example *Colpoda*, which are covered by a temporary, very thin wall entirely different from that of the resting cysts (Foissner, 1993).

5.4.3 *Sorogena stoianovitchae* (Colpodea)

This curious ciliate lives on rotting foliage of plants (Figs 5.20, 5.21). It has a size of 30–70 × 20–45 µm and belongs to the class Colpodea, possibly representing a distinct order (Foissner, 1993; Foissner and Stoeck, 2009).

Sorogena stoianovitchae is the only ciliate that undergoes fruiting body development, and thus was initially thought to be related to the slime molds (Bradbury and Olive, 1980). The development process can be classified into five stages (Olive

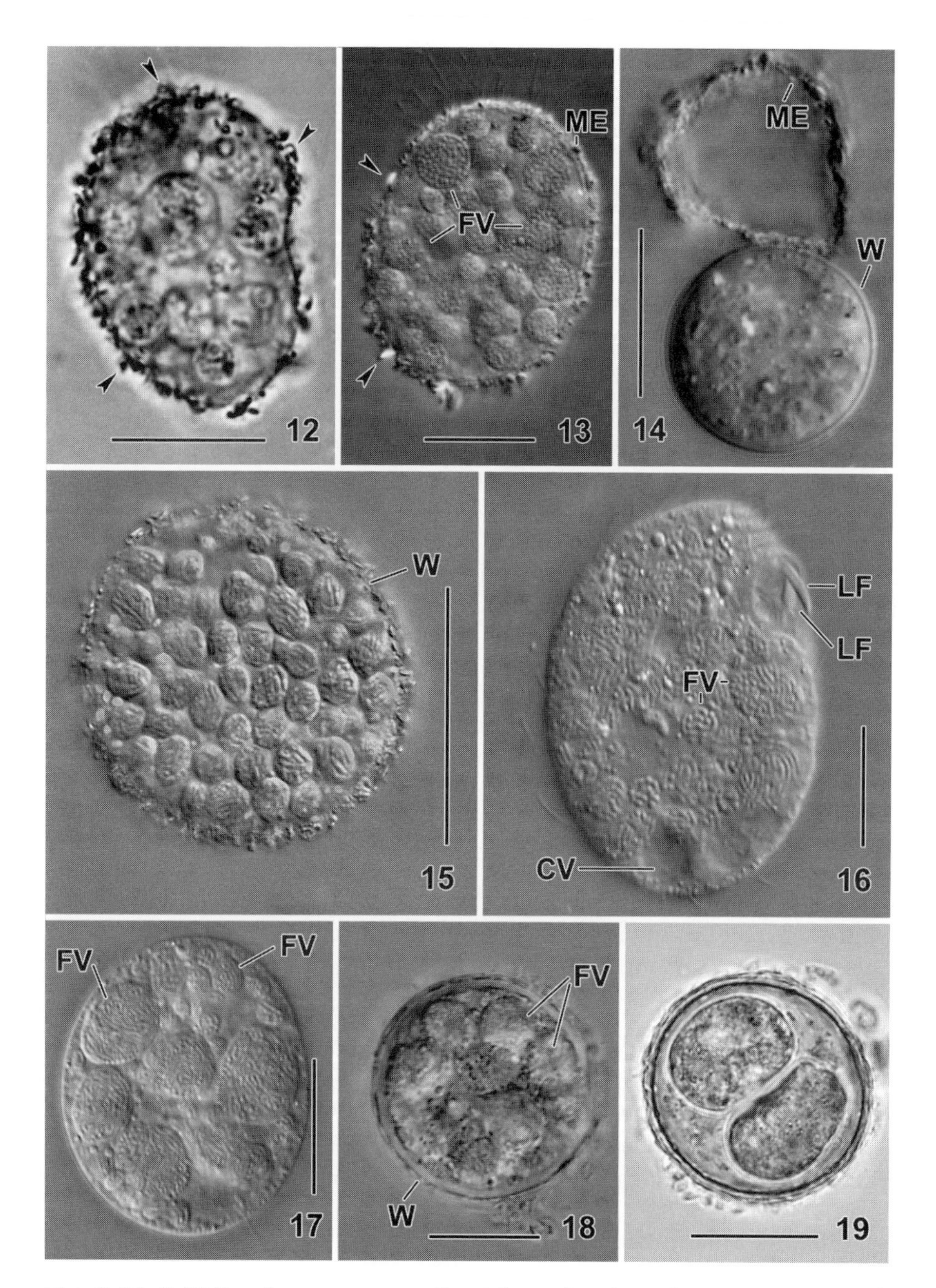

Figs 5.12–5.19 *Pseudomaryna australiensis* (**12–15**) and *Sandmanniella terricola* (**16–19**) from life. Both species collect bacteria, forming large, compact food vacuoles which are not digested in the trophic (**12, 13, 16, 17**) but in the cystic (**14, 15, 18**) stage, where they also divide (**19**). *P. australiensis* has a mineralic envelope (some particles marked by arrowheads), making it looking like a soil particle (**12–15**). CV – contractile vacuole, FV – food vacuoles, LF – left oral ciliary field, ME – mineralic envelope, W – cyst wall. Scale bars 20 μm. With permission from Foissner (2003) and Foissner and Stoeck (2009).

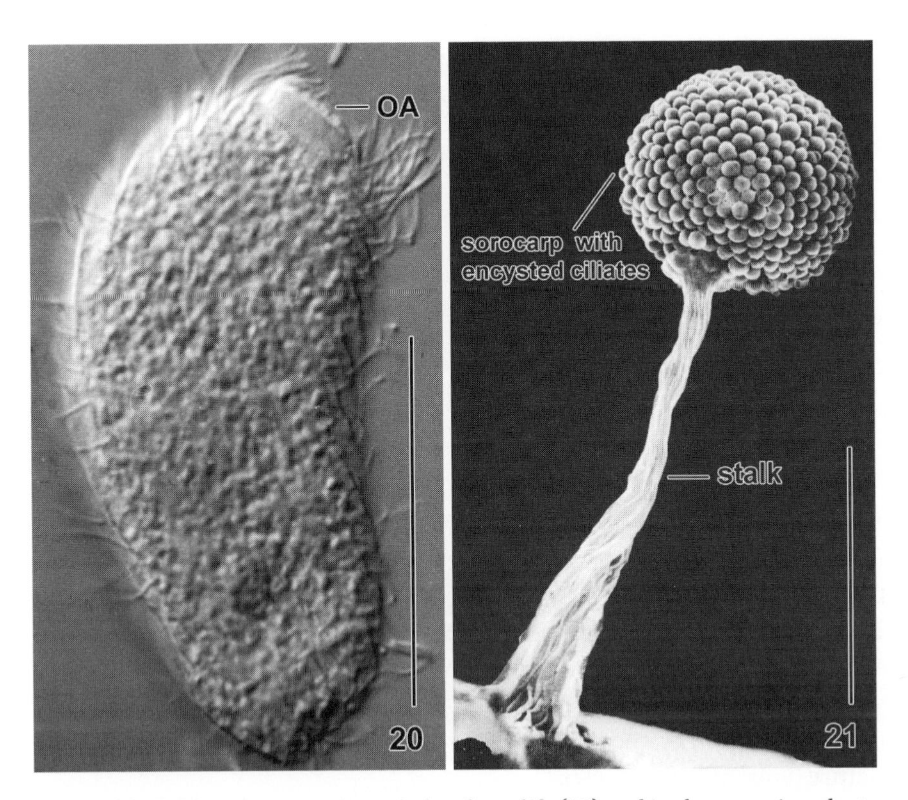

Figs 5.20, 5.21 *Sorogena stoianovitchae* from life (**20**) and in the scanning electron microscope (**21**). **20**: Right side overview of a trophont, showing the dome-shaped oral entrance (OA). **21**: Uniquely, *S. stoianovitchae* develops aerial sorocarps, quite similar to those of slime moulds. Scale bars 30 μm (**20**) and 200 μm (**21**). With permission from Bardele et al. (1991) and Olive and Blanton (1980).

and Blanton, 1980; Sugimoto and Endoh, 2008): aggregation, compact aggregation, secretion of a mucous matrix, stalk elongation and completion of the fruiting body. When *S. stoianovitchae* is mildly starved, several hundreds of cells aggregate beneath the water surface, and the aggregate develops into an aerial fruiting body, in which the individual cells encyst, forming a very thin wall (Blanton and Olive, 1983). Essential requirements for fruiting body development are high cell density, a light–dark cycle, and a dark period of more than 8 consecutive hours. In addition, the initial aggregation begins during the night and sunrise (light) triggers the subsequent development. The stalk of the sorocarp is composed of a matrix of complex protein-polysaccharide molecules (Blanton et al., 1983). Recently, Sugimoto and Endoh (2008) analysed the genes involved in fruiting body development. A BLASTX search revealed that sequences with high identity for extracellular proteins (mucin, proteophosphoglycan) or membrane proteins are likely candidates for aggregating material, mucous matrix and stalk material.

Table 5.2 State of 100 protargol-impregnated *Meseres corlissi* specimens from an exponentially growing culture.

State of specimens	Proportion (%)
Ordinary specimens	40
Ordinary dividing specimens	6
Dividing specimens with cyst wall precursors	2
Specimens with cyst wall precursors	23
Specimens with few food vacuoles	17
Specimens with few food vacuoles and with cyst wall precursors	12

5.4.4 *Meseres corlissi* and *Halteria grandinella* (oligotrichs)

Meseres (Figs 5.22–27) and *Halteria* (Figs 5.28–5.29) are closely related morphologically (Petz and Foissner, 1992) and genetically (Katz et al., 2005). This is sustained by their resting cysts, especially the occurrence and fine structure of the lepidosomes (extracellular, organic structures produced intracellularly by trophic and/or cystic protist species; Foissner et al., 2005). Further, the cysts share a considerable overall similarity, that is, the wall is composed of five layers with similar fine structure (Foissner et al., 2007).

However, there are also conspicuous differences: (i) the lepidosomes are spherical in *Meseres* (Figs 5.24–5.26), while conical in *Halteria* (Fig 5.29); (ii) the lepidosomes of *Meseres* are located in a slimy 'basal layer' (Fig 5.26), while those of *Halteria*, which lacks a basal layer, are attached to the ectocyst; (iii) *Meseres* has a bright (non-osmiophilic) zone between mesocyst and endocyst (Fig 5.27), while both are close together in *Halteria* (Foissner et al., 2007); (iv) *Halteria* lacks the chitin present in *Meseres*, which is unexpected considering the close morphologic and genetic relationship; (v) *Meseres* has five complex types of cyst wall precursors (Foissner and Pichler, 2006), while *Halteria* has possibly only three or four because it lacks the basal layer and the bright zone between mesocyst and endocyst (see items ii and iii); (vi) The 'curious structures', very likely reserve bodies produced by the autophagous vacuoles, have a different shape (Foissner, 2005; Foissner et al., 2007); and (vii) in contrast to *Halteria*, *Meseres* produces part of the cyst wall precursors in the morphostatic condition and even in dividing specimens (Table 5.2, Fig 5.23). This ability, which I term 'precursor stocking', may explain why *Meseres* is able to encyst within one hour, in spite of the complexity of the process (Foissner and Pichler, 2006). Precursor stocking is possibly more common than recognised, i.e. I observed it also in some haptorid ciliates (Foissner, unpublished).

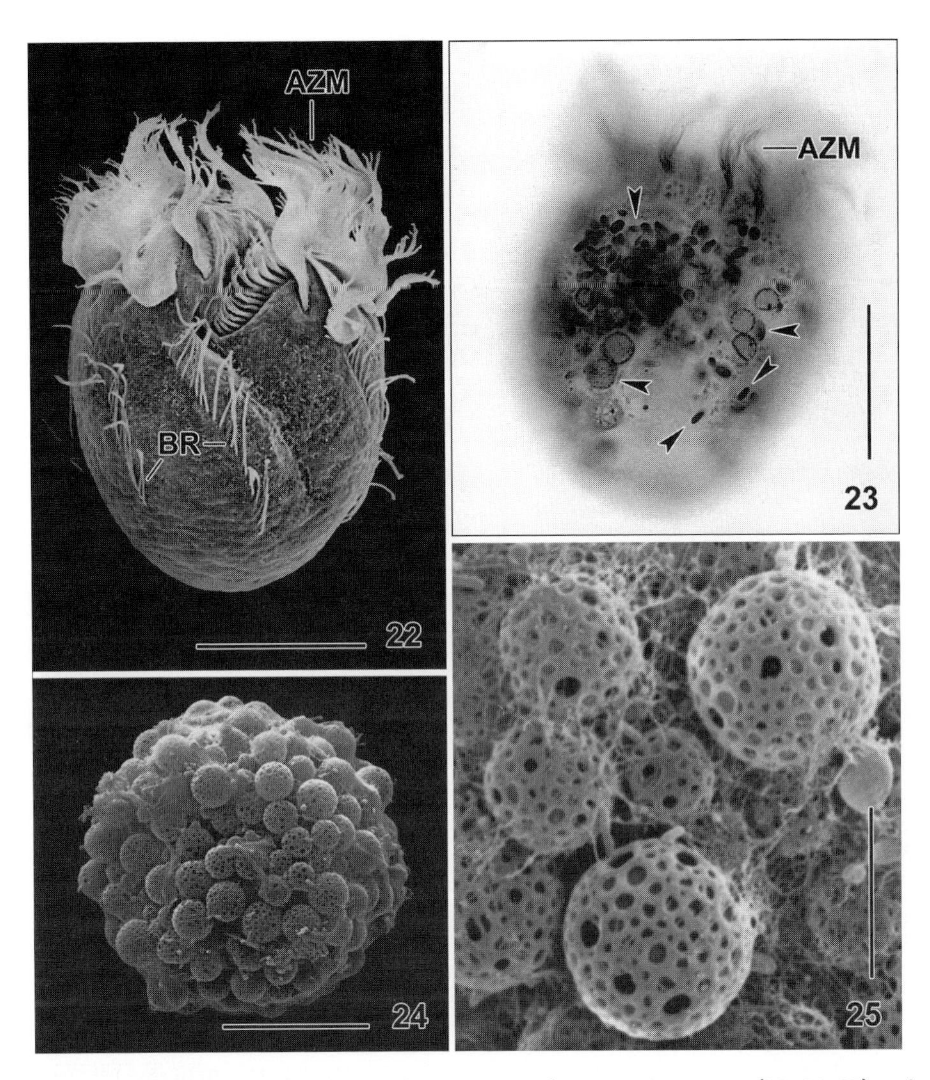

Figs 5.22–5.25 *Meseres corlissi* in the scanning electron microscope (**22**, **24**, **25**) and after silver (protargol) impregnation (**23**). **22**: Ventral overview, showing the conspicuous adoral zone of membranelles (AZM) and widely spaced somatic ciliary rows consisting of stiff bristles (BR). **23**: A morphostatic cell, as recognisable by the adoral zone of membranelles (AZM), which has numerous cyst wall precursors in the cytoplasm (arrowheads), including fully developed lepidosomes (see next figures). This phenomen is called 'precursor stocking'. **24, 25**: The globular resting cyst is covered by about 200 spherical lepidosomes, i.e. organic scales produced by the Golgi apparatus and present also in 23% of non-encysting specimens (Table 5.2; Fig 5.23, precursor stocking). The lepidosomes have an average diameter of 6 µm and have a reticular wall (**25**). Scale bars 5 µm (**25**), 20 µm (**24**), and 30 µm (**22, 23**). With permission from Petz and Foissner (1992), Foissner et al. (2005) and Foissner and Pichler (2006).

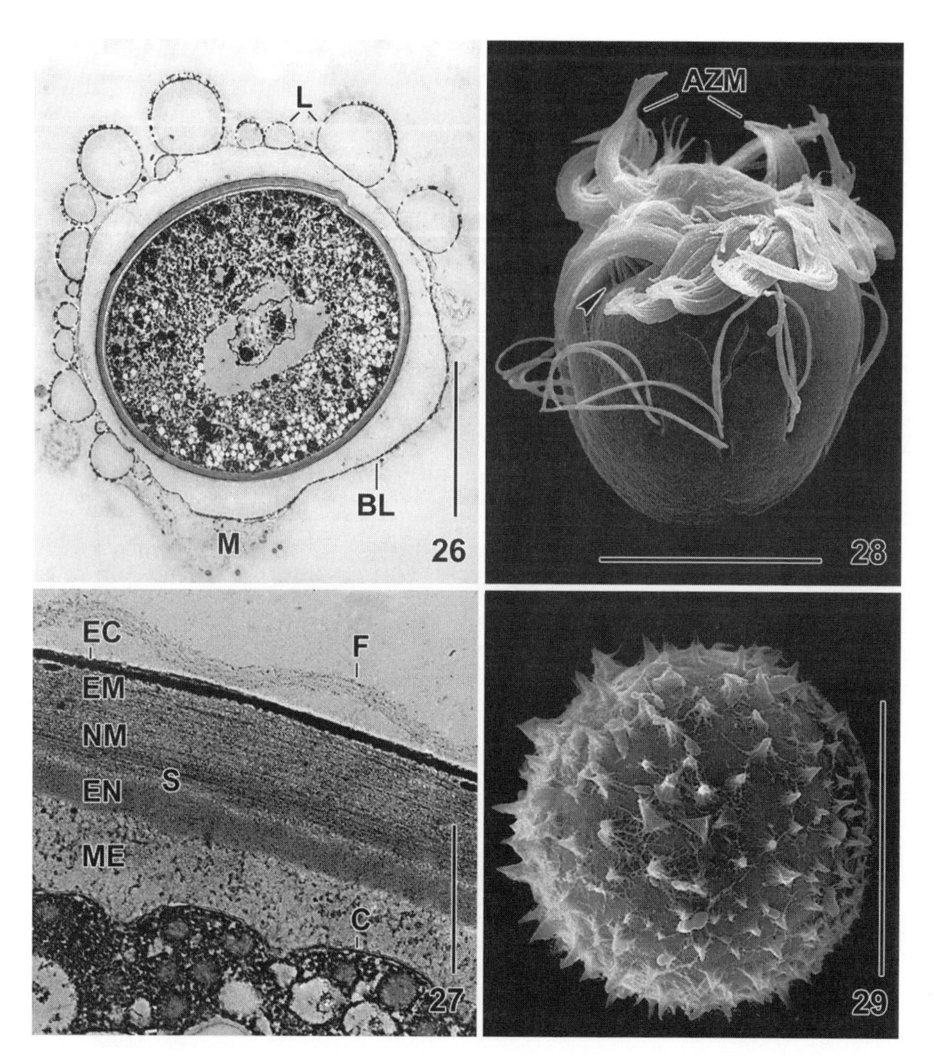

Figs 5.26–5.29 *Meseres corlissi* (**26, 27**) and *Halteria grandinella* (**28, 29**) in the transmission (**26, 27**) and scanning (**28, 29**) electron microscope. See also Figs 5.22–5.25. **26, 27**: *Meseres corlissi* has a complex cyst wall, consisting of (from outside to inside) lepidosomes (L) embedded in a slimy matrix (M), a basal layer (BL), a microfibrillar layer (F), an ectocyst (EC), an ectomesocyst (EM), an endomesocyst (NM), an endocyst (EN) and a metacyst (ME). The cortex (C) of the ciliate is maintained. **28**: Left side view with end of adoral zone of membranelles (AZM) marked by an arrowhead. Note the long jumping bristles. **29**: Like *Meseres, Halteria* has lepidosomes on the surface of the resting cyst. However, the lepidosomes are globular in *Meseres,* while conical in *Halteria*. Scale bars 1 µm (**27**), 15 µm (**28, 29**), and 20 µm (**26**). With permission from Foissner (2005) and Foissner et al. (2007).

The differences in the cyst structure of *Meseres* and *Halteria*, especially the complex lepidosomes and the presence of a chitin layer in the former, might at least partially explain their different ecology. Although both are cosmopolitan (Katz et al., 2005; Weisse et al., 2008), *Meseres* is very rare and possibly restricted to ephemeral freshwater habitats, while *Halteria* is one of the most common ciliates occurring in a wide variety of ephemeral and permanent limnetic environments (Foissner et al., 1991; Weisse et al., 2008).

As an inhabitant of ephemeral habitats, *Meseres* needs a 'stronger' cyst wall than *Halteria*. Indeed, the wall is twice as thick (1241 nm vs. 660 nm) and has a higher complexity (see above). While the chemical composition and the function of the lepidosomes, whose genesis and release takes a lot of energy, is still obscure, the chitin layer might be helpful in protecting the cell from mechanical and water stress as well as from bacterial decomposition because chitin is a very resistant matter. Finally, precursor stocking is an excellent way to use even short periods of good environmental conditions.

5.4.5 *Strombidium oculatum* (oligotrichs)

Strombidium oculatum is a tide-pool ciliate and an impressive example of circatidal encystment, first described by Fauré-Fremiet (1948) and later studied in detail by Jonsson (1994) and Montagnes et al. (2002). The ciliate, which has an obconical shape and is about 80 × 40 μm in size (Fig 5.30), is possibly restricted to the northern hemisphere (Agatha, S., pers. comm.). Usually, it is green due to sequestered chloroplasts and has a distinct, red eyespot composed of stigma obtained from chlorophyte prey. The cysts are flattened spheres, about 50 μm in diameter, and in the middle of the top surface there is a 10 μm-wide escape opening closed with a spumiform plug (Fig 5.31; Jonsson, 1994; Montagnes et al., 2002). Unfortunately, the fine structure and chemical composition of the cyst wall and the plug have not yet been investigated.

The circatidal behaviour runs as follows (Fig 5.32, Jonsson, 1994; Montagnes et al., 2002): for about 6 h, at low tide, *S. oculatum* is free-swimming in pools, and about 20–60 min before flushing of the pools it encysts on a substrate. Encystment lasts for about 19 h: two high tides and one intervening low tide. Excystment then occurs the next day about 30–40 min after the pools are isolated. Cells divide almost immediately after excysting, allowing the ciliate population to rapidly exploit potential food resources. Experiments and field observations revealed that *S. oculatum* responds phototaxically and exhibits seasonal trends in population dynamics with very low abundances in winter.

5.4.6 *Odontochlamys* spp. (Chilodonellidae)

These are small (~50 μm), bacterivorous ciliates living in terrestrial and limnetic habitats (Fig 5.33). They are remarkable in having the ability to change within a few

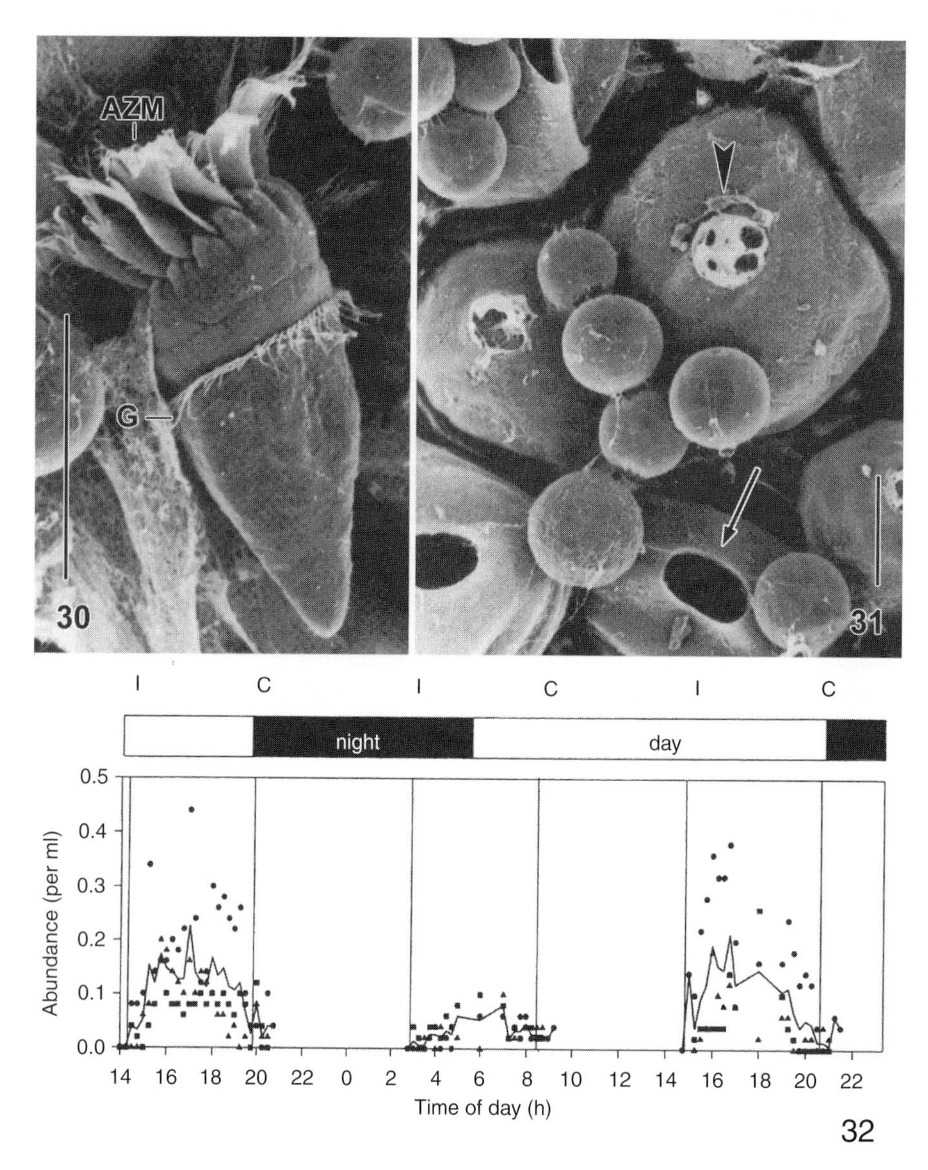

Figs 5.30–5.32 *Strombidium oculatum* in the scanning electron microscope. **30**: Lateral overview, showing the conspicuous adoral zone of membranelles (AZM) and the girdle ciliary row (G). **31**: Resting cysts are closed by a fibrous lid (arrowhead). When excysting, the lid disappears (arrow). **32**: Field observations of the change in abundance of *Strombidium oculatum* over the day–night cycle in three replicate tide pools (■, ●, ▲) over ~3 low and ~3 high tides. The solid line is the mean abundance of ciliates in the three pools. Vertical lines represent when pools were isolated by the outgoing tide (I) and covered by the incoming tide (C). Days and nights were delineated by sunrise and sunset. Scale bars 15 µm (**31**) and 25 µm (**30**). With permission from Jonsson (1994) and Montagnes et al. (2002).

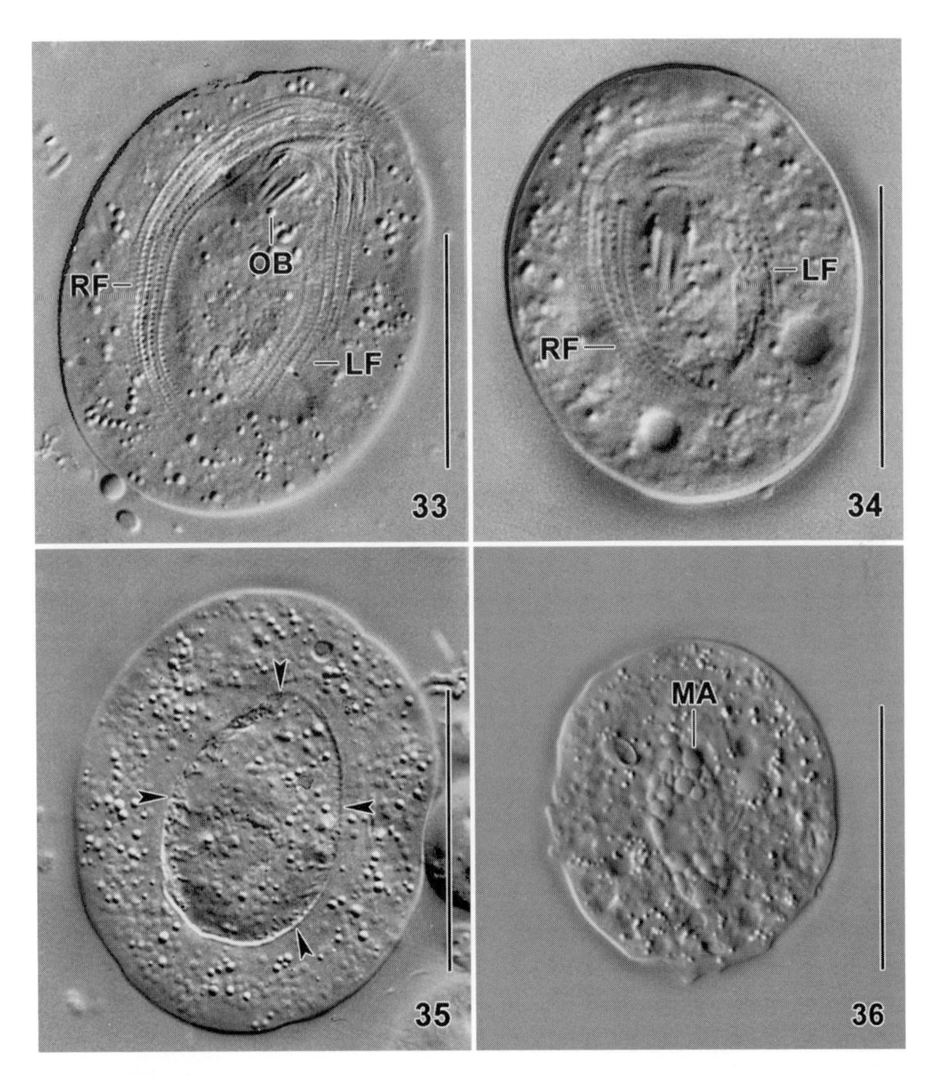

Figs 5.33–5.36 *Odontochlamys* spp. can encyst within 10 min. **33**: Ventral view of a trophic specimen, showing the oral basket (OB) and the right and left ciliary field (LF, RF). **34**: When encystment commences, the cell rounds up and the dorsal side begins to vault over the ventral one. **35**: Middle stage, showing that the dorsal side (margin marked by arrowheads) vaulted over most of the ventral side. **36**: Young resting cyst without distinct wall, showing the macronucleus (MA). Scale bars 25 μm. With permission from Blatterer and Foissner (1992) (**33**–**35**) and original (**36**).

minutes from the active into the cystic state. Thus, encystment can be observed under the microscope (Figs 5.34–5.36). For details, see figure captions.

Obviously, fast encystment is a strategy very helpful in ephemeral habitats, such as moss, leaf litter and small ponds, where these ciliates usually occur. Looking at the examples provided in this brief review, it becomes obvious that ciliates evolved several quite different strategies to survive in ephemeral habitats. It is likely that many more wait to be discovered.

5.5 Dispersal by humans

Biogeographic changes due to human activities have been largely ignored in the discussion of protist distribution, although a number of examples have been well known for a long time. For example, several tropical and Indopacific species of foraminifera entered the Mediterranean Sea via the Suez Canal (Lesseps' immigrants) and tropical aquaria. Moreover, it is likely that certain toxic dinoflagellates spread by human activities (Hallegraeff and Bolch, 1992). In rotifers, many of which have a similar size as ciliates, *Brachionus havanaensis* and *Keratella americana* have been introduced to southeast Asia by human activities (Segers, 2001). On the other hand, alpine zooplankton richness and genetic diversity have been only slightly influenced by anthropogenic stress and fish introduction, possibly due to the ability to produce long-lived resting stages withstanding unfavourable conditions (Winder et al., 2001).

Shipping (ballast water), the transport of goods and the construction of canals are three major reasons for the artificial dispersal of protists. Millions of tonnes of water and many thousands of tonnes of soil are transported across the world each year. Hallegraeff and Bolch (1992) and Hülsmann and Galil (2002) suppose that since the introduction of water as ballast in the middle of the nineteenth century, many protists may have spread globally, unheeded by protozoologists. The diatoms *Odontella sinensis* and *Coscinodiscus wailesii* entered the North Sea and the Baltic Sea rather recently, together with their parasites (Kühn, 1997; Hülsmann and Galil, 2002). Likewise, *Lagenophrys cochinensis*, an ectosymbiotic ciliate of wood-boring, marine isopods, has probably been transported from New Zealand to California in wooden ship hulls rather recently (Clamp, 2003), while the coccolithophore *Emiliania huxleyi* invaded the Black Sea about 1500 years ago (Winter et al., 1994). Elliott (1973) proposed that a species of the *Tetrahymena pyriformis* complex entered the Pacific Islands when man migrated westward from South and perhaps Central America. The same might have happened more recently with *Paramecium quadecaurelia*, a member of the *P. aurelia* sibling species complex. This species, which was known only from Australia, was recently reported from a pond of the city of Windhoek, the capital of Namibia (Przybós et al., 2003). Dispersal by ship's ballast water might also be responsible for the occurrence of four euryhaline psammobiontic (obligate sand-dwelling) testate amoeba species in the Great Lakes, Canada (Nicholls and MacIsaac, 2004), while marine dinoflagellates possibly cannot establish viable populations in these lakes (Fahnenstiel et al., 2009).

Another impressive example is the appearance of *Hydrodictyon* in New Zealand where this very distinctive alga had never been seen before. It was found in a pond belonging to a hatchery supplying fish and aquatic plants to aquarists. Obviously, *Hydrodictyon* had been imported together with fish or aquatic plants from East Asia (Kristiansen, 1996).

Freshwater diatoms show several impressive examples of human-mediated introductions. They have been reviewed by Vanormelingen et al. (2008), whose text I quote here: '*Asterionella formosa* is a widespread planktonic morphospecies and is frequently considered to be cosmopolitan; it is often seasonally dominant in eutrophic lakes. Detailed analysis of fossil material from 21 sediment cores (14 lakes) from New Zealand showed no trace of *A. formosa* in pre-European sediments, although it is now widespread, occurring in 45% of lakes for which phytoplankton records are available. The most likely vector for the introduction of *A. formosa* is the introduction of salmon ova into New Zealand lakes in the second half of the nineteenth century (Harper, 1994). It is highly unlikely that the species was extremely (i.e. not detectably) rare before European settlement as it is a species that can occur across a wide range of environmental conditions, from oligo- to eutrophic (Harper, 1994; Van Dam et al., 1994). Interestingly, *A. formosa* is also absent from other lake cores in Australasia, which might rule out environmental change due to the introduction of mammalian grazers as a cause for its sudden appearance. The recent spread in New Zealand of another exotic diatom, *Didymosphenia geminata* (Kilroy et al., 2007), is occurring long after the main human-induced environmental changes. Other convincing evidence for human mediated introduction of species (and hence previous dispersal limitation) among diatoms include the appearance of *Thalassiosira baltica* in the Laurentian Great Lakes (Edlund et al., 2000), and the North-American species *Gomphoneis minuta* and *Encyonema triangulum* in France (Coste and Ector, 2000).'

Wilkinson (2010) admonishes me for the lack of examples of human introductions in soil. There are none, unfortunately! And Wilkinson (2010) also did not provide any. However, he provided a solid discussion of soil introductions, some anecdotal evidence, and several suggestions to overcome the methodological problems which, indeed, are much more serious than in limnetic and marine environments.

5.6 Conclusions

Protist distribution is best described with the moderate endemicity model. But little information is available on the reasons why certain species are cosmopolitan and others are not. I argue that the break of Pangaea into Laurasia and Gondwana, the structure and physiology of the resting cysts, and human introductions are

Table 5.3 Percentages of dispersal routes of protists. Based on the calculation of Foissner (2008) that one-third of ciliates possibly have restricted distribution.

Dispersal routes	Amount (%)
Cosmopolitan distribution due to step-by-step dispersal and human introductions	35
Cosmopolitan distribution due to geological processes, euryoecious lifestyle and others	30
Restricted distribution due to morphological and physiological peculiarities of the resting cysts, break-up of Pangaea and insufficient time to disperse in young species	35

the most important factors for the dispersal of species (cosmopolitan or restricted distribution) and for their presence/absence in a certain habitat, at a certain time, and under certain environmental conditions. The same morphospecies may have different resting cysts, depending on the habitat in which the trophic cells live. Thus, long-range dispersal by air currents or animal vectors will not produce viable populations, for instance, ciliate species which evolved in rain forests will very likely not survive in Central Europe because their 'weak' cysts die during transport or do not withstand the different climate. This is substantiated by a comparative analysis of desiccation resistance of ciliate cysts from Namibia and rain forests in Borneo and the Malay Peninsula. Then, I present seven examples from ciliate resting cysts, showing their overwhelming morphological and physiological diversity, for instance, precursor stocking in *Meseres corlissi*, a glass cover in *Maryna umbrellata* and the circatidal cyst production of *Strombidium oculatum*. Step-by-step distribution and human introductions possibly also play a considerable role in the dispersal of species, especially at local, regional and continental scales. Several examples are provided. Table 5.3 shows rather speculative percentages on the contribution of several dispersal routes.

Acknowledgements

Financial support was provided by the Austrian Science Foundation (FWF, projects P20360-B17 and P19699-B17). The technical assistance of Mag. Barbara Harl is greatly acknowledged.

References

Bass, D., Richards, T.A., Matthai, L., Marsh, V., Cavalier-Smith, T. (2007). DNA evidence for global dispersal and probable endemicity of protozoa. *BMC Evolutionary Biology* **7**, 162.

Bardele, C.F., Foissner, W., Blanton, R.L. (1991). Morphology, morphogenesis and systematic position of the sorocarp forming ciliate *Sorogena stoianovitchae* Bradbury and Olive, 1980. *Journal of Protozoology* **38**, 7–17.

Blanton, R.L., Olive, L.S. (1983). Ultrastructure of aerial stalk formation by the ciliated protozoan *Sorogena stoianovitchae*. *Protoplasma* **116**, 125–135.

Blanton, R.L., Warner, S.A., Olive, L.S. (1983). The structure and composition of the stalk of the ciliated protozoan *Sorogena stoianovitchae*. *Journal of Protozoology* **30**, 617–624.

Blatterer, H., Foissner, W. (1992). Morphology and infraciliature of some cyrtophorid ciliates (Protozoa, Ciliophora) from freshwater and soil. *Archiv für Protistenkunde* **142**, 101–118.

Bradbury, P.C., Olive, L.S. (1980). Fine structure of the feeding stage of a sorogenic ciliate, *Sorogena stoianovitchae* gen. n., sp. n. *Journal of Protozoology* **27**, 267–277.

Caron, D.A. (2009). Past president's address: protistan biogeography: why all the fuss? *Journal of Eukaryotic Microbiology* **56**, 105–112.

Chao, A., Li, P.C., Agatha, S., Foissner, W. (2006). A statistical approach to estimate soil ciliate diversity and distribution based on data from five continents. *Oikos* **114**, 479–493.

Clamp, J.C. (2003). Ecology and geographic variation in *Lagenophrys cochinensis* (Ciliophora, Peritricha, Lagenophryidae), a widely distributed ectosymbiont of wood-boring, marine isopods. *Journal of Eukaryotic Microbiology* **50**, Abstract 82.

Corliss, J.O., Esser, S.C. (1974). Comments on the role of the cyst in the life cycle and survival of free-living protozoa. *Transactions of the American Microscopical Society* **93**, 578–593.

Coste, M., Ector, L. (2000). Diatomées invasives exotiques ou rares en France: principales observations effectuées au cours des dernières décennies. *Systematics and Geography of Plants* **70**, 373–400.

Cowling, A.J. (1994). Protozoan distribution and adaptation. In Darbyshire, J.F. (ed.), *Soil Protozoa, pp. 5–42*. Wallingford: CAB International.

Darling, K.F., Wade, C.M. (2008). The genetic diversity of planktic foraminifera and the global distribution of ribosomal RNA genotypes. *Marine Micropaleontology* **67**, 216–238.

Dolan, J.R. (2005). An introduction to the biogeography of aquatic microbes. *Aquatic Microbial Ecology* **41**, 39–48.

Edlund, M.B., Taylor, C.M., Schelske, C.L. et al. (2000). *Thalassiosira baltica* (Grunow) Ostenfeld (Bacillariophyta), a new exotic species in the Great Lakes. *Canadian Journal of Fisheries and Aquatic Sciences* **57**, 610–615.

Elliott, A.M. (1973). Life cycle and distribution of *Tetrahymena*. In Elliott, A.M. (ed.), *Biology of Tetrahymena, pp. 259–268*. Stroudsburg, PA: Hutchinson & Ross.

Epstein, S.S. (2009). Microbial awakenings. *Nature* **457**, 1083.

Fahnenstiel, G., Hong, Y., Millie, D., Doblin, M., Johengen, T., Reid, D. (2009). Marine dinoflagellate cysts in the ballast tank sediments of ships entering the Laurentian Great Lakes. *Verhandlungen der Internationalen Vereinigung für Limnologie* **30**, 1035–1038.

Fauré-Fremiet, E. (1948). Le rythme de marée du *Strombidium oculatum* Gruber. *Bulletin Biologique de la France et de la Belgique* **82**, 3–23.

Fenchel, T., Finlay, B.J. (2005). Cosmopolitanism and microbes. *SILnews* **44**, 5.

Finlay, B.J. (2002). Global dispersal of free-living microbial eukaryote species. *Science* **296**, 1061–1063.

Finlay, B.J., Esteban, G.F., Fenchel, T. (2004). Protist diversity is different? *Protist* **155**, 15–22.

Foissner, W. (1987). Soil protozoa: fundamental problems, ecological significance, adaptations in ciliates and testaceans, bioindicators, and guide to the literature. *Progress in Protistology* **2**, 69–212.

Foissner, W. (1993). Colpodea (Ciliophora). *Fischer, Stuttgart, Protozoenfauna* **4**, I–X + 798 pp.

Foissner, W. (1998). An updated compilation of world soil ciliates (Protozoa, Ciliophora), with ecological notes, new records, and descriptions of new species. *European Journal of Protistology* **34**, 195–235.

Foissner, W. (1999). Protist diversity: estimates of the near-imponderable. *Protist* **150**, 363–368.

Foissner, W. (2003). *Pseudomaryna australiensis* nov. gen., nov. spec. and *Colpoda brasiliensis* nov. spec., two new colpodids (Ciliophora, Colpodea) with a mineral envelope. *European Journal of Protistology* **39**, 199–212.

Foissner, W. (2004). Ubiquity and cosmopolitanism of protists questioned. *SILnews* **43**, 6–7.

Foissner, W. (2005). The unusual, lepidosome-coated resting cyst of *Meseres corlissi* (Ciliophora: Oligotrichea): transmission electron microscopy and phylogeny. *Acta Protozoologica* **44**, 217–230.

Foissner, W. (2006). Biogeography and dispersal of micro-organisms: a review emphasizing protists. *Acta Protozoologica* **45**, 111–136.

Foissner, W. (2008). Protist diversity and distribution: some basic considerations. *Biodiversity and Conservation* **17**, 235–242.

Foissner, W. (2009). The stunning, glass-covered resting cyst of *Maryna umbrellata* (Ciliophora, Colpodea). *Acta Protozoologica* **48**, 223–243.

Foissner, W., Pichler, M. (2006). The unusual, lepidosome-coated resting cyst of *Meseres corlissi* (Ciliophora: Oligotrichea): genesis of four complex types of wall precursors and assemblage of the cyst wall. *Acta Protozoologica* **45**, 339–366.

Foissner, W., Stoeck, T. (2009). Morphological and molecular characterization of a new protist family, Sandmanniellidae n. fam. (Ciliophora, Colpodea), with description of *Sandmanniella terricola* n. g., n. sp. from the Chobe floodplain in Botswana. *Journal of Eukaryotic Microbiology* **56**, 472–483.

Foissner, W., Blatterer, H., Berger, H., Kohmann, F. (1991). Taxonomische und ökologische Revision der Ciliaten des Saprobiensystems – Band I: Cyrtophorida, Oligotrichida, Hypotrichia, Colpodea. *Informationsberichte des Bayerischen Landesamtes für Wasserwirtschaft, München* **1/91**, 478 pp.

Foissner, W., Agatha, S., Berger, H. (2002). Soil ciliates (Protozoa, Ciliophora) from Namibia (Southwest Africa), with emphasis on two contrasting environments, the Etosha region and the Namib Desert. *Denisia* **5**, 1–1459.

Foissner, W., Müller H., Weisse, T. (2005). The unusual, lepidosome-coated resting cyst of *Meseres corlissi* (Ciliophora, Oligotrichea): light and scanning electron microscopy, cytochemistry. *Acta Protozoologica* **44**, 201–215.

Foissner, W., Müller, H., Agatha, S. (2007). A comparative fine structural and phylogenetic analysis of resting cysts in oligotrich and hypotrich Spirotrichea (Ciliophora). *European Journal of Protistology* **43**, 295–314.

Foissner, W., Weissenbacher, B., Krautgartner, W.-D., Lütz-Meindl, U. (2009). A cover of glass: first report of biomineralized silicon in a ciliate, *Maryna umbrellata* (Ciliophora: Colpodea). *Journal of Eukaryotic Microbiology* **56**, 519–530.

Green, J., Bohannan, B.J.M. (2006). Spatial scaling of microbial biodiversity. *Trends in Ecology and Evolution* **21**, 501–507.

Gutiérrez, J.C., Walker, G.K. (1983). Cystology: a new area in protozoology. *Proceedings of the 5th European Conference on Ciliate Biology*, Geneva (unpaged abstract).

Gutiérrez, J.C., Martin-González, A. (2002). Ciliate encystment-excystment cycle: A response to environmental stress. In Gutiérrez, J.C. (ed.), *Microbial Development Under Environmental Stress, pp. 29–49*. Research Signpost 37/611 (2), Fort P.O., Trivandrum-695 023, Kerala, India.

Gutiérrez, J.C., Díaz, S., Ortega, R., Martín-González, A. (2003). Ciliate resting cyst walls: a comparative review. *Recent Research Developments in Microbiology* **7**, 361–379.

Hallegraeff, G., Bolch, C.J. (1992). Transport of diatom and dinoflagellate resting spores in ships' ballast water: implications for plankton biogeography and aquaculture. *Journal of Plankton Research* **14**, 1067–1084.

Hamilton, W.D., Lenton, T.M. (1998). Spora and gaia: how microbes fly with their clouds. *Ethology Ecology and Evolution* **10**, 1–16.

Harper, M.A. (1994). Did Europeans introduce *Asterionella formosa* Hassall to New Zealand? In Kociolek, J.P. (ed.), *Proceedings of the 11th International Diatom Symposium 1990, pp. 479–484*. San Francisco, CA: California Academy of Sciences.

Hülsmann, N., Galil, B.S. (2002). Protists – a dominant component of the ballast-transported biota. In Leppäkoski, E. et al. (eds.), *Invasive Aquatic Species of Europe, pp. 20–26*. Dordrecht: Kluwer Academic.

Jonsson, P.R. (1994). Tidal rhythm of cyst formation in the rock pool ciliate *Strombidium oculatum* Gruber (Ciliophora, Oligotrichida): a description of the functional biology and an analysis of the tidal synchronization of encystment. *Journal of Experimental Marine Biology and Ecology* **175**, 77–103.

Katz, L.A., McManus, G.B., Snoeyenbos-West, L.O. *et al.* (2005). Reframing the "everything is everywhere" debate: evidence for high gene flow and diversity in ciliate morphospecies. *Aquatic Microbial Ecology* **41**, 55–65.

Kilroy, C., Biggs, B.J., Vieglais, C.C. (2007). *Didymosphenia geminata* in New Zealand: a science response to help manage an unwanted, invasive freshwater diatom. *Abstract book, ASLO Aquatic Sciences meeting*. Santa Fe (New Mexico), 4–9/02/2007.

Kristiansen, J. (1996). Dispersal of freshwater algae – a review. *Hydrobiologia* **336**, 151–157.

Kühn, S.F. (1997). *Victoriniella multiformis*, gen. et spec. nov. (incerta sedis), a polymorphic parasitoid protist infecting the marine diatom

Coscinodiscus wailesii Gran and Angst (North Sea, German Bight). *Archiv für Protistenkunde* **148**, 115–123.

Maguire, B., Jr. (1963). The passive dispersal of small aquatic organisms and their colonization of isolated bodies of water. *Ecological Monographs* **33**, 161–185.

Maguire, B., Jr. (1971). Community structure of protozoans and algae with particular emphasis on recently colonized bodies of water. In Cairns, J., Jr. (ed.), *The Structure and Function of Fresh-water Microbial Communities,* pp. 121–149. Research Division Monograph 3, Virginia Polytechnic Institute and State University, Blacksburg, Virginia 24061.

Martiny, J.B.H., Bohannan, B.J.M., Brown, J.H. et al. (2006). Microbial biogeography: putting microorganisms on the map. *Nature* **4**, 102–112.

Meier-Tackmann, D. (1982). Untersuchungen über die physiologische Funktion der Cystenhülle und die Resistenz der dünnwandigen Dauercysten von *Colpoda cucullus* O.F. Müller (Holotricha, Ciliata). *Zoologischer Anzeiger* **208**, 1–29.

Meier-Tackmann, D., Wenzel, F. (1988). Daten über Wettergeschehen und Reaktivierbarkeit der Dauercysten von *Colpoda cucullus* O.F. Müller (Holotricha, Ciliata). *Zoologischer Anzeiger* **220**, 277–290.

Meisterfeld, R. (1997). Thekamöben – ihr Potential für Ökosystemforschung und Bioindikation. *Abhandlungen und Berichte der Gesellschaft für Naturkunde, Görlitz* **69**, 87–95.

Meyer, E., Foissner, W., Aescht, E. (1989). Vielfalt und Leistung der Tiere im Waldboden. *Österreichische Forstzeitung* **3**, 15–18.

Montagnes, D.J.S., Wilson, D., Brooks, S.J., Lowe, C., Campey, M. (2002). Cyclical behaviour of the tide-pool ciliate *Stombidium oculatum*. *Aquatic Microbial Ecology* **28**, 55–68.

Müller, H., Foissner, W., Weisse, T. (2006). Role of soil in the life cycle of *Meseres corlissi* (Ciliophora: Oligotrichea): experiments with two clonal strains from the type locality, an astatic meadow pond. *Aquatic Microbial Ecology* **42**, 199–208.

Nicholls, K.H., MacIsaac, H.J. (2004). Euryhaline, sand-dwelling testate rhizopods in the Great Lakes. *Journal of Great Lakes Research* **30**, 123–132.

Nilsson, J.R. (2005). Ethanol affects endocytosis and proliferation of *Tetrahymena pyriformis* GL and promotes encystment. *Acta Protozoologica* **44**, 293–299.

Olive, L.S., Blanton, R.L. (1980). Aerial sorocarp development by the aggregative ciliate, *Sorogena stoianovitchae*. *Journal of Protozoology* **27**, 293–299.

Pawlowski, J., Holzman, M. (2008). Diversity and geographic distribution of benthic foraminifera: a molecular perspective. *Biodiversity and Conservation* **17**, 317–328.

Petz, W., Foissner, W. (1992). Morphology and morphogenesis of *Strobilidium caudatum* (Fromentel) *Meseres corlissi* n. sp., *Halteria grandinella* (Müller), and *Strombidium rehwaldi* n. sp., and a proposed phylogenetic system for oligotrich ciliates (Protozoa, Ciliophora). *Journal of Protozoology* **39**, 159–176.

Przybós, E., Hori, M., Fokin, S.I. (2003). Strains of *Paramecium quadecaurelia* from Namibia, Africa; genetic and molecular studies. *Acta Protozoologica* **42**, 357–360.

Segers, H. (2001). Zoogeography of the Southeast Asian Rotifera. *Hydrobiologia* **446/447**, 233–246.

Smith, H.G., Bobrov, A., Lara, E. (2008). Diversity and biogeography of testate amoebae. *Biodiversity and Conservation* **17**, 329–343.

Sugimoto, H., Endoh, H. (2008). Differentially expressed genes during fruiting body development in the aggregative ciliate *Sorogena stoianovitchae* (Ciliophora: Colpodea). *Journal of Eukaryotic Microbiology* **55**, 110–116.

Taylor, J.W., Turner, E., Townsend, J.P., Dettman, J.R., Jacobson, D. (2006). Eukaryotic microbes, species recognition and the geographic limits of species: examples from the kingdom Fungi. *Philosophical Transactions of the Royal Society B* **361**, 1947–1963.

Tyler, P.A. (1996). Endemism in freshwater algae with special reference to the Australian region. *Hydrobiologia* **336**, 1–9.

Van Dam, H., Mertens, A., Sinkeldam, J. (1994). A coded checklist and ecological indicator values of freshwater diatoms from The Netherlands. *Netherlands Journal of Aquatic Ecology* **28**, 117–133.

Vanormelingen, P., Verleyen, E., Vyverman, W. (2008). The diversity and distribution of diatoms: from cosmopolitanism to narrow endemism. *Biodiversity and Conservation* **17**, 393–405.

Wanner, M., Dunger, W. (2001). Biological activity of soils from reclaimed open-cast coal mining areas in Upper Lusatia using testate amoebae (protists) as indicators. *Ecological Engineering* **17**, 323–330.

Weisse, T., Strüder-Kypke, C., Berger, H., Foissner, W. (2008). Genetic, morphological, and ecological diversity of spatially separated clones of *Meseres corlissi* Petz & Foissner, 1992 (Ciliophora, Spirotrichea). *Journal of Eukaryotic Microbiology* **55**, 257–270.

Webster, J. (1983). *Pilze. Eine Einführung.* Berlin: Springer.

Wichterman, R. (1986). *The Biology of Paramecium*, 2nd Edition. New York: Plenum Press.

Wilkinson, D.M. (2001). What is the upper size limit for cosmopolitan distribution in free-living microorganisms? *Journal of Biogeography* **28**, 285–291.

Wilkinson, D.M. (2010). Have we underestimated the importance of humans in the biogeography of free-living terrestrial microorganisms? *Journal of Biogeography* **37**, 393–397.

Winder, M., Monaghan, M.T., Spaak, P. (2001). Have human impacts changed alpine zooplankton diversity over the past 100 years? *Arctic, Antarctic, and Alpine Research* **33**, 467–475.

Winter, A., Jordan, R.W., Roth, P.H. (1994). Biogeography of living coccolithophores in ocean waters. In Winter, A., Siesser, W.G. (eds.), *Coccolithophores*, pp. 161–177. Cambridge: Cambridge University Press.

Yang, S.H., Lee, K.-B., Kong, B., Kim, J.-H., Kim, H.-S., Choi, I.S. (2009). Biomimetic encapsulation of individual cells with silica. *Angewandte Chemie* **121**, 9324–9327.

6

Everything is everywhere: a twenty-first century de-/reconstruction with respect to protists

DAVID BASS[1] AND JENS BOENIGK[2]

[1] *Zoology Department, The Natural History Museum, London, UK*
[2] *General Botany, University Duisburg-Essen, Essen, Germany*

6.1 Introduction

The aphorism 'Everything is everywhere, but the environment selects' asserts that microbial taxa are found anywhere on earth that there is suitable habitat for them. It was crystallised in this form by Baas Becking (1934), who was in turn inspired by the ideas of Beijerinck (1913). The history of its establishment has been nicely summarised by O'Malley (2007) and its incorporation into contemporary thought described by de Wit and Bouvier (2006). It was further contextualised in terms of free-living protists by Fenchel and Finlay (2004). This concept (which we refer to hereafter as EiE) has recently been the focus of much heated debate in microbiology and protistology. This chapter will reassess the fundamental principles behind the EiE concept with respect to free-living protists in the light of recent findings and insights from twenty-first century molecular biology and microbial ecology. We do not intend to provide a survey of studies/taxa that apparently do or do not meet

Biogeography of Microscopic Organisms: Is Everything Small Everywhere?, ed. Diego Fontaneto. Published by Cambridge University Press.

its predictions; work in this field is proceeding rapidly and such a survey would soon become obsolete. The studies cited here were not selected because we agree or disagree particularly strongly with them, but because they illustrate points of our discussion. At the time of writing the most sensible view is that protist distribution is not fundamentally different to that of other organisms – the apparent differences being quantitative rather than qualitative – and overall fits the 'moderate endemicity model' (Foissner, 1999, 2004a, 2006, 2008; Hughes Martiny et al., 2006; Telford et al., 2006; Vyverman et al., 2007). Similar patterns and concepts continue to be discussed in other microbial and small multicellular groups (e.g. Finlay and Fenchel, 2004; Fierer, 2008). However, the means by which microbial distributions and dispersal can be measured and understood are necessarily different from those relating to larger organisms for a range of reasons – some more obvious than others.

Some of the main proponents of EiE in the protist world have been Finlay and colleagues (e.g. Finlay et al., 1996; Finlay, 2002; Fenchel and Finlay, 2004; Finlay and Fenchel, 2004), who argue that microbial eukaryotes are 'so abundant that continuous large-scale dispersal sustains their global distribution', and that 'sheer weight of [their] numbers might be expected to drive large-scale dispersal for purely statistical reasons' (Finlay, 2002 and references therein). It is important to bear in mind that the original formulation of EiE and many subsequent applications, including some contemporary ones, use morphology to define the operational taxoxomic unit (OTU) under consideration – i.e. the morphospecies (Finlay et al., 1996). It is apparently true that many or most protist morphospecies are globally distributed. There are myriad examples, particularly in groups of small, morphologically relatively characterless protists such as some flagellates and amoebae, but also in groups with more characters by which individual lineages can be distinguished by light microscopy, for example those with hardened and complex external features as in silica-scaled taxa and ciliates (Finlay et al., 1996; Finlay and Clarke, 1999; Wylezich et al., 2007; Kristiansen, 2008), and those with tests (Smith et al. 2008; Chapter 7). Conversely, there are some convincing examples of morphospecies with restricted distributions, particularly those found in poorly studied regions but not in apparently highly suitable habitats in much more intensively studied regions. Examples of this phenomenon have been cited in many groups including unicellular green algae (Coesel and Krienitz, 2008), planktonic foraminifera (Darling and Wade, 2008), testate amoebae (with particular reference to *Apodera vas* and *Certesella* spp., apparently occurring only south of the Tropic of Cancer desert belt: Mitchell and Meisterfeld, 2005; Smith and Wilkinson, 2007; Chapter 7), ciliates (Foissner et al., 2003; Foissner, 2004b; Stoeck et al., 2007a), diatoms (Vanormelingen et al., 2008) and chrysophytes (Kristiansen, 2008). However, the possibility cannot be ruled out that more intensive sampling, perhaps using different techniques, will prove at least some of these organisms to be more widely dispersed in viable form (even if they do not subsequently grow) and/or that some of these cases will prove to be driven by as yet unrecognised ecological factors.

A new set of opportunities and challenges for the EiE concept is coming from recent great advances in molecular biology – the increasing ease of probing environments using PCR primers at various levels of taxonomic specificity and obtaining sequence data which can be taxonomically assigned to protist groups and directly compared across sampling sites. The most striking outcome of studies using these techniques is that the molecular diversity of protists is much higher than that of plants and metazoa (e.g. Moreira and López-García, 2002; Berney et al., 2004; Cavalier-Smith, 2004; Guillou et al., 2004; Countway et al., 2005; Slapeta et al., 2005; Worden, 2006), and much greater than suggested by their morphological diversity as observed by light microscopy, even taking into account statistical extrapolations of morphospecies diversity from large, globally distributed data sets such as that for ciliates by Chao et al. (2006). Furthermore, many – if not all – protist morphospecies comprise many distinct lineages differing in ecological, physiological, behavioural, ultrastructural and/or other traits (e.g. Nanney et al., 1998; Habura et al., 2004; von der Heyden et al., 2004; Boenigk et al., 2005, 2006, 2007; Koch and Ekelund, 2005; Lowe et al., 2005; Slapeta et al., 2005, 2006; von der Heyden and Cavalier-Smith, 2005; Finlay et al., 2006; Stoeck et al., 2006; Scheckenbach et al., 2006; Bass et al., 2007, 2009a, 2009b, 2009c; Lilly et al., 2007; Hoef-Emden, 2007, 2008; Boenigk, 2008a, 2008b; Darling and Wade, 2008; Simon et al., 2008; Howe et al., 2009).

The most frequently used marker to measure this diversity is the small subunit rRNA gene (SSU rDNA) (Mindell and Honeycutt, 1990; Wuyts et al., 2000), which is generally well suited to this purpose: it is very strongly represented in online databases so provides a robust context for taxonomic assignation and phylogenetic analyses of newly obtained sequences, it is present in the genome in multiple copies so is relatively easy to detect even when cell numbers are low, and it is of sufficient length and encompasses a wide enough range of evolutionary rates for it to be an unusually good single gene for phylogenetic analyses. Therefore it is used as both an ecological, taxonomic and phylogenetic marker. Its use in the first two capacities has been criticised on the grounds that there is no direct link between variation in the SSU and species or ecotype boundaries; i.e. rDNA divergence is not the direct result of selective forces relevant to species or ecotype divergence. Such criticisms miss the point of marker genes, which simply offer 'calibration points' or 'barcodes' with which other biological information can be associated (biological species boundaries, ecotypes, phenotypic differences, etc.), and on the basis of which taxonomic unit differences can be categorised.

We will now examine the components of the EiE statement individually and in a contemporary context. We identify three key issues/considerations that must be addressed when assessing the biogeographic status of protists: (1) distinguishing between biogeographic patterns with ecological and historical bases; (2) determining an appropriate level of phylogenetic resolution at which to define taxonomic units that will be considered as potentially cosmopolitan or endemic; and (3) the implications of the term 'everywhere' in the EiE statement and the limitations of comprehensive sampling by both morphological and molecular survey methods.

6.2 The crucial but potentially elusive distinction between historical and ecological biogeography

EiE applies to biogeographic structures as a result of historical processes, *not* ecological tolerance/preference. This is emphasised by de Wit and Bouvier (2006), who point out that the Dutch 'maar' ('but') linking the two clauses 'everything is everywhere' and 'the environment selects' implies that there is potential conflict between them. EiE can be paraphrased by saying that protist taxa are dispersed globally, but only grow where conditions are suitable for them. Elsewhere they may be present in dormant forms or may be absent because they are unable to colonise and/or grow and have subsequently disappeared. They may or may not be detectable in non-suitable habitats by enrichment culturing (Finlay, 2002; Smirnov, 2003) and/or molecular techniques (see below). The only scenarios that would contradict the EiE assertion is (1) if a particular taxon had never reached some parts of the world because its members were in some way physically restricted from doing so, or (2) because the dispersal process took so long that the lineage had split into daughter lineages before the 'parent' taxon could achieve a cosmopolitan distribution, or (3) because dispersal necessarily takes so long without intervening suitable habitat that cells are unable to survive the journey in viable form. Otherwise, as long as individuals of a particular lineage are dispersed globally, whether they then grow or not (*as long as they have the potential to do so*), they can be said to have cosmopolitan distributions. This is a crucial distinction, and one that is often overlooked or misinterpreted by studies that claim restricted distributions for taxa in the sense that they provide evidence against EiE.

Therefore, the term 'ecological biogeography' describes only whether (or to what extent) a lineage proliferates at a particular site/region according to the suitability of conditions in that region, whereas 'historical biogeography' refers to where on earth lineages have reached or been transported to. The difference between these two concepts spotlights a crucial distinction between protist dispersal (and the ways and extent to which it can be detected) and that of many larger organisms. In most (all?) environments there exists a 'seedbank' of dormant protists, members of which can become active when conditions change to suit them. This seedbank probably at least partly explains the very long rank abundance tail of rare protist genotypes detected in both classical clone library and next generation sequencing (NGS) environmental diversity studies – as described in sections 4 and 5. (Such long tails also comprise lineages that may never be able to grow because conditions are vanishingly unlikely to change in their favour, and 'rare' sequences resulting from methodological errors and intra-lineage sequence variation, which artefactually inflate estimates of the diversity present.) This cryptic/dormant diversity is not present in organisms with larger dispersal stages, which are more easily seen and identified and are generally not able to persist in dormant forms for long or as indetectably. Such larger organisms have disproportionally influenced classical biogeographic theory; unlike protists, they have very small discrepancies between their potential (historically derived biogeographic range) and their realised (ecologically derived) ranges.

Plants and fungi are closer to protists in this respect – for example tropical tree fruits/seeds can be transported transcontinentally with ocean currents and may or may not become established, thus potentially expanding their true range. But such seeds washed up on northern European shores are not likely or expected to establish themselves; therefore it would be nonsense to say that their range includes northern Europe. However, the massive (and mostly unknown) diversity of protists and the difficulty of knowing whether they are active or not (or even present) means that such judgements are much harder to make for members of the microbial world. In plants with very small dispersal agents (e.g. orchid seeds) and sporophore-forming fungi dispersal is likely to be easier and wider per spore/seed than for tropical tree seeds, but because the lifecycle includes a conspicuous multicellular stage the distinction from protists stands. (Note: it is possible that many such dispersal agents are as easily transported as some protists, but orchids and many macro-fungi show strong historical biogeographic structuring. This is not a real challenge to the ubiquitous dispersal theory for protists as numbers of fungal and plant spores are many orders of magnitude lower than protist cells/cysts on a global scale.) Therefore, as well as a lack of a significant cryptic seedbank of multicellular organisms there is also a conceptual disjunction between the idea of 'distribution range' when applied to protists and multicells.

The relationship between geographic distance and ecological distance is far from understood in respect to protists, and the two concepts are often conflated in discussions of microbial biogeography, in many cases unavoidably as discussed further in section 6.4. All claims of endemism should ideally be clearly qualified by what the causes of such restricted ranges are, and the ecological conditions of sampling and detection methods reported in such studies closely scrutinised to ascertain whether an ecological explanation could in fact better explain a distribution pattern than a historical one. It is important that the cells are dispersed in a viable form, i.e. that they have the potential to grow and establish themselves wherever they reach. If they do not grow following introduction because of local conditions, including competitive exclusion from another strain, however closely related, then an ecological, not historical, explanation is required. If parts of the cell (for example an inorganic test or scale) are dispersed across greater distances than the organism can survive then a misleading historically mediated biogeographic distribution may result.

The means by which dispersal can be effected to conform to EiE are effectively unlimited. Purists are likely to cite such 'natural' mechanisms such as wind currents and precipitation and translocation by organismal vectors (internally or externally), but human-mediated transport, incidentally as a result of human travel around the globe by sea, land and air, and also deliberately (if such a thing were to occur) are valid dispersal mechanisms under the terms of EiE (e.g. Wilkinson, 2010).

6.3 Taxonomic units in protists and their relevance to the EiE debate

What level of phylogenetic resolution is appropriate for defining species/operational taxonomic units (OTUs) in protists? By extension, what can/does EiE mean by 'everything': how do we define the 'things'? It is increasingly apparent that this level differs significantly among groups of organisms (because genetic marker and trait evolution are not coupled). Thus it should be decided on a group-by-group basis, integrating molecular phylogenetic, morphological (ideally including ultrastructural), eco-physiological, functional and niche-inferred characteristics to distinguish between biologically informative taxonomic units.

Boenigk (2008b) uses the 'oldest' nanoflagellate genus *Monas* Müller 1773 to provide a historically far-reaching analysis of classification problems that are unique to or most pronounced in protists. These range from an initial uncertainty about how protists are related to all other life (and indeed whether they arose by abiogenesis or spontaneous generation, which would pose a completely different set of questions regarding biogeography!) to more modern concerns about reconstructing evolutionary histories, which is really only possible in protists using multigene molecular phylogenetics; a process that is not straightforward and in many respects still unresolved. As stated above, one of the major findings of molecular phylogenetics is that morphologically similar protists may be (and often are) unrelated: organisms originally affiliated with *Monas* are now known to belong to all major eukaryotic lineages. Levels of morphological conservation and convergent evolution in protists are generally very high (Nikolaev et al., 2004; Richards et al., 2006; Lilly et al., 2007; Boenigk, 2008a, 2008b).

Until molecular biology progressed to the point where it was relatively easy to obtain and analyse gene sequences from protist isolates the morphospecies was used as the basis taxonomic unit for protists (Finlay et al., 1996; Finlay, 2004). However, as explained above molecular analyses have shown that most if not all morphospecies harbour high levels of genetic variability – many comprising as much variation in the rRNA genes as in whole families, orders or classes of 'higher' organisms. These high levels of intra-morphospecies genetic diversity result from both morphological conservation and convergent evolution, rendering the morphospecies misleading and inadequate for the purpose of inferring evolutionary relationships (e.g. Lilly et al., 2007). Unsurprisingly, morphospecies also conflate ecophysiologically distinct lineages within a single 'taxon', thereby obscuring biologically meaningful differences in traits such as tolerance to different salinity and pH levels, temperatures, etc. and behavioural differences (e.g. Finlay, 2004; Boenigk et al., 2005, 2007; Scheckenbach et al., 2006; Bass et al., 2007; Foulon et al., 2008; Pfandl et al., 2009).

Recent work has shown that, across a wide range of sexual eukaryotes, biological species boundaries coincide with particular compensatory base changes in the ITS2 rDNA region (Amato et al., 2007; Coleman, 2007; Müller et al., 2007). The ITS (intergenic transcribed spacer) regions evolve much faster than the SSU and the majority of the LSU and therefore are able to resolve differences between closely related lineages more powerfully than the SSU alone (e.g. De Jonckheere, 1998; Amato et al., 2007; Coleman, 2007). Therefore, in sexual species it can be demonstrated that the SSU does not evolve fast enough to resolve species-level differences between lineages. For the vast majority of protist species sexuality is unknown, so the biological species concept cannot be applied. However, taxonomic units are still required in such organisms and it is arguably as valid to search for correspondences between marker genes and 'species'-level differences in asexuals as it is in sexuals, as long as the relationship between genotype and phenotype is consistently found. In asexual taxa (or those of unknown sexuality) consistent differences between lineages that are biologically informative such as salinity tolerance, growth vigour and propensity to form cysts can be found between different ITS-defined lineages (ITS-types) within a single SSU-type, e.g. Bass et al. (2007). This study also showed that some of these ITS-types had cosmopolitan distributions whereas others showed signs of regionally restricted distributions, although these may disappear with greater sampling effort. Whether it is agreed that these are separate 'species' or not is not the subject of this chapter. What is important is that we can consistently identify evolutionary units at a level of phylogenetic resolution that is appropriate for understanding the functional diversity of lineages in microbial communities.

At even higher levels of phylogenetic resolution it is increasingly apparent that real biological (morphological, ecological, etc.) differences can also be found *within* ITS-types, suggesting that even these hypervariable genetic markers are not resolving enough to distinguish between biological distinct entities in some organisms. Logares et al. (2007) provide the example of dinoflagellate morphospecies *Scrippsiella hangoei* and *Peridinium aciculiferum*, which show clear morphological differences but have identical ITS1, 5.8S, ITS2 and partial LSU rDNA sequences. Barth et al. (2006) showed that ITS regions were too conserved for intraspecific analyses of *Paramecium* spp., and that the faster-evolving mitochondrial cytochrome c oxidase I (COI) gene sequences were better suited to this purpose. Further examples include 'wrack' seaweeds (*Fucus* spp.), where morphological radiation in some species has occurred so rapidly that there are only low levels of genetic divergence among them, even in the quickly evolving ITS regions (Leclerc et al., 1998). In such cases it is possible that morphologically defined species could have historical biogeographies while their rDNA genotypes could be more widespread or even cosmopolitan.

This makes the cases of putative geographically restricted morphotypes mentioned in the introduction (e.g. *Apodera vas*) potentially all the more interesting. Do at least some of these morphotypes represent taxa in which morphological evolution is rapid in relation to genetic marker evolution and dispersal rate (as proposed by Logares et al. (2007) and as for the *Urocentrum turbo*-like isolate in an alpine anoxic lake (Stoeck et al., 2007a))? It is obvious that evolution does not proceed in a regular, clock-like manner. Acquisition of characters with strong selective advantages can be relatively fast (e.g. mitochondria, chloroplasts, secondary endosymbioses) and can lead to rapid radiation of the 'new' lineage. Different lineages appear to adapt to similar ecophysiological challenges at different rates relative to the evolution of a single marker gene. For instance, in some protist groups marine and non-marine strains are very similar in rDNA sequences (e.g. Hoef-Emden, 2008), whereas in others the rDNA sequence differences are much greater (von der Heyden et al., 2004; von der Heyden and Cavalier-Smith, 2005; Bass et al., 2009b). Cases of disproportionally rapid marker evolution are increasingly well known, for example in Foraminifera (Pawlowski and Berney, 2003), cyclotrichiid ciliates (Johnson et al., 2004; Bass et al., 2009c), and many parasitic lineages (Cavalier-Smith, 2004).

Even where higher phylogenetic resolution than that offered by rDNA is not required, the use of other markers is desirable at least to provide confirmation of rDNA results. Concordance between markers cannot be taken for granted: Przybos et al. (2009) showed that rDNA and COI sequences from individual strains in the *Paramecium aurelia* complex produced discordant phylogenetic tree topologies, which would obviously confound phylogeographic inferences. Multiple markers can also provide greater resolution, and a more genomic view of lineage radiation, adaptation and occurrence. Such considerations are fundamental to protist biogeography precisely because it is important to know in biologically functional terms *what* is where. Logares et al. (2009) analysed amplified fragment length polymorphism (AFLP) variation – a highly multi-locus, genomic fingerprinting approach – within dinoflagellate strains with identical rRNA cistron sequences. Some of these co-occurred within a single lake suggesting some form of sexual and/or ecological isolation between them; differential distribution patterns in others appeared associated with known ecological shifts. NGS technologies potentially provide sufficient sequencing capacity to associate such genomic data with biological function. However, as powerful as genomic approaches may be, they are not currently applicable to the culture-independent probing such as is possible with group-specific rDNA PCR primers, which relieved microscopists of the often impossible task of identifying a particular strain in sufficiently large numbers of environmental samples from around the world to be able to even begin to infer its distribution. The intensive genome-scale molecular approaches described above similarly require the isolation of individual cells or cultures from the environment. One limited solution to this issue is to analyse cDNA created from ribosomal RNA genes (Stoeck et al., 2007b), which, if analysed quantitatively using real time PCR could provide a measure of the relative activity of targeted lineages in different environments, thereby providing a more 'functional'/niche-directed perspective on their occurrence and distribution than the more activity-independent rDNA-based approach.

For biogeographic investigations it is important to have a choice of genetic markers and to know how best to employ them depending on the group of organisms and specific question being asked. We believe that ITS rDNA regions have a general value in this context because of their universality, methodological tractability and increasing database representation. We are not, by extension, advocating ITS rDNA as the best 'barcoding gene' for eukaryotes as a whole (although it may turn out to be). The use of accepted barcoding genes in protistan environmental studies would clearly be valuable, but in many cases is not possible or optimal. For example COI has been shown to be a very good barcoding gene in many groups (e.g. Frezal and Leblois, 2008; Ward et al., 2009), but in protists its general use is limited as knowledge and database representation of this gene is severely limited (or even absent) in many groups, and it is not universal as it is absent in amitochondrial taxa.

6.4 Is anything really 'everywhere' and how should we look?

The third key issue is the 'everywhere' element of the EiE statement. It is conceptually easy to show that a taxonomic unit (at whatever level) is globally distributed – i.e. is present in at least some suitable habitat in each biogeographic region of the world (e.g. Darling et al., 2004; Slapeta et al., 2006; Bass et al., 2007), although

SSU rDNA detection by PCR cannot distinguish between active and inactive or inviable cells, or even extracellular DNA. (The important distinction between cells being viable or inviable due to failure to survive the dispersal process was made earlier.) One way around this is to use rRNA-derived measures (as described above) to bias sampling towards metabolically active cells. However, for there to be a rDNA signal at all indicates that the genotype in question *has* been distributed to the site investigated, probably quite recently and very probably with the cell in a viable state if the DNA is still abundant and intact enough to detect repeatably in what is inevitably a minute proportion of the available sample-space (see below). Conversely, it is impossible to prove that a protist lineage is not/cannot be dispersed to a particular site, nothwithstanding the examples given above of morphospecies discovered in less well-sampled areas but never in apparently suitable habitats in much more intensively sampled regions. This is an obvious point (and is the reason why EiE is not a scientific hypothesis) but it highlights interesting aspects of protist ecology and the ways in which it is studied.

The necessary nature of environmental samples and their collection for either microscopic or molecular investigation imposes constraints on biogeographic studies, although these are difficult to quantify (ironically partly because of lack of information of the kind that such samples are intended to provide). For both molecular and morphological analyses individual samples need to be small (ideally and usually in the order of one gram of soil or sediment) and the intensity with which each sample can be investigated by both approaches is limited by time and financial constraints. As an example, imagine an apparently homogeneous meadow-like environment. How many samples (and what size of samples) are required to comprehensively survey the protist group(s) of interest within the meadow and to detect relatively rare lineages therein? An impossibly high number, for very rare taxa. For more abundant and diverse taxa, other problems arise. Howe et al. (2009) and Bass *et al.* (unpublished data) found that individual protist genera can be represented by at least tens of SSU genotypes in less than a gram of soil, that sets of genotypes from independent samples across a small site share only 30–40% of those genotypes, and that actual diversity even within the samples analysed may be several times greater than that actually detected in those studies. In addition there is often little overlap between SSU genotypes detected by molecular techniques and isolation of strains into culture from the same environmental sample, for a range of reasons (Bass et al., 2009a; Howe et al., 2009). The significance of these findings is almost certainly exaggerated by undersampling, but exactly the same difficulties of detection afflict biogeographic studies.

In any case there is very likely to be patchy distribution of protist taxa across a range of spatial scales caused by ecological heterogeneity and/or differences in community assembly processes. This means that it is very difficult to determine what is present even at a single site and within a relatively narrowly defined group of organisms, and that large amounts of screening may be necessary to detect some lineages in at least some of the habitats in which they occur (Foissner, 2008). This problem will be even greater for rare species, about which we know very little (see below). Compounding the ease with which a lineage can fail to be detected is temporal variation. Many, perhaps all protist communities, particularly in temperate zones, show a pronounced seasonality which means that very different sets of species are detectable at different points of the year. Over longer time scales such community shifts (a) may not show a regular annual pattern, (b) may be affected by ecological/climate changes on a range of different timescales, and (c) are likely to be subject to and influenced by changes in community assembly across time as ecosystems mature (Lawton, 2004; Rodriguez Zaragoza et al., 2005; Thackeray et al., 2008, Gilbert et al., 2009; Nolte et al., 2010). Theoretically, even for a single protist taxon, until the exact nature of its niche is known and can be measured and identified it will be extremely difficult to investigate potential niche-space globally to systematically determine its global distribution.

Futhermore, the conflation of geographic distance and ecological difference means that inferences about historical biogeography based on comparing protist communities in different parts of the world are usually unreliable. For example, it is impossible for us to identify ecologically equivalent freshwater ponds in the UK and Australia from the perspective of protist ecology, partly because we do not know enough about the relative importance of the environmental parameters at each site and therefore which ones to measure, but also because we know that the non-microbial biotic community in which the protists live will differ due to historical biogeographic structuring, even if the microbial community is potentially the same at both sites. There will obviously be differences in the higher plants that grow in and around the ponds and whose organic material decays within them, and in the animals that interact with the ponds, and even the planktonic meiofauna which graze on and are parasitised by protists. Therefore to some extent ecological distance increases *as a consequence of* geographic distance; the two cannot be decoupled. As we have seen, there is strong evidence that many protist genotypes (SSU- and ITS-types) are active on a global scale. Do these represent ecologically tolerant, generalist, weedy species analogous to generalist species in higher organisms? If so what proportion of protist species fall into this category?

Generally very little is known about the relative abundances of protist taxa on a global scale. Morphological and molecular analyses of protist communities (using both classical environmental gene cloning and NGS approaches) show concave taxon rank abundance curves familiar from community profiles in many organismal groups (e.g. Countway et al., 2005; Howe et al., 2009). Four hundred and fifty-four sequencing studies suggest that the tail of rare taxa (sometimes called the 'rare biosphere'; Pedrós-Alió, 2006; Sogin et al., 2006; Caron, 2009; Howe et al., 2009; Stoeck et al., 2010) is longer than previously thought (i.e. there are proportionally more rare taxa in all communities). Some of these lineages may be generally rare and/or ecologically specialised, in which case they would be interesting material for global distribution studies: according to the EiE hypothesis such lineages have a greater chance of showing historical biogeographic structuring than more abundant and/or generalist ones.

In fact, recent studies indicate that rather than 'everything', only generalist taxa may be ubiquitously distributed since they can achieve large population sizes and thereby high dispersal rates (Pither, 2007). Similarly, Nolte et al. (2010) suggest a systematic difference in the distribution patterns of abundant and rare taxa. Future analyses of the 'rare' biosphere using NGS methods will have the power to investigate this emerging pattern more intensively. Based on recent NGS data similar distribution patterns of micro- and macroorganisms are more strongly indicated than the capacity of earlier molecular techniques has allowed, i.e. global dispersal of abundant, generalist taxa contrasting with more restricted distributions of many less abundant taxa. Most clone library-based environmental DNA surveys relied on sample sizes below *c.* 1000 sequences (Kemp and Aller, 2004). Thus historic knowledge of protist distribution patterns, from the early nineteenth century to the recent molecular 'low-throughput' surveys, was therefore largely limited to the biogeography of relatively abundant taxa. Next generation sequencing approaches enable consistent detection of many more rare taxa than earlier molecular methods, and perhaps the detection of particularly rare taxa which were previously completely unknown.

6.5 Conclusions

In the light of the preceding discussion it is perhaps not surprising that most (although not all) known morphospecies have cosmopolitan distributions. It is also true that many SSU rDNA genotypes are also globally distributed (e.g. Finlay et al., 2006; Slapeta et al., 2006, Bass et al., 2007, 2009c; Darling and Wade, 2008) although we suggest that a higher proportion of SSU genotypes than of morphospecies show some degree of historical biogeographic structure. Faster-evolving markers have been less frequently used in biogeography studies, but cosmopolitan distributions of ITS rDNA genotypes have been demonstrated in freshwater bodies and soil (habitats which are arguably more likely to promote endemism due to their temporal transience and disjunct/ecologically heterogeneous nature) (Barth et al., 2006; Bass et al., 2007) and of COI and other protein-coding genes (beta-tubulin, chloroplast rbcL) in the marine picoplanktonic *Micromonas pusilla* (Slapeta et al., 2006). Therefore even the fastest-evolving markers so far employed in biogeographic studies can show cosmopolitan distributions of their host cells. Conversely, other genotypes of fast-evolving markers can show strong signs of restricted distributions (e.g. Bass et al., 2007; Darling et al., 2007; Darling and Wade, 2008).

A general principle emerges that is well illustrated by Fontaneto et al.'s (2008) analysis of COI sequences from global collections of the bdelloid rotifer genera *Adineta* and *Rotaria*. They related COI diversity to three different taxonomic units, in order from lowest to highest level of phylogenetic resolution: (1) traditional morphology-based species, (2) the most inclusive monophyletic clades (according to COI) containing a single morphospecies and (3) genetic clusters indicative of independently evolving lineages. Even at the highest levels of resolution (individual COI clusters) some lineages were cosmopolitan (although most were restricted to continents or smaller regions), but the general trend was of increasing endemicity with increasing phylogenetic resolution. This trend is generally true for all organisms but is more difficult to interpret in protists due to the relative lack of knowledge and variability of the relationship between molecular phylogeny and taxonomy than is the case for most multicellular organisms (Fig 6.1).

Furthermore, we are only just beginning to explore the 'rare' protistan biosphere. As many 'rare' protist taxa achieve population sizes of millions or even billions of individuals in a given habitat, these taxa cannot be considered rare in the classical sense, but appear so because we were not able to detect them with earlier methods due to the fact that some other protist taxa were even more abundant by orders of magnitude. These 'rare' – or more accurately 'less abundant' – protist taxa cannot be ignored in a biogeography debate and they probably account for the vast majority of protist lineage diversity (Pither, 2007; Bass et al., 2009a, 2009b; Nolte et al., 2010).

Thus we conclude that survey-based approaches to understanding protist biogeographic patterns are strongly and multiply confounded by difficulties of comprehensive sampling, taxonomic definitions and separating ecological from historical effects. We strongly assert that the total number of valid, free-living and extant protist taxonomic units, whether referred to as species or not, far exceeds the 90 000 estimated by Corliss (2000), the morphospecies-biased, depressed estimates of Finlay and Fenchel (1999), and even the revised figure of 300 000 suggested by Foissner (2008), by at least an order of magnitude. In this light (and its ecological consequences) the protist biogeography debate becomes immeasurably complex.

6.6 Prospects

Next generation sequencing technologies offer a huge increase in the depth to which environmental samples can be analysed (e.g. Sogin et al., 2006; Buée et al., 2009; Creer et al., 2010; Medinger et al., 2010; Nolte et al., 2010; Stoeck et al., 2010).

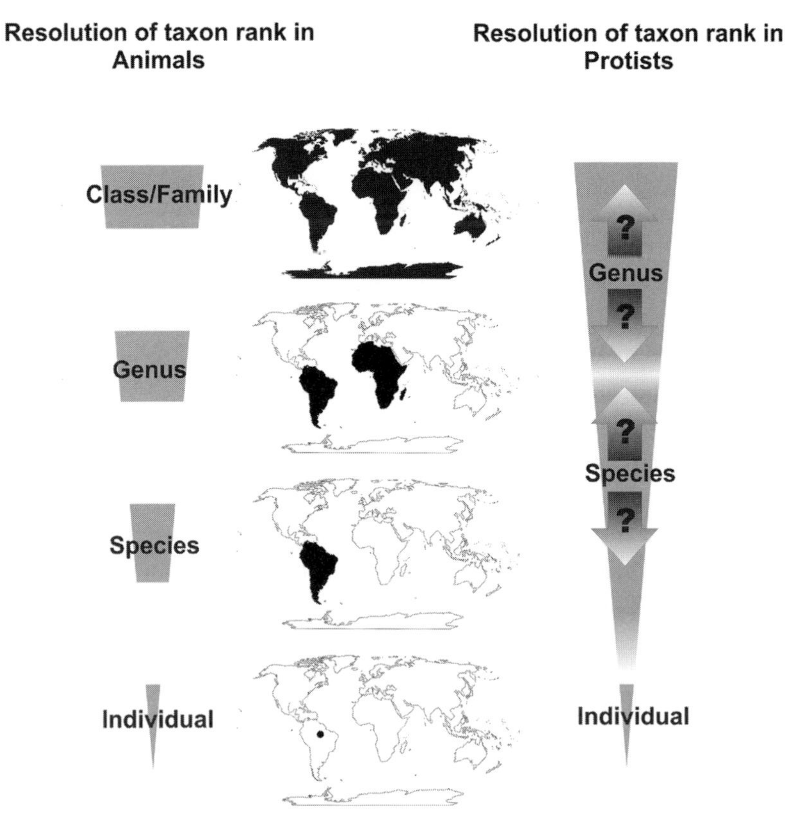

Fig 6.1 Biogeography is linked to taxon resolution. Higher-level taxa have wider distribution patterns. At the opposite extreme, an individual has a distinct occurrence at one point in space. Within a given taxon the distribution area decreases with finer taxon resolution. Species concepts in protists are unclear and often incomparable to the rank of species in higher eukaryotes. Therefore, for protists, distribution patterns are strongly interrelated with species delimitation. Higher eukaryotes with well-defined taxa and a high phylogenetic resolution of the species rank often have biogeographies. For protists the species rank is usually much broader and less well defined, thereby contributing to the perception of broad distribution.

However, this only ameliorates some of the problems associated with molecular detection of protist lineages on a global scale. Apart from the conceptual issues with defining protist OTUs there remain other methodological constraints. These constraints relate to both molecular biology and environmental sampling. In the former category, biases and artefacts in PCR primer activity and amplicon generation will still influence the likelihood of detecting particular genotypes and result in a skewed representation of the genotype profile of protist communities (e.g. Bass and Cavalier-Smith, 2004; Berney et al., 2004; Richards and Bass, 2005;

Stoeck et al., 2006, 2010 and references therein). In the latter category, choice of sampling sites, knowledge of protist niche parameters and the nature of habitat heterogeneity remain limiting as we currently know too little about them to be able to sample in a demonstrably consistent and systematic way. The *number* of independent samples required for adequate environmental surveying is at least as big a limitation for NGS as for 'classical' environmental gene cloning.

We suggest that a more experimental approach is required in the field of protist biogeography, testing specific ecological hypotheses. Much of the work that has been stimulated by EiE has attempted to identify lineages that show some degree of endemism, but as we have shown this is tackling the problem from a highly untractable perspective. It would instead be very enlightening and scientifically more robust to investigate which and how many lineages *do* conform to EiE, and to learn about their ecology and general biology. Questions to be addressed include: Are such lineages highly tolerant to stress and fluctuating environments (at least at population levels), and/or able to remain in a viable dormant form for long periods? Are they globally abundant, fast-growing, weedy species, with very small dispersal agents? Such factors together must contribute towards the 'dispersal pressure' of protist populations, but we know little about their relative importance or how long this list of factors is. Knowing more about the characteristics and conditions conducive to rapid and wide dispersal will provide a firmer baseline for describing other kinds of distributions and understanding the processes involved in creating them. It will also provide insight into which organisms (and therefore ecological agents) potentially provide the same ecosystem 'service' globally. Some studies (but currently a minority) have integrated community-level ecological processes with biogeographic considerations, for example Telford et al. (2006).

Almost nothing is known about the actual rates at which protist populations are dispersed on large scales. Rapid, long-distance dispersal of small particles is well demonstrated by the movement of weather systems around the globe, migrating animals and birds, human mobility, and very noticeably by the eruption of the volcano near the Eyjafjallajökull glacier (Iceland) in April 2010, which sent ash plumes visibly across Europe, dispersing innumerable particles (much larger than many protists) on continental scales and probably globally. Therefore the potential for equally rapid distribution of at least some protist lineages is huge. But what are the survival and successful colonisation rates (and associated biological processes) that ultimately determine the *effective* dispersal rate of widespread lineages? Some insight into dispersal rate may be gained by isolating into culture individual protist strains with apparently cosmopolitan, highly resolving genotypes (ITS rDNA, COI, etc.) collected from sites around the world, and then analysing them in terms of markers that evolve fast enough to be measured in the laboratory.

It would also be interesting to test examples of putatively geographically restricted morphospecies using highly specific molecular probes. Perhaps many of these taxa are more widespread than currently known but have not been detected because they are for some reason(s) relatively rare outside of a 'core' range. The examples given in the introduction of morphospecies apparently largely restricted to the southern hemisphere – i.e. not yet found in large areas of apparently suitable habitats in much more intensively sampled regions of the northern hemisphere – are intriguing as there are many vectors that could transport them from south to north. Unless they are very intolerant to removal from favourable conditions (as appears to be the case for some freshwater, desiccation-intolerant chrysophytes; Kristiansen, 2008) it would seem unlikely that they could not reach and grow in suitable conditions far beyond their current ranges. However, other factors come into play; for example it is increasingly clear that there is a strong effect of size on dispersal of testate amoebae: larger taxa (e.g. *Apodera* and *Certecella*) are more likely to show restricted global distributions than small taxa (Yang et al., 2010). Such cases offer potentially informative systems for studying biological factors resulting in limited dispersal and establishment, and also illustrate how non-trivial the separation of ecological and historical biogeographic processes can be when analysing protist distributions. For instance, is the *Urocentrum turbo*-like ciliate so far found only in an alpine anoxic lake by Stoeck et al. (2007a) more widespread in similar ecological conditions (assuming they can be found)?

Morphological and molecular analyses make very clear that most protistan diversity is currently unknown (e.g. Chao et al., 2006; Stoeck et al., 2010). However scientifically unfashionable it may become, it is important that we continue to discover and describe new protist taxa. There are not many reliable data about which free-living protists are globally rare, or highly specialised. Studying these would help to resolve to what extent morphology, abundance, population size and ecological generalism are drivers of global dispersal of protists. The overview in this chapter strongly reinforces the necessity of looking at each group of organisms individually and with fresh eyes. It is the differences between them, where these can be shown to be robust and then investigated in a hypothesis-driven framework, that could prove to be the most biologically interesting and informative sources of biogeographic insight.

References

Amato, A., Kooistra, W.H., Ghiron, J.H. et al. (2007). Reproductive isolation among sympatric cryptic species in marine diatoms. *Protist* **158**, 193–207.

Baas Becking, L.G.M. (1934). *Geobiologie of inleiding tot de milieukunde*. The Hague: Van Stockum and Zoon.

Barth, D., Krenek, S., Fokin, S.I., Berendonk, T.U. (2006). Intraspecific genetic variation in *Paramecium* revealed by mitochondrial cytochrome c oxidase I sequences. *Journal of Eukaryotic Microbiology* **53**, 20–25.

Bass, D., Cavalier-Smith, T. (2004). Phylum-specific environmental DNA analysis reveals remarkably high global biodiversity of Cercozoa (Protozoa). *International Journal of Systematic and Evolutionary Microbiology* **54**, 2393–2404.

Bass, D., Richards, T.A., Matthai, L., Marsh, V., Cavalier-Smith, T. (2007). DNA evidence for global dispersal and probable endemicity of protozoa. *BMC Evolutionary Biology* **7**, 162.

Bass, D., Howe, A.T., Mylnikov, A.P. et al. (2009a). Phylogeny and classification of Cercomonadida: *Cercomonas, Eocercomonas, Paracercomonas*, and *Cavernomonas* gen. n. *Protist* **160**, 483–521.

Bass, D., Chao, E.E., Nikolaev, S. et al. (2009b). Phylogeny of novel naked filose and reticulose Cercozoa: Granofilosea cl. n. and Proteomyxidea revised. *Protist* **160**, 75–109.

Bass, D., Brown, N., Mackenzie-Dodds, J. et al. (2009c). A molecular perspective on ecological differentiation and biogeography of cyclotrichiid ciliates. *Journal of Eukaryotic Microbiology* **56**, 559–567.

Beijerinck, M.W. (1913). De infusies en de ontdekking der backteriën. *Jaarboek van de Koninklijke Akademie voor Wetenschappen*. Amsterdam: Müller. (Preprinted in Verzamelde geschriften van M.W. Beijerinck, vijfde deel, pp. 119–140. Delft, 2921).

Berney, C., Fahrni, J., Pawlowski, J. (2004). How many novel eukaryotic 'kingdoms'? Pitfalls and limitations of environmental DNA surveys. *BMC Biology* **2**, 13.

Boenigk, J. (2008a). Nanoflagellates: functional groups and intraspecific variation. *Denisia* **23**, 331–335.

Boenigk, J. (2008b). The past and present classification problem with nanoflagellates exemplified by the genus *Monas*. *Protist* **159**, 319–337.

Boenigk, J., Pfandl, K., Stadler, P., Chatzinotas, A. (2005). High diversity of the '*Spumella*-like' flagellates: an investigation based on the SSU rRNA gene sequences of isolates from habitats located in six different geographic regions. *Environmental Microbiology* **7**, 685–697.

Boenigk, J., Pfandl, K., Garstecki, T. et al. (2006). Evidence for geographic isolation and signs of endemism within a protistan morphospecies. *Applied and Environmental Microbiology* **72**, 5159–5164.

Boenigk, J., Jost, S., Stoeck, T., Garstecki, T. (2007). Differential thermal adaptation of clonal strains of a protist morphospecies originating from different climatic zones. *Environmental Microbiology* **9**, 593–602.

Buée, M., Reich, M., Murat, C. et al. (2009). 454 pyrosequencing analyses of forest soils reveals an unexpectedly high fungal diversity. *New Phytologist* **184**, 449–456.

Caron, D.A. (2009). Past President's address: protistan biogeography: why all the fuss? *Journal of Eukaryotic Microbiology* **56**, 105–112.

Cavalier-Smith, T. (2004). Only six kingdoms of life. *Proceedings of the Royal Society of London B Biological Sciences* **271**, 1251–1262.

Chao, A., Li, P.C., Agatha, S. et al. (2006). A statistical approach to estimate soil ciliate diversity and distribution based on data from five continents. *Oikos* **114**, 479–493.

Coesel, P.F.M., Krienitz, L. (2008). Diversity and geographic distribution of desmids and other coccooid green algae. *Biodiversity and Conservation* **17**, 381–392.

Coleman, A.W. (2007). Pan-eukaryote ITS2 homologies revealed by RNA secondary structure. *Nucleic Acids Research* **35**, 3322–3329.

Corliss, J.O. (2000). Biodiversity, classification, and numbers of species of protists. In Raven, P.H., Williams, T. (eds.), *Nature and Human Society. The Quest for a Sustainable World,* pp. 130–155. Washington, DC: National Academy Press.

Countway, P.D., Gast, R.J., Savai, P., Caron, D.A. (2005). Protistan diversity estimates based on 18S rDNA from seawater incubations in the western North Atlantic. *Journal of Eukaryotic Microbiology* **52**, 95.

Creer, S., Fonseca, V.G., Porazinska, D.L. et al. (2010). Ultrasequencing of the meiofaunal biosphere: practice, pitfalls and promises. *Molecular Ecology* **19**, 4–20.

Darling, K.F., Wade, C.M., Stewart, I.A. et al. (2004). Molecular evidence for genetic mixing of Arctic and Antarctic subpolar populations of planktonic foraminifers. *Nature* **405**, 43–47.

Darling, K.F., Kucera, M., Wade, C.M. (2007). Global molecular phylogeography reveals persistent Arctic circumpolar isolation in marine planktonic protist. *Proceedings of the National Academy of Sciences USA* **104**, 5002–5007.

Darling, K.F., Wade, C.M. (2008). The genetic diversity of planktic foraminifera and the global distribution of ribosomal RNA genotypes. *Marine Micropaleontology* **67**, 216–238.

De Jonckheere, J.F. (1998). Sequence variation in the ribosomal internal transcribed spacers, including the 5.8S rDNA, of *Naegleria* spp. *Protist* **149**, 221–228.

De Wit, R., Bouvier, T. (2006). 'Everything is everywhere, but the environment selects'; what did Baas Becking and Beijerinck really say? *Environmental Microbiology* **8**, 755–758.

Fenchel, T., Finlay, B.J. (2004). The ubiquity of small species: patterns of local and global diversity. *BioScience* **54**, 777–784.

Fierer, N. (2008). Microbial biogeography: patterns in microbial diversity across space and time. In Zengler, K. (ed.), *Accessing Uncultivated Microorganisms: from the Environment to Organisms and Genomes and Back,* pp. 95–115. Washington, DC: ASM Press.

Finlay, B.J. (2002). Global dispersal of free-living microbial eukaryote species. *Science* **296**, 1061–1063.

Finlay, B.J. (2004). Protist taxonomy: an ecological perspective. *Philosophical Transactions of the Royal Society of London B* **359**, 599–610.

Finlay, B.J., Clarke, K.J. (1999). Apparent global ubiquity of species in the protist genus *Paraphysomonas. Protist* **150**, 419–430.

Finlay, B.J., Fenchel, T. (1999). Divergent perspectives on protist species richness. *Protist* **150**, 229–233.

Finlay, B.J., Fenchel, T. (2004). Cosmopolitan metapopulations of free-living microbial eukayotes. *Protist* **155**, 237–244.

Finlay, B.J., Corliss, J.O., Esteban, G., Fenchel, T. (1996). Biodiversity at the microbial level: the number of free-living ciliates in the biosphere. *Quaternary Review of Biology* **71**, 221–237.

Finlay, B.J., Esteban, G.F., Brown, S., Fenchel, T., Hoef-Emden, K. (2006). Multiple cosmopolitan ecotypes within a microbial eukaryote morphospecies. *Protist* **157**, 377–390.

Foissner, W. (1999). Protist diversity: estimates of the near-imponderable. *Protist* **150**, 363–368.

Foissner, W. (2004a). Ubiquity and cosmopolitanism of protists questioned. *SIL News* **43**, 6–7.

Foissner, W. (2004b). Two new "flagship" ciliates (Protozoa. Ciliophora) from Venezuela: *Sleighophrys pustulata* and *Luporinophrys micelae*. *European Journal of Protistology* **41**, 99–117.

Foissner, W. (2006). Biogeography and dispersal of micro-organisms: a review emphasizing protists. *Acta Protozoologica* **45**, 111–136.

Foissner, W. (2008). Protist diversity and distribution: some basic considerations. *Biodiversity and Conservation* **17**, 235–242.

Foissner, W., Strüde-Kypke, M., van der Staay, G.W.M., Moon-van der Staay, S.-Y., Hackstein, J.H.P. (2003). Endemic ciliates (Protozoa, Ciliophora) from tank bromeliads (Bromeliaceae): a combined morphological, molecular, and ecological study. *European Journal of Protistology* **39**, 365–372.

Fontaneto, D., Barraclough, T.G., Chen, K., Ricci, C., Herniou, E.A. (2008). Molecular evidence for broad-scale distributions in bdelloid rotifers: everything is not everywhere but most things are very widespread. *Molecular Ecology* **17**, 3136–3146.

Foulon, E., Not, F., Jalabert, F. et al. (2008). Ecological niche partitioning in the picoplanktonic green alga Micromonas pusilla: evidence from environmental surveys using phylogenetic probes. *Environmental Microbiology* **10**, 2433–2443.

Frezal, L., Leblois, R. (2008). Four years of DNA barcoding: current advances and prospects. *Infection Genetics and Evolution* **8**, 727–736.

Gilbert, J.A., Field, D., Swift, P. et al. (2009). The seasonal structure of microbial communities in the Western English Channel. *Environmental Microbiology* **11**, 3132–3139.

Guillou, L., Eikrem, W., Chrétiennot-Dinet, M.J. et al. (2004) Diversity of picoplanktonic prasinophytes assessed by direct nuclear SSU rDNA sequencing of environmental samples and novel isolates retrieved from oceanic and coastal marine ecosystems. *Protist* **155**, 193–214.

Habura, A., Pawlowski, J., Hanes, S.D., Bowser, S.S. (2004). Unexpected foraminiferal diversity revealed by small-subunit rDNA analysis of Antarctic sediment. *Journal of Eukaryotic Microbiology* **51**, 173–179.

Hoef-Emden, K. (2007). Revision of the genus *Cryptomonas* (Cryptophyceae) II: incongruences between the classical morphospecies concept and molecular phylogeny in smaller pyrenoid-less cells. *Phycologia* **46**, 402–428.

Hoef-Emden, K. (2008). Molecular phylogeny of phycocyanin-containing cryptophytes: evolution of biliproteins and geographical distribution. *Journal of Phycology* **44**, 985–993.

Howe, A.T., Bass, D., Vickerman, K., Chao, E.E., Cavalier-Smith, T. (2009). Phylogeny, taxonomy, and astounding genetic diversity of

Glissomonadida ord. nov., the dominant gliding zooflagellates in soil (Protozoa: Cercozoa). *Protist* **160**, 159–189.

Hughes Martiny, J.B., Bohannan, B.J.M., Brown, J.H. et al. (2006). Microbial biogeography: putting microorganisms on the map. *Nature Reviews Microbiology* **4**, 102–112.

Johnson, M.D., Tengs, T., Oldach, D.W., Delwiche, C.F., Stoecker, D.K. (2004). Highly divergent SSU rRNA genes found in the marine ciliates *Myrionecta rubra* and *Mesodinium pulex*. *Protist* **155**, 347–359.

Kemp, P.F., Aller, J.Y. (2004). Estimating prokaryotic diversity: when are 16S rDNA libraries large enough? *Limnology and Oceanography: Methods* **2**, 114–125.

Koch, T.A., Ekelund, F. (2005). Strains of the heterotrophic flagellate *Bodo designis* from different environments vary considerably with respect to salinity preference and SSU rRNA gene composition. *Protist* **156**, 97–112.

Kristiansen, J. (2008). Dispersal and biogeography of silica-scaled chrysophytes. *Biological Conservation* **17**, 419–426.

Lawton, J.H. (2004). Japan Prize commemorative lecture: biodiversity, conservation and sustainability. *Notes and Records of the Royal Society of London* **58**, 321–333.

Leclerc, M.C., Barriel, V., Lecointre, G., de Reviers, B. (1998). Low divergence in rDNA ITS sequences among five species of *Fucus* (Phaeophyceae) suggests a very recent radiation. *Journal of Molecular Evolution* **46**, 115–120.

Lilly, E.L., Halanych, K.M., Anderson, D.M. (2007). Species boundaries and global biogeography of the *Alexandrium tamarense* complex (Dinophyceae). *Journal of Phycology* **43**, 1329–1338.

Logares, R., Rengefors, K., Kremp, A. et al. (2007). Phenotypically different microalgal morphospecies with identical ribosomal DNA: a case of rapid adaptive evolution? *Microbial Ecology* **53**, 549–561.

Logares, R., Boltovskoy, A., Bensch, S., Laybourn-Parry, J., Rengefors, K. (2009). Genetic diversity patterns in five protist species occurring in lakes. *Protist* **160**, 301–317.

Lowe, C.D., Day, A., Kemp, S.J., Montagnes, D.J.S. (2005). There are high levels of functional and genetic diversity in *Oxyrrhis marina*. *Journal of Eukaryotic Microbiology* **52**, 250–257.

Medinger, R., Nolte, V., Vinay Pandey, R. et al. (2010). Diversity in a hidden world: potential and limitation of next-generation sequencing for surveys of molecular diversity of eukaryotic microorganisms. *Molecular Ecology* **19**, 32–40.

Mindell, D.P., Honeycutt, R.L. (1990). Ribosomal RNA in vertebrates: evolution and phylogenetic applications. *Annual Review of Ecology and Systematics* **21**, 541–566.

Mitchell, E.A.D., Meisterfeld, C. (2005). Taxonomic confusion blurs the debate on cosmopolitanism versus local endemism of free-living protists. *Protist* **156**, 263–267.

Moreira, D., López-García, P. (2002). The molecular ecology of microbial eukaryotes unveils a hidden world. *Trends in Microbiology* **10**, 31–38.

Müller, T., Philippi, N., Dandekar, T., Schultz, J., Wolf, M. (2007). Distinguishing species. *RNA* **13**, 1469–1472.

Nanney, D.L., Park, C., Preparata, R., Simon, E.M. (1998). Comparison of sequence differences in a variable 23S rRNA domain among sets of cryptic species of ciliated protozoa. *Journal of Eukaryotic Microbiology* **45**, 91–100.

Nikolaev, S.I., Berney, C., Fahrni, J.F. et al. (2004). The twilight of Heliozoa and rise of Rhizaria, an emerging supergroup of amoeboid eukaryotes. *Proceedings of the National Academy of Sciences USA* **101**, 8066–8071.

Nolte, V., Pandey, R.V., Jost, S. et al. (2010). Contrasting seasonal niche separation between rare and abundant taxa conceals the extent of protist diversity. *Molecular Ecology* **79**, 2908–2915.

O'Malley, M.A. (2007). The nineteenth-century roots of 'everything is everywhere'. *Nature Reviews Microbiology* **5**, 647–651.

Pawlowski, J., Berney, C. (2003). Episodic evolution of nuclear small subunit ribosomal RNA gene in the stem-lineage of Foraminifera. In Donoghue, P.C.J., Smith, M.P. (eds.), *Telling the Evolutionary Time: Molecular Clocks and The Fossil Record*, pp. 107–118. Systematics Association special volume 66. London: Taylor and Francis.

Pedrós-Alió, C. (2006). Marine microbial diversity: can it be determined? *Trends in Microbiology* **14**, 257–263.

Pfandl, K., Chatzinotas, A., Dyal, P., Beonigk, J. (2009). SSU rRNA gene variation resolves population heterogeneity and ecophysiological differentiation within a morphospecies (Stramenopiles, Chrysophyceae). *Limnology and Oceanography* **54**, 171–181.

Pither, J. (2007). Comment on "Dispersal limitations matter for microbial morphospecies". *Science* **316**, 1124. Author reply 1124.

Przybos, E., Tarcz, S., Fokin, S. (2009). Molecular polymorphism of *Paramecium tetraurelia* (Ciliophora, Protozoa) in strains originating from difference continents. *Folia Biologica* **57**, 57–63.

Richards, T.A., Bass, D. (2005). Molecular screening of free-living microbial eukaryotes: diversity and distribution using a meta-analysis. *Current Opinion in Microbiology* **8**, 240–253.

Richards, T.A., Dacks, J.B., Jenkinson, J.M., Thornton, C.R., Talbot, N.J. (2006). Evolution of filamentous plant pathogens: Gene exchange across eukaryotic kingdoms. *Current Biology* **16**, 1857–1864.

Rodriguez Zaragoza, S., Mayzlish, E., Steinberger, Y. (2005). Seasonal changes in free-living amoeba species in the root canopy of *Zygophyllum dumosum* in the Negev Desert, Israel. *Microbial Ecology* **49**, 134–141.

Scheckenbach, F., Wylezich, C., Mylnikov, A.P., Weitere, M., Arndt, H. (2006). Molecular comparisons of freshwater and marine isolates of the same morphospecies of heterotrophic flagellates. *Applied and Environmental Microbiology* **72**, 6638–6643.

Simon, E.M., Nanney, D.L., Doerder, F.P. (2008). The "*Tetrahymena pyriformis*" complex of cryptic species. *Biodiversity and Conservation* **17**, 365–380.

Slapeta, J., Moreira, D., López-García, P. (2005). The extent of protist diversity: insights from molecular ecology of freshwater eukaryotes. *Proceedings of the Royal Society of London B* **272**, 2073–2081.

Slapeta, J., López-García, P., Moreira, D. (2006). Global dispersal and ancient cryptic species in the smallest marine eukaryotes. *Molecular Biology and Evolution* **23**, 23–29.

Smirnov, A.V. (2003). Optimizing methods of the recovery of gymnamoebae from environmental samples: a test of ten popular enrichment media, with some observations on the development of cultures. *Protistology* **3**, 47–57.

Smith, H.G., Wilkinson, D.M. (2007). Not all free-living microorganisms have cosmopolitan distributions – the case of *Nebela* (*Apodera*) *vas* Certes (Protozoa: Amoebozoa: Arcellinida). *Journal of Biogeography* **34**, 1822–1831.

Smith, H.G., Bobrov, A., Lara, E., (2008). Diversity and biogeography of testate amoebae. *Biodiversity and Conservation* **17**, 329–343.

Sogin, M., Morrison, H.G., Huber, J.A. et al. (2006). Microbial diversity in the deep sea and underexplored "rare biosphere". *Proceedings of the National Academy of Sciences USA* **103**, 12115–12120.

Stoeck, T., Hayward, B., Taylor, G.T., Varela, R., Epstein, S.S. (2006). A multiple PCR primer approach to access the microeukaryotic diversity in environmental samples. *Protist* **157**, 31–43.

Stoeck, T., Bruemmer, F., Foissner, W. (2007a). Evidence for local ciliate endemism in an Alpine anoxic lake. *Microbial Ecology* **54**, 478–486.

Stoeck, T., Zuendorf, A., Breiner, H.-W., Behnke, A. (2007b). A molecular approach to identify active microbes in environmental eukaryote clone libraries. *Microbial Ecology* **53**, 328–339.

Stoeck, T., Bass, D., Nebel, M. et al. (2010). Multiple marker parallel tag environmental DNA sequencing reveals a highly complex eukaryotic community in marine anoxic water. *Molecular Ecology* **19 (s1)**, 21–31.

Telford, R.J., Vandvik, V., Birks, H.J. (2006). Dispersal limitations matter for microbial morphospecies. *Science* **312**, 1015.

Thackeray, S.J., Jones, I.D., Maberly, S.C. (2008). Long-term change in the phenology of spring phytoplankton: species-specific responses to nutrient enrichment and climate change. *Journal of Ecology* **96**, 523–535.

Vanormelingen, P., Verleyen, E., Vyverman, W. (2008). The diversity and distribution of diatoms: from cosmopolitanism to narrow endemism. *Biodiversity and Conservation* **17**, 393–405.

von der Heyden, S., Chao, E.E., Cavalier-Smith, T. (2004). Genetic diversity of goniomonads: an ancient divergence between marine and freshwater species. *European Journal of Phycology* **39**, 343–350.

von der Heyden, S., Cavalier-Smith, T. (2005). Culturing and environmental DNA sequencing uncover hidden kinetoplastid biodiversity and a major marine clade within ancestrally freshwater *Neobodo designis*. *International Journal of Systematics and Evolutionary Microbiology* **55**, 2605–2621.

Vyverman, W., Verleyen, E., Sabbe, K. et al. (2007). Historical processes constrain patterns in global diatom diversity. *Ecology* **88**, 1924–1931.

Ward, R.D., Hanner, R., Hebert, P.D.N. (2009). The campaign to DNA barcode all fishes. *Journal of Fish Biology* **74**, 329–356.

Wilkinson, D.M. (2010). Have we underestimated the importance of humans in the biogeography of free-living terrestrial microorganisms? *Journal of Biogeography* **37**, 393–397.

Worden, A.Z. (2006). Picoeukaryote diversity in coastal waters of the Pacific Ocean. *Aquatic Microbial Ecology* **43**, 165–175.

Wuyts, J., De Rijk, P., Van de Peer, Y. et al. (2000). Comparative analysis of more than 3000 sequences reveals the existence of two pseudoknots in area V4 of eukaryotic small subunit ribosomal RNA. *Nucleic Acids Research* **28**, 4698–4708.

Wylezich, C., Mylnikov, A.P., Weitere, M., Arndt, H. (2007). Distribution and phylogenetic relationships of freshwater thaumatomonads with a description of the new species *Thaumatomonas coloniensis* n. sp. *Journal of Eukaryotic Microbiology* **54**, 347–357.

Yang, J., Smith, H.G., Sherratt, T.N., Wilkinson, D.M. (2010). Is there a size limit for cosmopolitan distribution in free-living microorganisms? A biogeographical analysis of testate amoebae from polar areas. *Microbial Ecology* **59**, 635–645.

7

Arcellinida testate amoebae (Amoebozoa: Arcellinida): model of organisms for assessing microbial biogeography

THIERRY J. HEGER[1], ENRIQUE LARA[2] AND EDWARD A.D. MITCHELL[2]

[1] *WSL, Swiss Federal Institute for Forest, Snow and Landscape Research, Ecosystem Boundaries Research Unit, Wetlands Research Group, Lausanne, Switzerland; Laboratory of Ecological Systems, École Polytechnique Fédérale de Lausanne (EPFL), Lausanne, Switzerland; Department of Zoology and Animal Biology, University of Geneva, Switzerland; and Biodiversity Research Center, University of British Columbia, Vancouver, Canada*

[2] *Institute of Biology, Laboratory of Soil Biology, University of Neuchâtel, Neuchâtel, Switzerland*

7.1 Introduction

Although widely recognised as essential participants in ecosystem processes and representing a significant part of the Earth's biodiversity (Clarholm, 1985; Corliss, 2002; Schröter et al., 2003; Falkowski et al., 2004), eukaryotic microorganisms

Biogeography of Microscopic Organisms: Is Everything Small Everywhere?, ed. Diego Fontaneto. Published by Cambridge University Press.

are very poorly understood from evolutionary and biogeographic points of view. Major questions concerning the diversity and the distribution of protists remain completely unresolved. Arcellinida testate amoebae are an excellent group from which to get insights into these questions because they are easy to collect, present in different habitats and they build a shell of characteristic morphology that remains even after the organism's death. In this group, both cosmopolitan and restricted distribution patterns have been documented. Some morphospecies such as *Apodera vas(=Nebela vas)*, *Alocodera cockayni* or the whole genus *Certesella* have been reported as one of the most convincing examples of heterotrophic protists with restricted distributions (Foissner, 2006; Smith and Wilkinson, 2007; Smith et al., 2008). Arcellinida testate amoebae belong to the eukaryotic supergroup Amoebozoa (Nikolaev et al., 2005) and are morphologically characterised by the presence of lobose pseudopodia and a shell (test) composed from proteinaceous, calcareous or siliceous material. It can be either self-secreted or composed of recovered and agglutinated material. The Arcellinida covers a relatively broad range of sizes (mostly between 20 and 250 μm). At least some species have the ability to form a resting stage allowing their persistence under unfavourable conditions and a relatively unlimited dispersal capacity (Corliss and Esser, 1974; Foissner, 1987).

Arcellinida diversity is estimated at about 1500 species, mostly belonging to the genera *Centropyxis*, *Difflugia* and *Nebela* (Meisterfeld, 2002). They are diverse and abundant in virtually all terrestrial and freshwater aquatic habitats on Earth from the tropics to the poles (Meisterfeld, 2002) but they were not reported from truly marine environments. A few species have however successfully colonised brackish water ecosystems such as the marine supralittoral zone (Golemansky, 2007; Todorov and Golemansky, 2007), as well as the saline soils (Bonnet, 1959). The Arcellinida feed on bacteria, plant cells, protists, fungi or small metazoans (Foissner, 1987; Yeates and Foissner, 1995; Gilbert et al., 2000). Moisture conditions and pH are major environmental variables controlling the occurrence of testate amoebae (Charman and Warner, 1992; Charman, 1997) and the response of testate amoebae to different ecological gradients and pollutions make them a useful tool for palaeoecological studies and pollution monitoring (Charman et al., 2004; Nguyen-Viet et al., 2007, 2008; Laggoun-Defarge et al., 2008; Lamentowicz et al., 2008; Mitchell et al., 2008; Kokfelt et al., 2009).

7.2 Biogeography of Arcellinida: historical views

The first investigations on the biogeography of microorganisms date back to the mid nineteenth century when Christian Gottfried Ehrenberg (1795–1876) claimed that mountain ranges separated divergent populations of 'infusoria' (ciliates) (Ehrenberg, 1838, 1850). In his famous voyage on the Beagle, Darwin collected some dust fallen after a storm in the Cape Verde Islands and sent this sample to

Ehrenberg for analysis of the ciliate populations. Surprisingly for them, they found two species observed only in South America and none observed in Africa, in spite of the fact that the wind was blowing westwards and that Africa was closer than America. This apparent contradiction, in addition to the observation that relatively large particles could be transported ('above the thousandth of an inch square'), incited Darwin to think that small organisms have a huge dispersal potential.

The paradigm of the ubiquity of microorganisms came later with the emergence of microbiology/bacteriology as a recognised scientific discipline (O'Malley, 2007). At that time, the study of environmental samples was based on the retrieval and characterisation of pure cultures, and it was observed that identical organisms could be found whenever identical nutritional and physical conditions were provided. Thus, microorganisms had to be ubiquitous, and all environments were provided with a constant input of a 'seed rain' of microbes awaiting the adequate conditions to prosper. This idea was developed mainly by Beijerinck, the founder of the Delft School in Microbiology, who showed that it was possible to predict the composition of a ciliate community knowing the parameters of the environment (Beijerinck, 1913). The famous sentence 'Everything is everywhere, but the environment selects' was later formulated by one of his followers, Baas Becking (Baas Becking, 1934; de Wit and Bouvier, 2006).

This paradigm seemed to rule the viewpoints of the scientific community on the biogeography and distribution of protists well into the twentieth century. Eugène Penard, a famous pioneer on the study of testate amoebae, was convinced that the objects of his studies could be found everywhere when suitable conditions were met (Penard, 1902). In his 1902 monograph, Penard recorded 92% of all Arcellinida and Euglyphida testate amoebae species described to that date in the Lake Geneva area alone. However, with hindsight, he had described many of the species in that monograph or in earlier studies, so this finding may not be very surprising. Indeed, in the following decades, other researchers such as Heinis (1914) and Deflandre (1928, 1936) analysed samples from other biogeographic regions and observed significantly different faunas. Later, Penard himself eventually revised his opinion of the cosmopolitanism of testate amoebae (Penard, 1938).

These pioneering studies had a major influence on later research. By the end of the twentieth century and the beginning of the twenty-first, the existence of limited geographic ranges in testate amoebae was admitted by almost all specialists. Diatom taxonomists also reached a similar conclusion (Kilroy et al., 2003; Van de Vijver et al., 2005; Vanormelingen et al., 2008).

7.3 Endemic Arcellinida morphospecies

Arcellinida testate amoebae comprise some of the most convincing illustrations of non-cosmopolitan heterotrophic protists. The conspicuous testate amoebae

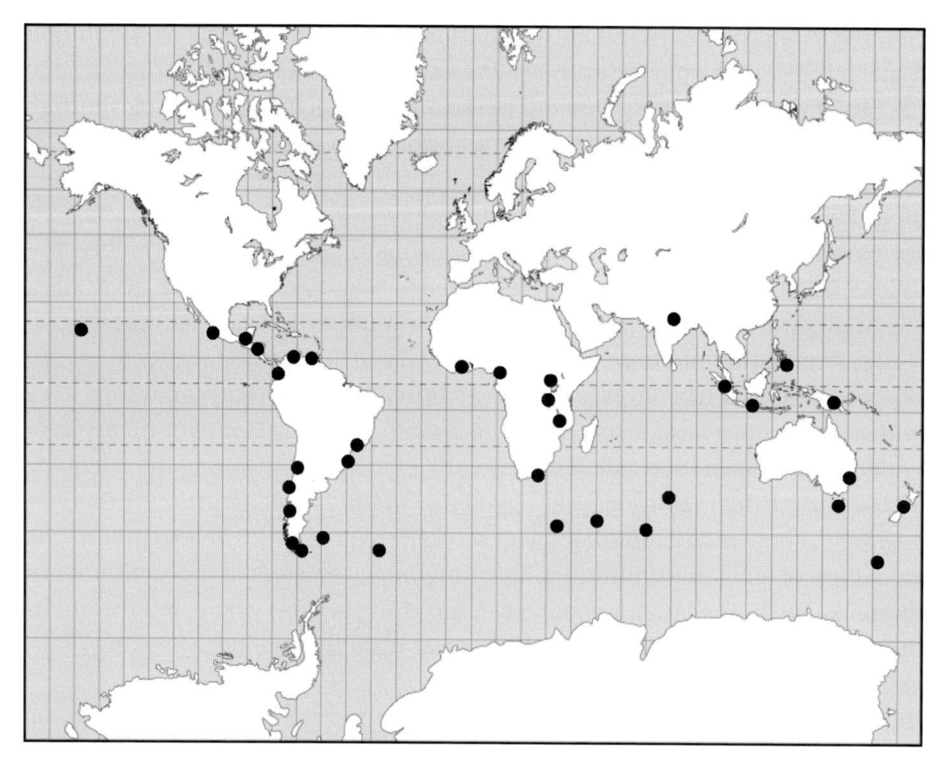

Fig 7.1 Sketch map showing *Apodera vas (=Nebela vas)* records in the southern hemisphere and the tropics (after Smith and Wilkinson, 2007).

Apodera vas and *Certesella* spp., which occur only to the south of the Tropic of Cancer desert belt (Fig 7.1) (Smith and Wilkinson, 2007; Smith et al., 2008), represent classical examples of morphospecies with restricted distributions.

The geographic distributions of these organisms contradict Baas Becking's 'Everything is everywhere, but the environment selects' tenet because in spite of the fact that they were found frequently in the southern hemisphere and low latitudinal zones of the northern hemisphere, they were never encountered in similar Holarctic habitats where most studies on testate amoebae from distinct habitats such as mosses, soils or aquatic environments took place. These genera also stand out by their very distinct morphologies which allow unambiguous identification, making them good examples of 'flagship species' (*sensu* Foissner, 2006). This contrasts with many other taxa of testate amoebae and other free-living protists whose geographic distribution cannot be established with any degree of certainty because of taxonomic uncertainties (Smith et al., 2008; Heger et al., 2009). It is therefore quite likely that besides these flagship species, many more taxa also have limited distributions (see further).

7.4 Assessing the distribution of *Apodera vas* and *Certesella* morphospecies in Mesoamerica

Mesoamerica has been the place of contact between South American and North American fauna and flora, an event that took place 3 Mya when the Panamá isthmus was closed allowing macroorganisms to disperse between the two American continents (Webb, 1991). However, in comparison with what is known with macroorganisms the potential impact of the rise of the isthmus on the dispersion of microorganisms between North and South America (and/or vice versa) remains completely unknown. Mesoamerica represents an interesting test case for the ongoing debate over microbial biogeography: if long-distance dispersal is easy for microorganisms then the existence or absence of this isthmus should not affect distribution patterns. However, if microorganisms do not disperse easily over long distances, then the presence of an isthmus will affect distribution patterns by allowing southern taxa to migrate northwards or vice versa.

Because of the possible existence of distinct faunas of free-living protists in South and North America, Mesoamerica is a key region for the study of protistan biogeography. However, although the biogeographic distributions of *Apodera vas* and *Certesella* spp. are reasonably well documented at a global scale, their distributions in Mesoamerica have so far not been investigated systematically (Heinis, 1911; Kufferath, 1929; Bonnet, 1977a; Madrazo-Garibay and López-Ochoterena, 1982; Smith and Wilkinson, 2007; Smith et al., 2008). To our knowledge, only two studies based on a very small number of samples reported the presence of *Apodera vas* in Costa Rica, Guatemala and Mexico (Golemansky, 1967; Laminger, 1973) while *Certesella* spp. were only reported in Guatemala (Laminger, 1973). To this day, the second northernmost reported continental occurrence of *Apodera vas* is Mexico (Laminger, 1973); Nepal is the northernmost occurrence (Bonnet, 1977b). The aim of this study is to estimate the distribution of the 'southern hemisphere and tropical endemic Hyalospheniids' (SHTEH; i.e. *Apodera vas* and members of genus *Certesella*) along a south-north latitudinal Mesoamerican transect extending from Panama to the south of Mexico.

7.5 Methods and sampling area

To assess the biogeographic distributions of *Apodera vas* and *Certesella* morphospecies, over 200 moss and litter samples were collected along a transect from Panama to Mexico. In addition to the latitudinal gradient, this region is characterised by altitudinal gradients, resulting in a broad diversity of biomes from wet lowland rainforest and cloud forest to dry scrublands and desert. In this study, wet moss and wet litter samples were collected in cloud and mesophilous forests between April and May 2007 (Table 7.1) because such moist substrates were described as suitable habitats for *A. vas* and *Certesella* spp. (SHTEH) (Smith and Wilkinson, 2007 and references therein). To extract testate amoebae, moss and litter samples were shaken energetically in water and then sieved and back-sieved using appropriate mesh sizes (250 and 20 μm). The occurrence of *Apodera vas* and *Certesella* spp. was checked using a light microscope under 200× magnification. Cells from *Apodera vas* specimens from Monte Cristo (El Salvador) were isolated and documented by scanning electron microscopy (SEM). The SEM was performed as described in Heger et al. (2009). Up to now, three *Certesella* species have been described (Meisterfeld, 2002). However, given that the taxonomy of genus *Certesella* is still relatively unclear, we adopted a conservative approach and did not distinguish among *Certesella* species.

7.6 Results and discussion

This survey confirmed the widespread occurrence of SHTEH in Mesoamerica. For the first time, *Apodera vas* was recorded in Panama, Salvador and Nicaragua while *Certesella* morphospecies were recorded for the first time in Panama, Costa Rica, Nicaragua, Salvador and Mexico (Table 7.1, Figs 7.2 and 7.3). We found *Certesella* spp. and *Apodera vas* specimens in most samples collected from favourable forest habitats (i.e. permanently wet moss and litter samples). The exact northernmost distribution limit of SHTEH in Mexico remains unclear. The region located between Mexico City and the USA was indeed poorly or not investigated. All samples with *Apodera vas* or *Certesella* spp. were collected at elevations between about 1000 and 3100 m a.s.l. *Apodera vas* and *Certesella* spp. co-occurred in seven of the 35 samples (Table 7.1), confirming that these species have relatively similar ecological tolerances (Charman, 1997). Our data also confirm that as far as suitable microhabitats occur, these SHTEH can live in relatively distinct ecosystems. For instance, we found *Apodera vas* and *Certesella* spp. in the Monteverde cloud forest where mean annual precipitation and temperature were 2500 mm and 18.8 °C (Clark et al., 2000) as well as in the Volcan Poás forest where precipitation exceeded 3400 mm and mean annual temperature was lower than 13 °C (Rowe et al., 1992; Martinez et al., 2000). In the literature, these species with restricted distributions were also reported in South American peatlands (Zapata et al., 2008) and New Zealand peatlands (Charman, 1997). Interestingly, some moss species harbouring SHTEH in the south of the Tropic of Cancer desert belt are also present in the Holarctic habitats where SHTEH were never reported. For instance, *Apodera vas*

Table 7.1 Locations and characteristics of all *Apodera vas* and *Certesella* spp. records in Mesoamerica.

Apodera vas	*Certesella* sp.	Sampling location	Country	Coordinates	Sampling date	Substrate	Ecosystem	Altitude (m)	Reference
	*	Parque La Amistad	Panama	08°53′58.5″N 82°37′11″W	30.3.2007	moss	Cloud forest	~2455	new record
*		Parque La Amistad	Panama	08°54′04.9″N 82°37′13″W	30.3.2007	moss	Cloud forest	~2450	new record
	*	Parque La Amistad	Panama	08°54′05″N 82°37′05″W	30.3.2007	moss	Cloud forest	~2380	new record
*		Parque Nacional Chirripo	Costa Rica	09°27′N 83°34′W	11.4.2007	moss	Cloud forest	~2350	new record
*		Parque Nacional Chirripo	Costa Rica	09°27′53″N 83°34′03″W	11.4.2007	moss	Cloud forest	2066	new record
*		Parque Nacional Chirripo	Costa Rica	09°27′15.6″N 83°32′55.1″W	11.4.2007	moss	Cloud forest	~2400	new record
*		N.A	Costa Rica	c. 9–10°N 82–84°W	1966	N.A	N.A	2000–4000	Laminger (1973)
*	*	Volcan Barva	Costa Rica	10°08′N 84°06′W	15.4.2007	moss	Cloud forest	~2830	new record
*		Volcan Barva	Costa Rica	10°08′N 84°06′W	15.4.2007	moss	Cloud forest	~2830	new record
*	*	Volcan Poás	Costa Rica	10°11′27″N 84°13′58.6″W	13.4.2007	moss	Cloud forest	~2575	new record

Table 7.1 (*cont.*)

Apodera vas	*Certesella* sp.	Sampling location	Country	Coordinates	Sampling date	Substrate	Ecosystem	Altitude (m)	Reference
*	*	Volcan Poás	Costa Rica	10°11′27″N 84°13′58.6″W	13.4.2007	moss	Cloud forest	~2575	new record
*	*	Volcan Poás	Costa Rica	10°11′27″N 84°13′58.6″W	13.4.2007	moss	Cloud forest	~2575	new record
	*	Volcan Poás	Costa Rica	10°11′27″N 84°13′58.6″W	13.4.2007	moss	Cloud forest	~2575	new record
*		Monteverde reserva	Costa Rica	10°18′N 84°47′W	18.4.2007	moss	Cloud forest	~1570	new record
*		Monteverde reserva	Costa Rica	10°18′N 84°47′W	18.4.2007	moss	Cloud forest	~1570	new record
*		Monteverde reserva	Costa Rica	10°18′N 84°47′W	18.4.2007	moss	Cloud forest	~1570	new record
*		Monteverde reserva	Costa Rica	10°18′N 84°47′W	18.4.2007	moss	Cloud forest	~1570	new record
*		Santa Elena reserva	Costa Rica	10°20′N 84°47′W	20.04.2007	moss	Cloud forest	~1550	new record
*		Volcan Mombacho	Nicaragua	11°50′07″N 85°58′46″W	20.04.2007	moss	Cloud forest	1123	new record
*		Volcan Mombacho	Nicaragua	11°50′03″N 85°58′48″W	20.04.2007	moss	Cloud forest	1143	new record

*		Parque Montecristo	Salvador	14°25′N 89°21′W	24.4.2007	moss	Cloud forest	2140	new record
*		Parque Montecristo	Salvador	14°25′N 89°21′W	24.4.2007	moss	Cloud forest	~2000	new record
*		Parque Montecristo	Salvador	14°25′N 89°21′W	24.4.2007	moss	Cloud forest	~2000	new record
*		Parque Montecristo	Salvador	14°25′N 89°21′W	24.4.2007	moss	Cloud forest	~2000	new record
*		Parque Montecristo	Salvador	14°25′N 89°21′W	24.4.2007	moss	Cloud forest	~2000	new record
*	*	N.A	Guatemala	c. 14–15°N, 89–92°W	1966/1970	N.A	N.A	2000–4000	Laminger (1973)
*	*	Biotopo del Quetzal	Guatemala	15°12′N 90°13′ W	27.04.2007	moss	Cloud forest	~1650	new record
	*	Biotopo del Quetzal	Guatemala	15°12′N 90°13′W	27.04.2007	moss	Cloud forest	~1650	new record
	*	Biotopo del Quetzal	Guatemala	15°12′N 90°13′W	27.04.2007	moss	Cloud forest	~1650	new record
	*	Biotopo del Quetzal	Guatemala	15°12′N 90°13′W	27.04.2007	moss	Cloud forest	~1650	new record
	*	Biotopo del Quetzal	Guatemala	15°12′N 90°13′W	27.04.2007	moss	Cloud forest	~1650	new record
*		Biotopo del Quetzal	Guatemala	15°12′N 90°13′W	27.04.2007	moss	Cloud forest	1750–1900	new record

Table 7.1 (*cont.*)

Apodera vas	*Certesella* sp.	Sampling location	Country	Coordinates	Sampling date	Substrate	Ecosystem	Altitude (m)	Reference
	*	Biotopo del Quetzal	Guatemala	15°12′N 90°13′W	27.04.2007	moss	Cloud forest	1750–1900	new record
*		Ixtlán	Mexico	17°17′45″N 96°23′ 53″W	5.5.2007	moss	Cloud forest	2350	new record
*	*	Ixtlán de Juárez	Mexico	17°22′49″N 96°26′ 51″W	6.5.2007	moss	Mesophilous forest	3050	new record
*		Tequila	Mexico	18°43′32″N 97°04′ 37′ ′W	8.5.2007	litter	Forest	1770	new record
*		Tequila	Mexico	18°43′32″N 97°04′ 37″W	8.5.2007	litter	Forest	1770	new record
*		N.A	Mexico	c. 19°N, 97°W	1966	N.A	N.A	2000ñ4000	Laminger (1973)
*		Between Desierto de los Leones and Cruz Blanca	Mexico	19°16′N, 99°19′W	1967	moss	Temperate forest	3100–3700	Golemansky (1967)

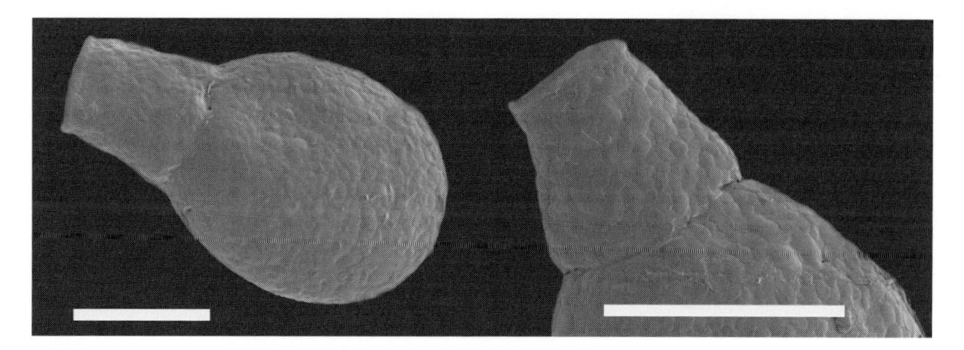

Fig 7.2 Scanning electron pictures illustrating *Apodera vas* from Salvador. Scale bars 50 μm.

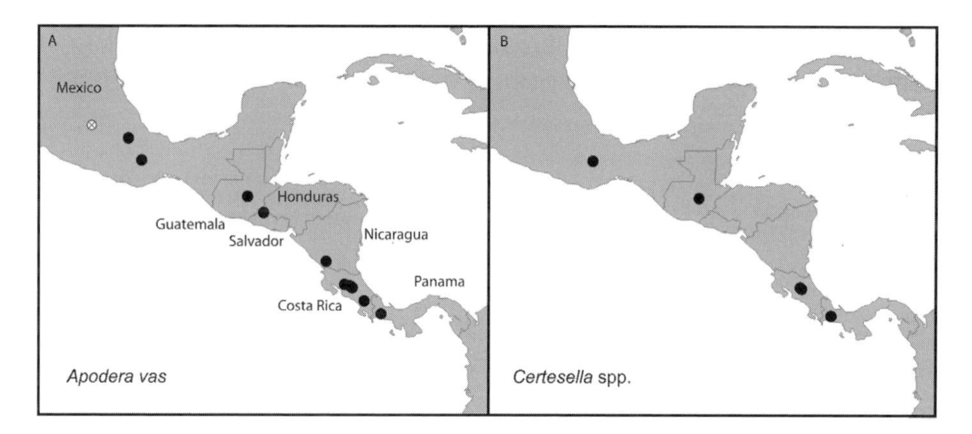

Fig 7.3 Sketch maps showing (A) the occurrences of the testate amoebae *Apodera vas* and (B) *Certesella* spp. along a Mesoamerican transect between Panama and the south of Mexico. Black dots indicate new records while the white dot represents the *Apodera vas* record published by Golemansky (1967). The records published by Laminger (1973) were not reported on this map because of the lack of accurate coordinates. Details of sites are given in Table 7.1.

were found in a *Sphagnum magellanicum* sample collected in Laguna Esmeralda (near Ushuaia, Tierra del Fuego, Argentina; E. Lara, unpublished data) but were never reported from holarctic *Sphagnum magellanicum* samples. Altogether, these data support the interpretation that the absence of these species in Holarctic habitats is not explained by the lack of a specific habitat in the northern hemisphere. Abundant favourable moist soil and moist litter habitats are indeed present in Europe and North America from the Boreal tundra to temperate rain forests. Thus, the SHTEH example is clearly inconsistent with the 'Everything is everywhere, but the environment selects' tenet.

7.7 Evaluating the consequences of potential cryptic species within endemic Arcellinida morphospecies

Cryptic diversity has been commonly reported in several free-living protist morphospecies (de Vargas et al., 1999; Slapeta et al., 2005; Heger et al., 2010). Within the endemic *Certesella* and *Apodera vas* morphospecies, the presence of a hidden diversity is therefore also possible. In the case of *Apodera vas*, several indices such as its polymorphic shell (Smith and Wilkinson, 2007; Zapata and Fernández, 2008) and its relatively wide environmental tolerances indeed suggest the presence of cryptic species (species which cannot be discriminated by morphology alone) or pseudocryptic species (species with subtle morphological dissimilarities, possibly visible only by scanning electron microscopy) within this morphospecies. Each of these potential hidden species can have either a distribution corresponding to the actual *Apodera vas*/*Certesella* morphospecies distributions or a more restricted distribution. In order to evaluate the genetic diversity within these endemic morphospecies, molecular-based studies are needed.

7.8 Potential factors governing the distribution of Arcellinida

The debate over cosmopolitan vs. limited endemism in free-living microorganisms has mostly focused on the question of taxonomy: defenders of cosmopolitanism are usually 'lumpers' while defenders of endemism are usually 'splitters' (see Finlay et al., 2004 and answer by Mitchell and Meisterfeld, 2005 for an example). Recent developments in taxonomy and especially in molecular taxonomy have provided several examples showing that the splitters may be right and that we have even underestimated the true diversity of testate amoebae (Lara et al., 2008). However, this debate is only one part of the whole story. Several other important factors influence the distribution of free-living microorganisms and all of these suffer from a clear lack of data or conceptual framework:

1. Dispersal by wind. Free-living microorganisms are assumed to be small enough to be easily transported by wind (Finlay, 2002). However, although several studies have revealed the presence of microorganisms (mostly bacteria and fungi) in the atmosphere and highlight the importance of atmospheric transport for microbial large scale dispersion (Darwin, 1846; Griffin et al., 2002; Kellogg and Griffin, 2006; Gorbushina et al., 2007; Hervas et al., 2009; Pearce et al., 2009), no data are available for testate amoebae. Atmospheric circulation models could be used to model potential long-distance

dispersal of microorganisms under different scenarios (e.g. size of the organisms) (Wilkinson et al., In prep).

2. Dispersal by animals. The main potential factor for natural dispersal of free-living microorganisms other than wind is migratory animals, mainly birds. The feathers of birds constitute ideal sampling and transportation devices for microorganisms. This will be most efficient for birds living in wetlands and ground-nesting and/or feeding taxa. Although several studies have shown that birds can transport high numbers of microorganisms (Wuthrich and Matthey, 1980; Bisson et al., 2009), we are far from having a clear image of the local and global potential dispersal of free-living microorganisms by birds. A potential limitation of birds as vectors of free-living microorganisms is that the nesting habitat of many long-distance migratory birds is very different from their wintering habitats. For example, many shorebirds nest in freshwater wetlands and bogs in the boreal and arctic regions but winter in brackish water wetlands.
3. Dispersal by humans. Wilkinson (2010) discussed the potential role of humans in dispersing terrestrial free-living microorganisms. Given the exponential increase in travel across continents, this is currently a very significant mechanism for long-distance dispersal. Humans may have contributed to the dispersal of microorganisms for millennia through their own migrations, the spread of agriculture and the development of long-distance ocean travel.
4. Survival during transportation. Long-distance transport, especially on birds' feathers or by wind exposes microorganisms to unfavourable conditions such as drought, freezing and UV radiation. Therefore it may be that, although many microorganisms can be transported over long distances, they do not survive and cannot colonise a new habitat. It would therefore be useful to test experimentally the resistance of various microorganisms to the range of conditions they are likely to experience during transport in order to determine if survival plays a role in shaping biogeographic patterns. Many free-living microorganisms are able to encyst and when encysted would presumably survive extended periods of time in the air. However, the encystment and excystment capacities of most microorganisms are not well documented (Corliss and Esser, 1974; Foissner, 1987).
5. Establishment in an existing community. The potential for viable microorganisms falling on soil or into an aquatic environment to establish new populations will depend on several factors such as the ecological conditions of the new environment (do microclimatic and physico-chemical conditions match its own requirements?) and biotic interactions (are potential prey organisms present in the case of a predator, could local predators wipe out the newly

developed population?). The importance of competition in a community assembly of testate amoebae is currently unclear. For example Wanner and Xylander (2005) found no evidence for species turnover in a primary colonisation sequence of sand dunes. Experiments could quite easily be done to test the potential for alien species to become established in existing communities but such experiments have to our knowledge not been performed with Arcellinida.

Taken together these factors show that the dispersal potential of free-living microorganisms is clearly not only a matter of size.

7.9 Perspectives

Palaeoecology could potentially provide useful information on the dispersal potential of Arcellinida. For example, peat deposits and lake sediments could be studied to determine how Arcellinids (re-)colonised those habitats after the last glacial retreat or if their diversity on isolated islands increased with the arrival of humans. However, a potential confounding factor of this latter possible study is that humans usually strongly modify the vegetation of such islands, either directly by clearing forests for agriculture or indirectly by introducing animals such as goats and cows that strongly alter the vegetation. Such changes in turn may affect the ecology of wetlands and lakes, thus causing shifts in communities of microorganisms that are not necessarily related to the establishment of new taxa on the island but simply the colonisation of habitats that became favourable to taxa that already lived on the island but in other locations.

An international network of suction traps has been established since the nineteenth century for the study of plant pests such as aphids (Klueken et al., 2009). These traps could easily be used to sample microorganisms from the air and obtain a quantitative estimate of the colonisation rate and diversity of air-transported microbes.

In order to better understand the Arcellinida gene flow among regions and continents, it would be also highly relevant to assess the phylogeography of selected Arcellinida morphospecies. Indeed, a purely morphological approach does not provide any information on the genetic structure of the populations and the level of gene flow among populations. If a highly variable gene was studied and little or no genetic differentiation among populations of a species was observed, this would provide evidence for a high level of dispersal, consistent with the 'ubiquity theory'. In contrast, a high degree of genetic differentiation among populations would indicate that limited geographic distributions exist in protists, providing evidence against the 'ubiquity theory'. In addition, such a study would also contribute to resolving taxonomic uncertainties.

Finally, process-based modelling and sensitivity tests of all model parameters would be useful to assess which factors may play the biggest role in determining the dispersal potential of free-living microorganisms. Models cannot be expected to provide definitive answers, at least not with the current lack of data and understanding on critical parameters that will need to be included in the model, but they would be useful in clarifying which of the many open questions matters most.

Acknowledgements

This work was funded by Swiss NSF projects n° 205321-109709 (to E. Mitchell) and (Ambizione fellowship, E. Lara) and the Swiss Academy of Sciences (travel/ PZ00P2_122042 grant). Additional funding to EM by CCES projects RECORD and BigLink is kindly acknowledged. We thank the authorities of the Smithsonian Tropical Research Institution, especially Yves Basset, the coordinator of the Canopy Crane Access Systems for allowing fieldwork and providing logistical support. We are also grateful for the fieldwork assistance of Tanja Schwander. The authors wish to thank Elena Rossel for technical support, Kathryn Lannas for GIS assistance and Humphrey Smith as well as two anonymous referees for fruitful comments on the manuscript. SEM at the EPFL was made possible through the Interdisciplinary Centre for Electron Microscopy (CIME).

References

Baas Becking, L.G.M. (1934). *Geobiologie of inleiding tot de milieukunde.* The Hague: Van Stockum and Zoon.

Beijerinck, M.W. (1913). *De infusies en de ontdekking der backteriën.* Amsterdam: Müller.

Bisson, I.A., Marra, P.P., Burtt, E.H., Sikaroodi, M., Gillevet, P.M. (2009). Variation in plumage microbiota depends on season and migration. *Microbial Ecology* **58**, 212–220.

Bonnet, L. (1959). Quelques aspects des populations Thécamoebiennes endogées. *Bulletin de la Société d'Histoire Naturelle de Toulouse* **94**, 413–428.

Bonnet, L. (1977a). Faunistique et biogéographie des Thécamoebiens. I. Thécamoebiens des sols du Mexique. *Bulletin de la Société d'Histoire Naturelle de Toulouse* **113**, 40–44.

Bonnet, L. (1977b). Le peuplement thécamoebien des sols du Népal et son intérêt biogéographique. *Bulletin de la Société d'Histoire Naturelle de Toulouse* **113**, 331–348.

Charman, D.J. (1997). Modelling hydrological relationships of testate amoebae (Protozoa: Rhizopoda) on New Zealand peatlands. *Journal of the Royal Society of New Zealand* **27**, 465–483.

Charman, D.J., Warner, B.G. (1992). Relationship between testate amebas (Protozoa, Rhizopoda) and microenvironmental parameters on a forested peatland in Northeastern

Ontario. *Canadian Journal of Zoology – Revue Canadienne de Zoologie* **70**, 2474–2482.

Charman, D.J., Brown, A.D., Hendon, D., Karofeld, E. (2004). Testing the relationship between Holocene peatland palaeoclimate reconstructions and instrumental data at two European sites. *Quaternary Science Reviews* **23**, 137–143.

Clarholm, M. (1985). Interactions of bacteria, protozoa and plants leading to mineralization of soil-nitrogen. *Soil Biology and Biochemistry* **17**, 181–187.

Clark, K.L., Lawton, R.O., Butler, P.R. (2000). The physical environment. In Nadkarni NM, Wheelwright NT (eds.), *Monteverde, Ecology and Conservation of a Tropical Cloud Forest*, pp 16–38. New York: Cambridge University Press.

Corliss, J.O. (2002). Biodiversity and biocomplexity of the protists and an overview of their signifiant roles in maintenance of our biosphere. *Acta Protozoologica* **41**, 199–219.

Corliss, J.O., Esser, S.C. (1974). Role of cyst in life-cycle and survival of free-living protozoa. *Transactions of the American Microscopical Society* **93**, 578–593.

Darwin, C. (1846). An account of the fine dust which often falls on vessels in the Atlantic Ocean. *Quarterly Journal of the Geological Society* **2**, 26–30.

Deflandre, G. (1928). Le genre *Arcella* Ehrenberg. Morphologie-Biologie. Essai phylogénétique et systématiqe. *Archiv für Protistenkunde* **64**, 152–287.

Deflandre, G. (1936). Etude monographique sur le genre *Nebela* Leidy. *Annales de Protistologie* **5**, 201–286.

de Vargas, C., Norris, R., Zaninetti, L., Gibb, S.W., Pawlowski, J. (1999). Molecular evidence of cryptic speciation in planktonic foraminifers and their relation to oceanic provinces. *Proceedings of the National Academy of Sciences USA* **96**, 2864–2868.

de Wit, R., Bouvier, T. (2006). 'Everything is everywhere, but the environment selects'; what did Baas Becking and Beijerinck really say? *Environmental Microbiology* **8**, 755–758.

Ehrenberg, C.G. (1838). Ein Blick in das tiefere organische Leben der Natur. In Voss VvL (ed.), *Die Infusionstheirchen als vollkommene Organismen*. Leipzig.

Ehrenberg, C.G. (1850). On infusorial deposits on the River Chutes in Oregon. *American Journal of Science* **9**, 140.

Falkowski, P.G., Katz, M.E., Knoll, A.H. et al. (2004). The evolution of modern eukaryotic phytoplankton. *Science* **305**, 354–360.

Finlay, B.J. (2002). Global dispersal of free-living microbial eukaryote species. *Science* **296**, 1061–1063.

Finlay, B.J., Esteban, G.F., Fenchel, T. (2004). Protist diversity is different? *Protist* **155**, 15–22.

Foissner, W. (1987). Soil protozoa: fundamental problems, ecological significance, adaptation in ciliates and testaceans, bioindicators, and guide to the literature. *Progress in Protozoology* **2**, 69–212.

Foissner, W. (2006). Biogeography and dispersal of micro-organisms: a review emphasizing protists. *Acta Protozoologica* **45**, 111–136.

Gilbert, D., Amblard, C., Bourdier, G., Francez, A.-J., Mitchell, E.A.D. (2000). Le régime alimentaire des thécamoebiens. *L'Année Biologique* **39**, 57–68.

Golemansky, V. (1967). Tecamebianos Muscicolas (Rhizopoda, Testacea) de Mexico. *Revista de la Sociedad Mexicana de Historia Natural* **18**, 73–77.

Golemansky, V. (2007). Testate amoebas and monothalamous foraminifera (Protozoa) from the Bulgarian Black Sea Coast. In Fet V., Popov, A. (eds.), *Biogeography and Ecology of Bulgaria*, pp. 555–570. Dordrecht: Springer.

Gorbushina, A.A., Kort, R., Schulte, A. et al. (2007). Life in Darwin's dust: intercontinental transport and survival of microbes in the nineteenth century. *Environmental Microbiology* **9**, 2911–2922.

Griffin, D.W., Kellogg, C.A., Garrison, V.H., Shinn, E.A. (2002). The global transport of dust – an intercontinental river of dust, microorganisms and toxic chemicals flows through the Earth's atmosphere. *American Scientist* **90**, 228–235.

Heger, T.J., Mitchell, E.A.D., Ledeganck, P. et al. (2009). The curse of taxonomic uncertainty in biogeographical studies of free-living terrestrial protists: a case study of testate amoebae from Amsterdam Island. *Journal of Biogeography* **36**, 1551–1560.

Heger, T.J., Mitchell, E.A.D., Golemansky, V. et al. (2010). Molecular phylogeny of euglyphid testate amoebae (Cercozoa: Euglyphida) suggests transitions between marine supralittoral and freshwater/terrestrial environments are infrequent. *Molecular Phylogenetics and Evolution* **55**, 113–122.

Heinis, F. (1911). Beiträge zur Kenntnis der Centralamerikanischen Moosfauna. *Revue Suisse de Zoologie* **19**, 253–256.

Heinis, F. (1914). Die Moosfauna Columbiens: In Voyage d'exploration scientifique en Colombie. *Mémoires de la Société Neuchâteloise des Sciences Naturelles* **5**, 675–730.

Hervas, A., Camarero, L., Reche, I., Casamayor, E.O. (2009). Viability and potential for immigration of airborne bacteria from Africa that reach high mountain lakes in Europe. *Environmental Microbiology* **11**, 1612–1623.

Kellogg, C.A., Griffin, D.W. (2006). Aerobiology and the global transport of desert dust. *Trends in Ecology and Evolution* **21**, 638–644.

Kilroy, C., Sabbe, K., Bergey, E.A., Vyverman, W., Lowe, R. (2003). New species of Fragilariforma (Bacillariophyceae) from New Zealand and Australia. *New Zealand Journal of Botany* **41**, 535–554.

Klueken, A.M., Hau, B., Ulber, B., Poehling, H.M. (2009). Forecasting migration of cereal aphids (Hemiptera: Aphididae) in autumn and spring. *Journal of Applied Entomology* **133**, 328–344.

Kokfelt, U., Rosen, P., Schoning, K. et al. (2009). Ecosystem responses to increased precipitation and permafrost decay in subarctic Sweden inferred from peat and lake sediments. *Global Change Biology* **15**, 1652–1663.

Kufferath, H. (1929). Algues et protistes muscicoles, corticoles et terrestres récoltés sur la montagne de Barba (Costa Rica). *Annales de Cryptogamie Exotique* **2**, 23–52.

Laggoun-Defarge, F., Mitchell, E., Gilbert, D. et al. (2008). Cut-over peatland regeneration assessment using organic matter and microbial indicators (bacteria and testate amoebae). *Journal of Applied Ecology* **45**, 716–727.

Lamentowicz, M., Obremska, M., Mitchell, E.A.D. (2008). Autogenic succession, land-use change, and climatic influences on the Holocene development of a kettle-hole mire in Northern Poland. *Review of Palaeobotany and Palynology* **151**, 21–40.

Laminger, H. (1973). Die Testaceen (Protozoa, Rhizopoda) einiger Hochgebirgsgewässer von Mexiko, Costa Rica und Guatemala. *Internationale Revue der gesamten Hydrobiologie und Hydrographie* **58**, 273–305.

Lara, E., Heger, T.J., Ekelund, F., Lamentowicz, M., Mitchell, E.A.D. (2008). Ribosomal RNA genes challenge the monophyly of the Hyalospheniidae (Amoebozoa: Arcellinida). *Protist* **159**, 165–176.

Madrazo-Garibay, M., López-Ochoterena, E. (1982). Segunda lista taxonómica comentada de protozoarios de vida libre de México. *Revista latinoamericana de microbiología* **24**, 281–295.

Martinez, M., Fernandez, E., Valdes, J. et al. (2000). Chemical evolution and volcanic activity of the active crater lake of Poas volcano, Costa Rica, 1993–1997. *Journal of Volcanology and Geothermal Research* **97**, 127–141.

Meisterfeld, R. (2002). Order Arcellinida Kent, 1880. In Lee JJ, Leedale GF, Bradbury P (eds.), *The Illustrated Guide To The Protozoa*, pp. 827–860. Lawrence, KS: Society of Protozoologists.

Mitchell, E.A.D., Meisterfeld, R. (2005). Taxonomic confusion blurs the debate on cosmopolitanism versus local endemism of free-living protists. *Protist* **156**, 263–267.

Mitchell, E.A.D., Charman, D.J., Warner, B.G. (2008). Testate amoebae analysis in ecological and paleoecological studies of wetlands: past, present and future. *Biodiversity and Conservation* **17**, 2115–2137.

Nguyen-Viet, H., Gilbert, D., Mitchell, E.A.D., Badot, P.M., Bernard, N. (2007). Effects of experimental lead pollution on the microbial communities associated with *Sphagnum fallax* (Bryophyta). *Microbial Ecology* **54**, 232–241.

Nguyen-Viet, H., Bernard, N., Mitchell, E.A.D., Badot, P.M., Gilbert, D. (2008). Effect of lead pollution on testate amoebae communities living in *Sphagnum fallax*: an experimental study. *Ecotoxicology and Environmental Safety* **69**, 130–138.

Nikolaev, S.I., Mitchell, E.A.D., Petrov, N.B. et al. (2005). The testate lobose amoebae (order Arcellinida Kent, 1880) finally find their home within Amoebozoa. *Protist* **156**, 191–202.

O'Malley, M.A. (2007). The nineteenth century roots of 'everything is everywhere'. *Nature Reviews Microbiology* **5**, 647–651.

Pearce, D.A., Bridge, P.D., Hughes, K.A. et al. (2009). Microorganisms in the atmosphere over Antarctica. *FEMS Microbiology Ecology* **69**, 143–157.

Penard, E. (1902). *Faune rhizopodique du bassin du Léman*. Genève: Kündig.

Penard, E. (1938). *Les infiniment petits dans leurs manifestations vitales*, pp 1–212. Genève: Georg & Cie.

Rowe, G.L., Brantley, S.L., Fernandez, M. et al. (1992). Fluid-volcano interaction in an active stratovolcano – the crater lake system of Poás volcano, Costa Rica. *Journal of Volcanology and Geothermal Research* **49**, 23–51.

Schröter, D., Wolters, V., De Ruiter, P.C. (2003). C and N mineralisation in the decomposer food webs of a European forest transect. *Oikos* **102**, 294–308.

Slapeta, J., Moreira, D., López-García, P. (2005). The extent of protist diversity: insights from molecular ecology of freshwater eukaryotes. *Proceedings of the Royal Society B – Biological Sciences* **272**, 2073–2081.

Smith, H.G., Bobrov, A., Lara, E. (2008). Diversity and biogeography of testate amoebae. *Biodiversity and Conservation* **17**, 329–343.

Smith, H.G., Wilkinson, D.M. (2007). Not all free-living microorganisms have cosmopolitan distributions – the case of *Nebela* (*Apodera*) *vas* Certes (Protozoa: Amoebozoa: Arcellinida). *Journal of Biogeography* **34**, 1822–1831.

Todorov, M., Golemansky, V. (2007). Seasonal dynamics of the diversity and abundance of the marine interstitial testate amoebae (Rhizopoda: Testacealobosia and Testaceafilosia) in the Black Sea supralittoral. *Acta Protozoologica* **46**, 169–181.

Van de Vijver, B., Gremmen, N.J.M., Beyens, L. (2005). The genus *Stauroneis* (Bacillariophyceae) in the Antarctic region. *Journal of Biogeography* **32**, 1791–1798.

Vanormelingen, P., Verleyen, E., Vyverman, W. (2008). The diversity and distribution of diatoms: from cosmopolitanism to narrow endemism. *Biodiversity and Conservation* **17**, 393–405.

Wanner, M., Xylander, W.E.R. (2005). Biodiversity development of terrestrial testate amoebae: is there any succession at all? *Biology and Fertility of Soils* **41**, 428–438.

Webb, S.D. (1991) Ecogeography and the Great American Interchange. *Paleobiology* **17**, 266–280.

Wilkinson, D.M. (2010). Have we underestimated the importance of humans in the biogeography of free-living terrestrial microorganisms? *Journal of Biogeography* **37**, 393–397.

Wilkinson, D.M., Koumoutsaris, S., Mitchell, E.A.D., Bey, I. (In prep) Investigating the aerial distribution of microorganisms with a model of the global atmospheric circulation.

Wuthrich, M., Matthey, W. (1980). Les diatomées de la Tourbière du Cachot (Jura Neuchâtelois) III. Etude de l'apport éolien et du transport par les oiseaux et insectes aquatiques. *Swiss Journal of Hydrology* **42**, 269–284.

Yeates, G.W., Foissner, W. (1995). Testate amebas as predators of nematodes. *Biology and Fertility of Soils* **20**, 1–7.

Zapata, J., Fernández, L. (2008). Morphology and morphometry of *Apodera vas* (Certes, 1889) (Protozoa: Testacea) from two peatlands in Southern Chile. *Acta Protozoologica* **47**, 389–395.

Zapata, J., Yáñez, M., Rudolph, E. (2008). Thecamoebians (Protozoa: Rhizopoda) of the peatland from Puyehue National Park (40° 45'S; 72° 19'W), Chile. *Gayana* **72**, 9–17.

8

Everything is not everywhere: the distribution of cactophilic yeast

PHILIP F. GANTER

Department of Biological Sciences, Tennessee State University, Nashville, TN, USA

8.1 Introduction

Cactophilic yeast form a community of fungi confined to the necrotic tissues of certain species of cacti. Although there is much we do not know about this community, our current understanding has implications for the generality of the 'Everything is everywhere, but the environment selects' (EiE) hypothesis (Finlay and Clarke, 1999; Fenchel and Finlay, 2004a; de Wit and Bouvier, 2006). The hypothesis of ubiquitous distributions for free-living microbial species is attractive because it solves a conceptual problem for biologists studying such small organisms: how do they discover new resource patches when their motion is passive? EiE provides an answer. Microbes reach such large population sizes that passive dispersal is sufficient to discover a new resource as it becomes available. It is a positive feedback loop. The larger its global population, the more likely a microbe is to find new resources and the larger its global population will become, increasing its likelihood of discovering more resources. Microbes that live in a patchy environment can quickly overexploit the patches and must often undergo difficult migrations across inhospitable territory in order to reach the next patch. This is true if

Biogeography of Microscopic Organisms: Is Everything Small Everywhere?, ed. Diego Fontaneto. Published by Cambridge University Press.

distance, time or both separate patches. As passive agents, they must be resistant to the stresses inherent in dispersal. For microbes in a patchy habitat, ubiquity is not only an outcome of large population sizes. It also rests on the assumption of passive dispersal of resistant life forms, often in the form of resistant spores.

Here, I argue that this life history is not the only one available to free-living microbes in a patchy environment. I argue that cactophilic yeast distributions are not global and cactophilic yeast populations do not need to attain large sizes because these yeast use a different mode of dispersal: active vectoring by animals. But before I attempt to discuss the relevance of the association between cactophilic yeast and insects to assessing the generality of the EiE hypothesis, it will be necessary to describe this yeast community and to establish the (remarkable) level of endemism found there.

8.2 Yeast

Yeast is the collective name given to fungi that have abandoned hyphal growth and adopted unicellular growth (I include here both true unicellularity and pseudohyphal growth – growth in short chains of rounded cells, not the elongated tubes characteristic of true hyphal growth; Crampin et al., 2005) (Fig 8.1). The difference between yeast and hyphal fungal lifestyles is a continuum and many examples of dimorphic fungi, those that can switch between hyphal and yeast growth, are known (Sudbery et al., 2004). Most of the yeast species thus far described are not dimorphic but are fully committed to life as a microbe. The two largest fungal clades (sometimes called the Dikarya) are the Basidiomycota and the Ascomycota, each of which appears to be monophyletic. Although there is, as of yet, no consensus on the number, membership or phylogeny of the subclades within each of these large clades (Hibbett, 2004; Lutzoni et al., 2004; James et al., 2006; Spatafora et al., 2006; Marcet-Houben and Gabaldón, 2009), enough is known about fungal evolutionary history to allow some observations relevant to yeast. First, yeast lifestyles are found in both clades of the Dikarya. Second, all current phylogenies of these two groups call for multiple origins of the yeast lifestyle and, third, the Saccharomycotina (= Hemiascomycetes without the Taphrinomycotina) is the largest monophyletic yeast clade by far. Most of the yeast analysed here come from Saccharomycotina with the addition of a few genera belonging to the Basidiomycota.

The second point above is of particular relevance to the EiE hypothesis. If one accepts that hyphal fungi are macroorganisms (Ferguson et al., 2003; Cairney, 2005), then each of those clades in which the yeast lifestyle has been derived from the ancestral, hyphal lifestyle is an independent origin of unicellularity and is potentially a test of the generality of the EiE hypothesis. If EiE is a general feature of small organisms, then it should be adopted upon making the transition from macro to micro size. Fungi can also be seen as preadapted for conformity with

Fig 8.1 Images of the cactophilic system. (A) A saguaro (*Carnegiea gigantea*) with a rot that is leaking liquid down the side of the cactus in the northern Sonoran Desert; (B) an *Opuntia* fruit rot; (C) a disappointingly healthy *Opuntia stricta* (*inermis*) in the Flinders Range, South Australia; (D) a rotting arm of *Stenocereus griseus* on the island of Curaçao, Netherland Antilles; (E) hat-shaped spores (two dehisced asci can be seen) and vegetative cells of *Pichia insulana*; (F) *P. insulana* colonies growing on agar; (G) vegetative cells of *Pichia cactophila*, the most common cactophilic yeast. Note: only E and G are at the same scale.

the EiE hypothesis. Both the Ascomycota and Basidiomycota normally produce many minute, stress-resistant spores that are often wind or water dispersed. Thus, finding clades of non-conforming yeast (with respect to EiE) within the fungi may further our understanding of those situations to which the principle of EiE might be applied.

8.3 Cacti

Although I will focus on yeast confined to cacti, this does not mean confinement to narrow geographic limits. Cacti are perhaps the most familiar succulent xerophytes. The stem-succulent members of the family are exclusively native to the Nearctic and Neotropic biogeographic realms and range from southern Canada to

Patagonia and from coastal dunes to the Altiplano wherever soils are well drained (Anderson, 2001). In addition, humans have carried them to Asia, Africa, Europe and Australia. Some cacti have established large viable populations in these non-native habitats and some populations have become invasive (Murray, 1982). Yeast are found in three places in succulent cacti: in the nectaries of their flowers, in rotting fruits, and in stem necroses following a breach in the tough outer cuticle. Stem necroses, often referred to as rot pockets or rots, are active sites of bacterial and yeast activity (Fogleman and Foster, 1989). They are not found in all species of cacti. Plant size and stem morphology seem to be the most important determinants of susceptibility to necrosis formation but no definitive studies have been done on this topic. Stem necroses may quickly heal by isolation of the damaged tissue through callusing, may persist for long periods if the stem in which they occur is large, or may even cause the death of the cactus, although this is not common. In some giant Pachycereinae cacti of the Sonoran Desert, stem necroses may persist so long that liquefied stem tissue drains down the side of the stem and moistens the soil at the base of the cactus, forming a fourth yeast habitat associated with cacti.

The flowers, fruit rots and stem necroses are visited by many insects (Starmer et al., 1988b), some of which breed in them (Heed, 1977a). With the exception of some stem necroses caused by lepidopteran larvae, insects do not produce the breaches that initiate stem rots, the habitat that is the focus of this paper. The origin of stem necroses is not well documented but mammal and bird activities, impacts due to objects carried by wind and frost damage have been implicated. Often, the first insects to arrive at rots, whether in stems or fruits, are *Drosophila* (Fogleman and Foster, 1989) and there are many species of *Drosophila* that breed only in cactus stem necroses (Heed, 1982, 1989). One species, *Drosophila mettleri*, has even colonised the soil soaked by liquid flowing from Pachycereinae necroses (Heed, 1977b). Cactophilic *Drosophila* adults deposit cactophilic yeast onto their feeding substrates through both regurgitation and defecation (Starmer, 1982; Ganter et al., 1986; Ganter, 1988; Fogleman and Foster, 1989).

Cacti are potential oases in a dry landscape. To avoid the exploitation this attracts, cacti have a range of physical attributes that make it difficult for other organisms to feed on them. One attribute relevant to the yeast community is the accumulation of secondary metabolites in their somatic tissues. For those cacti that form rots, these metabolites range from mucoid polysaccharides to more toxic compounds such as alkaloids and triterpene glycosides, in concentrations from trace amounts to 25–40% of a stem's dry weight for some of the saponins (Kircher, 1982). This variation contributes greatly to heterogeneity in the cactophilic yeast habitat. Another contributing factor is variation in the amount of free simple sugars. These are abundant in nectar and fruits and are common in some cactus stems. However, other cacti bind the majority of simple sugars in their stems to other molecules and the sugars are available only if secreted microbial enzymes cleave them free (Kircher, 1982; Fogleman and Abril, 1990). Thus, the habitat for yeast associated with cacti consists of relatively small, temporary patches that vary in chemical makeup, persistence and species composition but that have, in total, a global distribution, especially since cacti have been exported to so many parts of the world. Patches of suitable habitat are not initiated by the activity of the microbial community but, once found, are quickly colonised and exploited until the patch is exhausted. The ecology of the microbial community found in cactus rots is consistent with the ecology of other free-living microbes. However, microbial colonisation occurs via both passive dispersal (for at least some bacteria) and vectoring by insects (Fogleman and Foster, 1989).

Before I proceed to a discussion of endemism in cactophilic yeast, it is useful to examine the degree to which cacti represent a separate yeast habitat and the evidence for differences among the yeast communities within cacti. Although I have collected from cacti in many regions, I have not often collected from other neighbouring habitats at the same time as the cactus collection was done. In many instances, cactus fruits and flowers were not found on the cacti or were not there in sufficient abundance that they were collected. However, enough data have been collected to give some indication of the degree of overlap between cactus stem necrosis yeast (referred to hence as cactophilic yeast) communities and neighbouring yeast communities (Table 8.1). The tree flux and flower communities in the table qualify as neighbouring communities for several reasons. They were all collected within 30 kilometres and two days of the cactus communities to which they are compared. In addition, all three habitats have insect communities that feed on and vector the yeast found in them (Ganter et al., 1986). The insects can act as a means of isolating (Ganter, 1988) or connecting the habitats (Ganter et al., 1986), depending on their behaviours. The metric chosen for the comparison is proportional overlap (Bloom, 1981; Vegelius et al., 1986).

There is no objective criterion for comparing proportional overlap measures, but Table 8.1 shows that there was no overlap between the yeast communities collected from cacti and *Ipomoea* flowers in Peru and very little overlap between the yeast collected from cacti and tree fluxes in either Texas or Arizona. Comparison metrics of the three communities associated with cacti are more variable and range from no overlap to values as high as 0.38 for the comparison of stem rots and fruit rots sampled in Brazil. Starmer et al. (1988a) collected yeast from *Opuntia* and fruit rots in Florida in 1986. The overlap between the stem and fruit communities in their data was 0.16, smaller than the 0.28 in my data. In order to evaluate these overlap values, I compare the values in Table 8.1 to overlap values calculated from within the stem community using collections done at different times and places. The average of the eight comparisons between cactus-yeast communities (stem, fruit and flower) in Table 8.1 is 0.16. The average overlap between collections of stem rots

Table 8.1 Proportional overlap between cactus and neighbouring habitats.

		Cactus fruits	**Cactus flowers**	***Ipomoea* flowers**	**Tree flux**
Peru (2002)					
	Cactus stem rots	0.06	0.32	0	
	Cactus fruits		0	0	
	Cactus flowers			0	
Texas, USA (1998)					
	Cactus stem rots	0.04			0.06
	Cactus fruits				0.02
Northeastern Brazil (2002)					
	Cactus stem rots	0.38	0.01		
	Cactus fruits		0.19		
Florida, USA (1990–1996)					
	Cactus stem rots	0.28			
Arizona, USA (1986)					
	Cactus stem rots				0.02

from different species of *Opuntia* in Texas was 0.57 (n = 8, data not shown). The average overlap for comparisons of yeast collections from rots in different genera of cacti (from several regions) was 0.42 (n = 9, data not shown). In evaluating these numbers, one should keep in mind that the eight comparisons of stem, fruit and flower collections in Table 8.1 are based on data collected at the same time and in the same place (often from the same cactus) yet the communities show little evidence of overlap. The comparisons of stem rots were done from collections made years and many miles apart but yielded much higher similarities. Thus, I conclude that the cactus yeast community (whether stem, fruit or flower) is isolated from neighbouring, non-cactus yeast communities and that the cactus stem community is also isolated from cactus fruit and cactus flower communities, although to a lesser degree than from neighbouring non-cactus yeast communities.

8.4 Cactophilic yeast endemism

In this section, I will examine endemism within the cactus stem necrosis community, which has been sampled far more extensively than either the cactus flower or cactus fruit necrosis communities. The first evidence for endemism I present is the taxonomic makeup of the community. The basis for this overview is a compilation of data from two sources: 3451 strains isolated by me (with the help acknowledged below) and 5159 strains isolated by Dr William T. Starmer and his co-workers (Starmer et al., 1990). The total of 8610 strains includes only strains that belong to described taxa (species and varieties) or to known taxa that have not been published as of yet. Almost all come from North and South America, the Caribbean Islands and Australia.

First, I must establish that it is appropriate to combine these two data sets. There are 106 taxa (102 species and four varieties) in the combined data. Of the 106, 44 occur in both data sets, 26 are found only in my data and 36 are found only in the Starmer data. However, the 44 species found in common represent 95% of all 8610 strains. The proportional overlap between the two collections is 0.74, a large value considering that there are differences in the locations and species of cacti sampled. Based on the degree of overlap and numerical importance of the species common to both data sets, it seems reasonable to combine the data sets.

Not all yeast found in cactus rots are members of the cactophilic community. If one assumes that the strains isolated represent both community members and 'accidentals', strains that do not normally grow in stem necroses, then elimination of rare isolates is justified. For this reason, Table 8.2, which presents the combined data set, includes those 30 taxa with proportional representation over a half per cent (0.005) of the total. However, this arbitrary criterion is not sufficient, as there are rare species that must be considered as cactus species in spite of their rareness because they have never been collected from other habitats. The last eight species in Table 8.2 are added under this second criterion. The elimination of accidentals reduces the total number of isolates in the combined data set by less than 5%.

8.4.1 Taxonomic analysis of endemism

Of the 38 species in Table 8.2, 25 are found only in the cactophilic habitat, ten are listed as possible cactophilic species, and three are widespread, occurring in many habitats. The cactophilic isolates are 82% of the total (excluding the 'Possible' category) while the widespread species comprise only 3.5%. Species for which there is evidence that they are native only to the cactophilic habitat were placed in the 'Possible' group. *Galactomyces geotrichum* (Naumova et al., 2001), *Candida guilliermondii* (Vaughan-Martini et al., 2005), and *Cryptococcus laurentii* (Takashima et al., 2003; Pohl et al., 2006) are known to be complexes of species and the cactus isolates show some physiological differences from the standard description (and, for *Cr. laurentii*, some rDNA sequence difference has

Table 8.2 Taxonomic analysis of 8244 strains isolated from the cactophilic habitat. 'Cactus' refers to taxa restricted to the cactophilic habitat, 'Possible' to taxa for which there is evidence that the cactophilic isolates represent separate undescribed species, and 'Wide' to taxa collected from many habitats. The species from the *Sporopachydermia cereana* complex are aggregated. *P.* 'brazilensis' is an undescribed variety of *P. cactophila*.

Taxon	Type	Isolates	Proportion of total	Cumulative proportion
Total		8244		
Pichia cactophila	Cactus	1914	0.2322	0.232
P. 'brazilensis'	-			
Candida sonorensis	Cactus	1329	0.1612	0.393
Sporopachydermia cereana complex	Cactus	1049	0.1272	0.521
S. cereana	-			
S. australis	-			
S. brasilensis	-			
S. centralis	-			
S. obscura	-			
S. opuntiana	-			
S. oaxacaensis	-			
S. pachycereana	-			
S. stenocereana	-			
S. trichocereana	-			
Clavispora opuntiae	Cactus	643	0.0780	0.599
Pichia kluyveri	Possible	277	0.0336	0.632
Myxozyma mucilagina	Cactus	248	0.0301	0.662
Geotrichum clavatum	Possible	207	0.0251	0.687
Dipodascus starmeri	Cactus	194	0.0235	0.711
Pichia deserticola	Cactus	191	0.0232	0.734
Starmera caribaea	Cactus	166	0.0201	0.754
Pichia angusta-like	Cactus	152	0.0184	0.773

Table 8.2 (*cont.*)

Taxon	Type	Isolates	Proportion of total	Cumulative proportion
Starmera amethionina	Cactus	146	0.0177	0.790
Pichia eremophila	Cactus	131	0.0159	0.806
Cryptococcus albidus	Widespread	125	0.0152	0.821
Candida boidinii	Widespread	124	0.0150	0.836
Pichia mexicana	Possible	121	0.0147	0.851
Rhodotorula mucilaginosa	Possible	116	0.0141	0.865
Pichia insulana	Cactus	99	0.0120	0.877
Rhodotorula minuta	Possible	98	0.0119	0.899
Galactomyces geotrichum	Possible	96	0.0116	0.901
Pichia heedii	Cactus	94	0.0114	0.912
Phaffomyces opuntiae	Cactus	83	0.0101	0.922
Pichia norvegensis	Possible	80	0.0097	0.932
Cryptococcus laurentii	Possible	78	0.0095	0.941
Starmera pachycereana	Cactus	70	0.0085	0.950
Candida guilliermondii	Possible	68	0.0082	0.958
Candida caseinolytica	Cactus	62	0.0075	0.966
Pichia barkeri	Cactus	60	0.0073	0.973
Debaryomyces hansenii	Possible	50	0.0061	0.979
Kloeckera apis	Widespread	48	0.0058	0.985
Phaffomyces thermotolerans	Cactus	29	0.0035	0.988
Pichia pseudocactophila	Cactus	25	0.0030	0.991

Table 8.2 (*cont.*)

Starmera 'australensis'	Cactus	24	0.0029	0.994
Phaffomyces antillensis	Cactus	20	0.0024	0.997
Candida orba	Cactus	12	0.0015	0.998
Starmera 'curaçao'	Cactus	9	0.0011	0.999
Pichia cephalocereana	Cactus	5	0.0006	0.999
Starmera 'guatemala'	Cactus	1	0.0001	1.000

also been found; Ganter, unpublished data). *Pichia norvegensis* is a rare human pathogen (Leask and Yarrow, 1976) in its anamorphic state but is common in *Opuntia* rots in some locales and is very likely a cactophilic species (Starmer et al., 1990). Anamorphs of *Pichia mexicana* have been isolated rarely from soil, fruit and insect-associated habitats (Kurtzman, 1998), but the majority of isolations are from cacti. Variation among the *P. mexicana* phenotypes found in cacti suggests that this might be a complex of sibling species (Ganter, unpublished data). *Geotrichum clavatum, Rhodotorula minuta* and *Rhodotorula mucilaginosa* are widely distributed anamorphs with variable physiological profiles. In all three cases, the strains collected from cacti show consistent differences with the standard description (Ganter, unpublished data). Although each species is variable within the cactus habitat, the variations represent separate clusters and provide evidence that the cactus strains are distinct taxa. One sexual species, *Debaryomyces hansenii,* is widespread but is noted for phenotypic variation (Groenewald et al., 2008; Jacques et al., 2009; Nguyen et al., 2009). Differences between cactophilic isolates and isolates from other habitats and the description of strains from neighbouring mesquite fluxes as a new species (Phaff et al., 1998) suggest that the cactophilic strains of *D. hansenii* may be a separate species as well. Of the widely distributed species, both *Candida boidinii* and *Cryptococcus albidus* might have been classified as 'Possible' in Table 8.2 but the evidence is not as clear as for the others. *Kloeckera apis* is very common in cactus fruit (and acidic fruits in general) and is probably an accidental from that habitat. *Pichia kluyveri* has long been known from both fruit and cactus sources and will be discussed later as it is a particularly striking example of endemism. Naumov et al. (1997a) presented evidence that the cactophilic strains of *Pichia angusta* (once *Hansenula polymorpha,* now *Ogataea polymorpha*) were a separate species but did not formally describe the new species and Table 8.2 lists them as cactus yeast.

There are two cactophilic species not in the Starmer–Ganter combined data. *Dipodascus australiensis* and *D. spicifer* are known from only a few specimens but both appear to be cactophilic (von Arx, 1977; de Hoog et al., 1986). The latter species was described from a single isolate from an *Opuntia* rot in Arizona and the former from three isolates found in *Opuntia* rots from Australia and South Africa plus one isolate from a rot in *Euphorbia ingens*, also from South Africa. It is not known why they are absent from the combined collection. Ongoing analysis of strains from this group (Dipodascus is closely related to Galactomyces and Geotrichum; Kurtzman and Robnett, 1995) collected in South America has uncovered several taxa related to but distinctly separate from both *D. australiensis* and *D. spicifer* (Antonielli, Cardinali and Ganter, unpublished data), which reinforces the claim that *D. australiensis* and *D. spicifer* are cactophilic species.

The conclusion from this taxonomic analysis is clear: the cactophilic habitat is dominated by species found only in cactus rots, although it is not closed and yeast from other habitats occur regularly but rarely. The dominance of cactus-specific species will probably become more complete as more taxonomically relevant sequences are examined in those taxa listed as 'Possible' in Table 8.2 and as isolates currently listed as 'unknowns' (332 in both data sets) are examined for new taxa. EiE posits that microbial habitats should be species poor (at the global scale) compared with macrobial habitats (Fenchel and Finlay, 2004b), but the cactophilic habitat is rich in endemic species, as rich as the insect community associated with the resource. A closer examination of the cactophilic community (sections 8.4.2 and 8.4.3 below) will show that narrow endemism is very common. Here, narrow endemism means geographic distributions restricted to a subset of locales in which suitable cacti are found. Thus, narrow endemism means that a species is absent from a portion of its niche (the niche according to Grinnell). Narrow endemism is a violation of EiE because EiE permits no geographic restrictions to a species' distribution. Wherever the habitat is found so should those microbes able to exploit it ('the habitat selects'). Of the 25 cactophilic taxa, only three have ever been considered ubiquitous within the habitat: *Pichia cactophila, Candida sonorensis* and *Sporopachydermia cereana. Sporopachydermia cereana* has already been shown to be a complex of similar, closely related species (Lachance et al., 2001b), all with restricted geographic ranges. In Table 8.2, all of the cactophilic *Sporopachydermia* strains have been lumped together. I could not separate the strains that I had not isolated into their proper species because I did not have access to the original physiological profiles and rDNA sequences are not available. For those strains in my collection, rDNA sequence information is only available for a few strains and those few sequences are not close matches to previously published *S. cereana* complex sequences. In addition, many unsequenced strains in my collection have physiological profiles that do not match those described by Lachance et al. (2001b), indicating that there may be more species in the complex than presently described. The *Sporopachydermia cereana* complex, once thought to be a single species, is an outstanding example of narrow geographic endemism within the cactophilic habitat. This leaves two 'ubiquitous' species and I will discuss the ubiquity of *P. cactophila* and *C. sonorensis* in the last portion of this section.

8.4.2 Narrow endemism within the cactophilic habitat

Initially, the hat-spored, cactophilic yeast species (which includes almost all of the sexual cactophilic yeast species) were described as *Pichia membranifaciens*, a polymorphic member of the once large genus *Pichia* (eight of the 11 positive assimilation abilities vary among strains in the description of *P. membranifaciens*). Correlations between habitat, geography and physiology led to the description of a set of new cactophilic species. The newly described cactophilic *Pichia* species had an unusually large range of mol G+C% values (from 27% to 48%) and Starmer et al. (1986) found that the mol G+C content of their DNA had a pattern: cactophilic *Pichia* species were clustered into groups spaced at intervals of 3% (the 30%, 33% and 36% groups were the most speciose). Ribosomal DNA sequence data has since been used to separate most of the members of the 33% group into two endemic cactophilic genera (Yamada et al., 1997, 1999). Although the new genera were not closely related to *Pichia* or to one another, the rDNA data confirmed that within-clade relationships were close. I will examine three of these lineages as examples of very narrow geographic endemism common among cactophilic yeast.

One observation is germane here. Since their description, no strain of any of the species that were originally thought to be *P. membranifaciens* has been isolated from a habitat that was not necrotic plant tissue and physically close to a population of cacti containing the species. For instance, some endemic cactophilic species have been isolated (always rarely) from tree sap fluxes in Arizona but sap fluxes in Canada do not contain any accidental cactophilic species. There is one possible exception and its distribution is a bit of a mystery. *Pichia norvegensis* was first identified in its anamorphic state as a cause of vaginitis (most cases reported are from Europe; Leask and Yarrow, 1976). Extensive sampling of humans shows that we are a very rare host for this species. One strain from a pig's gut and several strains from cheese (also European) widen its habitat somewhat. Although it was never mistaken for *P. membranifaciens*, it has been isolated from cacti (mostly *Opuntia* and members of the Cereeae) over 80 times in both North and South America. There are many more isolates from cacti than from all other habitats combined. Because of its odd distribution, it is listed as a possible cactophilic species in Table 8.2. Genetic variation among the isolates has not been investigated but might shed light on a very disjunct distribution.

Starmera. There are six taxa in the *Starmera* group (Table 8.3). The genus *Starmera* was proposed in 1997 for a cactophilic *Pichia* species, *P. amethionina*, with two varieties and a divergent rDNA sequence (Yamada et al., 1997). A second

Table 8.3 Relatedness within the cactophilic G+C groups for *Starmera* (a), *Phaffomices* (b) and *Pichia* (c). The upper triangle has DNA-DNA reassociation values and the lower triangle has hybrid spore viabilities. a, b, 33 mol% G+C Species; c, 30 mol% G+C Species. Sources are listed at the bottom of the table. Spaces indicate that no data are available. *Pichia heedii* has 33 mol% G+C, but shows only background homology with either *Starmera* or *Phaffomyces* species. *Pichia barkeri* has 36 mol% G+C, but has a remarkably high DNA-DNA homology of 20% with *P. kluyveri*.

(a)

Starmera	**S. am**	**S. pa**	**S. ca**	**S. au**	**S. cu**	**S. gu**
S. amethionina	-	65–85%	40%	87%		
S. pachycereana	51%	-	37%	64%		
S. caribaea	0%	0%	-			
S. 'australensis'				-		
S. 'curaçao'					-	
S. 'guatemala'						-

(b)

Phaffomyces	**Ph. op.**	**Ph. th.**	**Ph. an**	**C. or.**
Ph. opuntiae	-	28–34%	55%	27%
Ph. thermotolerans		-	26%	52%
Ph. antillensis			-	24%
Candida orba				-

(c)

Pichia	**P. kluyveri**	**P. eremophila**	**P. cephalocereana**	**C. eremophila**
P. kluyveri	-	66%	72%	64%
P. eremophila	30%	-	69%	94%
P. cephalocereana	81%	39%	-	70%
Candida eremophila				-

Starmera sources: Starmer et al. (1978, 1980), Holzschu and Phaff (1982), Phaff et al. (1992); *Phaffomyces* sources: Starmer et al. (1979, 1984, 2001), Holzschu et al. (1985); *Pichia* sources: Phaff et al. (1987a, 1987b), Kurtzman et al. (2008).

cactophilic *Pichia* species, now *S. caribaea*, was transferred to the genus later (Yamada et al., 1999). Kurtzman et al. (2008), using concatenated gene sequences, has placed two other *Pichia* species (*P. quercuum* and *P. dryadoides*) in the genus, but the bootstrap support for this expanded clade is low while the support for the cactophilic subclade is 100%. For our purposes, *Starmera* refers to just the cactophilic members of the clade. Three of the six members are described species and three are not. In the case of *S.* 'australensis', evidence for it being a valid species is found in Holzschu et al. (1985). For the remaining two, support for their inclusion as separate species is based on differences between them and the other *Starmera* taxa in both physiological profile and sequence of the D1/D2 region of the LSU rDNA genes (Ganter, unpublished data).

Starmera is found in almost every region and cactus type (Table 8.4). All of the taxa within the genus are distributed almost without overlap and are, for this reason, narrowly endemic species. Instances where only one strain in a locale was found on an unexpected cactus type have been ignored because there is variation in some of the key taxonomic physiological abilities and, in no case, had sequencing been done to corroborate the unexpected occurrence. The only exception to the no-overlap rule is that both *S. pachycereana* and *S. amethionina* are found on *Opuntia* in the Sonoran Desert and in West Texas. This overlap may occur on *Opuntia* because it is not the preferred host cactus for either species. Although more *Opuntia* rots have been sampled in the Sonoran Desert and West Texas than any other cactus type, *S. amethionina* has been isolated more often from Stenocereinae than *Opuntia* rots and *S. pachycereana* more often from Pachycereinae than *Opuntia* rots, so the overlap on *Opuntia* is not on the primary resource of either species.

Phaffomyces. Four species are currently recognised in the *Phaffomyces* clade (Table 8.3). Average DNA–DNA reassociation values are lower among *Phaffomyces* taxa than among *Starmera* taxa. Mating between *Phaffomyces* species is complicated by the reluctance of *Ph. thermotolerans* to mate with itself. For this reason, no spore viability data are available (Table 8.3). Starmer et al. (1979) concluded that *Ph. thermotolerans* is normally haploid, a situation found in other cactophilic yeast. A haplontic lifestyle may also be the reason that *Candida orba* is a member of *Candida* rather than *Phaffomyces*, although clearly in the *Phaffomyces* clade (evidence from similarity of rDNA gene sequences and physiologies). While no strain of *C. orba* will mate with another *C. orba* strain or with the mating types of any *Phaffomyces* species, all *C. orba* strains show initial mating reactions and conjugate with the h+ mating type of *P. antillensis* (no spores are produced), although there is only 24% DNA–DNA homology between them (Table 8.3), a value so low as to preclude conspecificity. As only a handful of *C. orba* strains have ever been isolated, we may simply not have isolated the h– mating type as of yet. This problem may be exacerbated if most strains belong to a single mating type, as can

Table 8.4 The distribution of species in the genera *Starmera* (a) and *Phaffomyces* (b). X indicates that the cactus type either does not occur in the region or that it never or very rarely hosts a stem necrosis and N indicates that the cactus type has not been investigated in the region. Spaces are locales where no member of the yeast group has been found, although the host plant is present. The host cacti have been divided into chemically distinct types where possible: *Opuntia*, Pachycereinae and Stenocereinae – the last two are divisions of the Pachycereae (Gibson, 1982). The 'Other Col.' category includes columnar cacti in several families including Cereeae, Cacteae, Trichocereae and Notocactaceae (taxonomy according to Anderson, 2001). In addition, the 'Other Col.' category includes hosts from the Pachycereae for which no biochemical analysis has been done. The category 'Caribbean Islands' does not include the island of Curaçao.

(a) *Starmera*

Locale	Opuntia	Stenocereinae	Pachycereinae	Other Col.
Sonoran Desert	*S. amethionina*	*S. amethionina*	*S. pachycereana*	X
West Texas	*S. amethionina*	X	X	X
East Texas	*S. caribaea*	X	X	X
Florida	*S. caribaea*	X	X	X
Caribbean Islands	*S. caribaea*	*S. caribaea*	*S. caribaea*	*S. caribaea*
Curaçao	S. 'curaçao'	S. 'curaçao'	S. 'curaçao'	X
Guatemala		N	S. 'guatemala'	N
Brazil		*S. amethionina*	*S. amethionina*	*S. amethionina*
Peru	*S. amethionina*	N		
Australia	S. 'australensis'	X	X	X

(b) *Phaffomyces*

Locale	Opuntia	Stenocereinae	Pachycereinae	Other Col.
Sonoran Desert			*Ph. thermotolerans*	X
West Texas		X	X	X
East Texas		X	X	X
Florida		X	X	X
Caribbean Islands				*Ph. antillensis*

Table 8.4 (*cont.*)

Curaçao				X
Guatemala				N
Brazil				
Peru		N		*Ph.* 'peru'
Australia	*Ph. opuntiae*, *Candida orba*	X	X	X

be the case for *Clavispora opuntiae* (Rosa et al., 1992; Lachance et al., 1994) and *Kurtzmaniella cleridarum* (Lachance and Starmer, 2008).

The distribution of *Phaffomyces* species is even narrower than the distribution of *Starmera* species (Table 8.4). *Phaffomyces* is absent from many regions and cactus types within regions. *Ph. thermotolerans* is found only in Sonoran Pachycereinae rots, *Ph. antillensis* only in Cereeae rots from islands in the northern Caribbean region. The only case of overlap occurs between *Ph. opuntiae* and *C. orba* on *Opuntia* in Australia, where cacti are not native and have been there less than 230 years (Dodd, 1940). The most obvious explanation is that both species of yeast were imported from either the Nearctic or Neotropical realms, although their regions of origin are not known. Invasive cacti can be a severe economic problem and the Australian government scoured native cactus habitats for potential agents of biological control and shipped many cactus specimens to Australia (Dodd, 1940). So far, sampling from the new world has produced no strains with the standard *Ph. opuntiae* physiological profile. Several strains similar to *Ph. opuntiae* have been collected from Peru and the Sonoran Desert with single strains isolated in the Caribbean and Venezuela. All of these strains are phenotypically distinct from the *Ph. opuntiae* found in Australia and sequencing of the Peruvian strains indicates they are a new, undescribed species. Although *Ph. opuntiae* occurs widely in Australia and is often common in Australian cactus rots, no strains of *C. orba* have been found anywhere other than within 100 km of Brisbane, Australia. However, much of South America is either not yet sampled or is undersampled. Until the origins of these two species are understood, it is impossible to tell if their overlap is due to their recent import into Australia or occurs in the species' native range. No matter how this is resolved, this cactophilic genus consists of narrow endemics that are most often rare where they do occur. Many regions have no member of this group, although suitable host plants are present. EiE has no explanation for the absence of *Phaffomyces* from so many cacti and locations.

Pichia kluyveri. Many of the cactophilic yeast are still in the genus *Pichia*, although many species have been transferred out of the genus (Kurtzman and

Robnett, 2010). Within the cactophilic *Pichia*, two cases of narrow endemism occur. One case is that of *Pichia kluyveri* and its relatives. This species is, at this time, unique in that there is no doubt that it occurs as a regular member of the cactophilic community but is not confined to it. *Pichia kluyveri* is well known from rotting acidic fruits worldwide. Starmer et al. (1992) collected *P. kluyveri* from halved tomatoes exposed to insects at 15 locations across the USA. Only the northernmost locations (in Maine and northern New York state) failed to yield multiple strains of *P. kluyveri*. Enough strains of *P. kluyveri* from the cactus community have been sequenced to establish that at least some of the strains in cacti are identical to *P. kluyveri* from acidic fruit collected outside of the geographic range of cacti (Ganter, unpublished data). This is the only case where sequence data have confirmed the identity of cactophilic and non-cactophilic strains. Such investigations have so often yielded evidence for separate cactophilic taxa that it is reasonable to be cautious of relying on the similarity of physiological profiles as presumptive evidence for conspecificity when the strains come from both cactophilic and non-cactophilic habitats (von Arx, 1977; Naumov et al., 1997a; Phaff et al., 1997, 1998).

The distribution of *P. kluyveri* within the cactophilic habitat is only partially known. The problem arises from the likelihood that some strains identified as *P. kluyveri* may actually be distinct taxa. This possibility was first noticed when collections in North America and the Caribbean yielded locale- or host-related phenotypic variation (Phaff et al., 1987a). From this observation, two varieties of *P. kluyveri* were described (recently elevated to species status by Kurtzman et al., 2008): *P. cephalocereana*, from Caribbean columnar cacti, and *P. eremophila*, from *Opuntia* cladodes in the northern Sonoran and Chihuahuan Deserts. Anamorphic strains of *P. eremophila* were found in Sonoran Stenocereinae rots. These strictly cactophilic taxa are narrowly endemic species related to (and, perhaps, derived from) the cosmopolitan species *P. kluyveri*. However, isolations of *P. kluyveri*-like strains from cacti since the description of *P. cephalocereana* and *P. eremophila* have complicated the situation. The diagnostic differences between the taxa (ability to ferment glucose, ability to kill a standard strain of *Candida glabrata*, host preference and geographic origin) are sufficient to separate these taxa when collecting from the same region and species of cacti from which *P. eremophila* and *P. cephalocereana* were described but will not suffice to identify strains collected from new host plants or from new locales. Almost all possible combinations of the diagnostic characters have appeared in collections. The recent sequence analysis that confirmed the presence of cosmopolitan *P. kluyveri* strains in cacti also revealed two divergent strains that are probably new species, although all of the strains tested had the physiological profile of standard *P. kluyveri* (Ganter, unpublished data). Thus, identification requires sequencing in this group, something not regularly done when the data presented here were collected. The only conclusion possible is that the *P. kluyveri* clade has a cosmopolitan distribution within the cactophilic habitat but, until more sequence data are collected, it is not possible to separate what may be several individual, narrowly endemic taxa.

This scenario is consistent with the only study of genetic variation in the *P. kluyveri* group (Ganter and de Barros Lopes, 2000). Between-strain random amplified polymorphic DNA (RAPD) and amplified fragment length polymorphism (AFLP) variation was so great that few monomorphic bands were detected by either method. Strains analysed included members of each taxon (now species, then varieties) and some strains not classifiable by phenotypic determination. The results showed that some strains with the same physiological profile were not closely related. The most important predictor of relatedness was geographic origin.

Kurtzman et al. (2008) placed *P. kluyveri, P. cephalocereana* and *P. eremophila* in a well-supported clade with *P. barkeri, P. heedii* and *P. nakasei*. All of these species are associated either with rotting cacti, rotting fruit or both. Only *P. nakasei* has not been isolated from a cactus rot. *Pichia barkeri*, listed as a cactophilic species, has never been a dominant yeast in any collection and has been isolated from cactus fruit rots, which might indicate that it is not strictly cactophilic. All six members of the clade are oligotrophic and have similar physiological profiles. *Pichia nakasei*, the only species not associated with cacti, is most divergent from the 'average' phenotype for the clade and Kurtzman et al. (2008) place it basal to the rest of the clade. If, as more loci and more cactophilic strains are sequenced, *P. nakasei* remains basal, then it is reasonable to hypothesise that there is a cactophilic 'kluyveri' niche for strains having the correct phenotype and that the niche is filled in different geographic regions by independently evolving lineages from this clade. The alternative to this hypothesis is that the similarity of phenotypes is a consequence of phylogenetic inertia. However, there is some evidence for the former hypothesis. *Pichia barkeri* has a physiological profile as close to *P. kluyveri* as either *P. cephalocereana* or *P. eremophila*, yet it's DNA–DNA reassociation value is only 21% with *P. kluyveri* (Phaff et al., 1987b), a value much lower than the value for either *P. eremophila* or *P. cephalocereana* (Table 8.3). Moreover, *P. barkeri* has a mol% G+C DNA composition of 36%, far from the 30% values for *P. kluyveri, P. cephalocereana* and *P. eremophila. Pichia heedii* and *P. nakasei* have intermediate values of 33%. Thus, the *P. kluyveri*-like taxa have physiological profiles more similar than their relatedness would predict, which indicates adaptation to a niche in the cactophilic environment, not phylogenetic inertia. Genome-wide comparisons of these lineages will be necessary to elucidate the reason or reasons for the magnitude of the variation in proportional nucleotide composition for such closely related organisms.

8.4.3 Geographic variation within ubiquitous cactophilic species

The only yeast in the cactophilic habitat that might merit description as ubiquitous are *Pichia cactophila* and *Candida sonorensis*, which account for about 40% of all cactophilic isolates (Table 8.2). Both are found on all host types and from all locales, with one notable exception, discussed below. Both are amictic, although the details differ. No spores have been detected in any *C. sonorensis* isolates, although each of the 1300+ strains collected so far has been examined for spores as a routine part of the identification procedure. *Pichia cactophila* strains often produce spores but almost always two large spores per ascus. Although the genetics have not been investigated, the assumption is that meiosis occurs in *P. cactophila* but mating takes place within the ascus, producing two diploid spores and preventing outcrossing. Here, I will demonstrate that these two species and another widespread but far less common asexual species, *Myxozyma mucilagina*, show significant levels of genetic variation linked to geographic origin of the strain. In these amictic species with less phenotypic variation than genetic variation, the existence of local lineages is evidence for narrow endemism.

Endemism at the species level has already been described for *P. cactophila*. In two locales (the Sonoran Desert and Curaçao, N.A.) strains were found with four spores instead of the expected two. The Sonoran four-spored strains have been described as *P. pseudocactophila* (Holzschu et al., 1983) and the Caribbean four-spored strains as *P. insulana* (Ganter et al., 2010). In neither case has a reliable physiological test been found that will distinguish *P. pseudocactophila* or *P. insulana* from *P. cactophila*. In the Sonoran Desert, *P. cactophila* and *P. pseudocactophila* divide the habitat by host type. Sporogenous *Pichia pseudocactophila* are found on Pachycereinae cacti. Although only a few sequences have been obtained, it appears that the asporogenous strains from that host type are also *P. pseudocactophila* (M.-A. Lachance, unpublished data). In the Caribbean, *P. cactophila* and *P. insulana* divide the habitat by geography. On Curaçao, sequences from both sporogenous and asporogenous strains are *P. insulana* and occur on Stenocereinae, Pachycereinae and *Opuntia* hosts. A few asporogenous *P. insulana* strains have been identified from other Caribbean islands but they appear rare outside of Curaçao and are not confined to a single host type. Unfortunately, no strains from Venezuela have been sequenced, so it is not known if Curaçao is the only location dominated by *P. insulana*. So, two narrowly endemic species related to *P. cactophila* are already known and separate collections in eastern South America by W.T. Starmer (personal communication) and me contain a physiological variant of *P. cactophila* that may represent a third sibling species. The distributions of *P. pseudocactophila* and *P. insulana* are not consistent with EiE. Each is restricted within very narrow geographic limits (*P. insulana* common only on Curaçao and *P. pseudocactophila* only in the northern Sonoran Desert) and *P. pseudocactophila* is found on only a subset of suitable hosts within its limited geographic range.

The large number of *P. cactophila* without unusual phenotypes still constitutes a potentially ubiquitous cactophilic species. However, RAPD variation was extensive within the species in a study of 32 strains from different regions and host types (Ganter and Quarles, 1997). There were 133 polymorphic bands from eight primers and each band was found in an average of nine of the 32 *P. cactophila* strains. Principal component analysis of this variation produced strong support for geographic location as the most important predictor of RAPD variation. Not all regions were represented in this study (strains came from Florida, Antigua, Argentina, Australia and the Sonoran Desert), but the distance effect was important enough that it was possible to separate strains from north and south Florida. Sequence data are not available for these strains, but many produced two-spored asci and these can be assumed to be *P. cactophila*. As there was no evidence in the RAPD data that the two-spored strains differed from the asporogenous strains, it is probable that these were also *P. cactophila*. Although *P. cactophila* is a cosmopolitan cactophilic yeast, its population genetic structure is consistent with narrow endemism, not EiE.

Proponents of EiE contend that there is no significance to geographically linked neutral genetic variation. If the phenotypes are identical, what does it matter that neutral variation is linked to geography? The regional differences in *P. cactophila* strains are not simply neutral variation but have phenotypic and ecological consequences. Many yeast secrete proteins that are toxic to other yeast. This phenomenon is labelled killing and the proteins are collectively referred to as killer factors (Magliani et al., 1997). Because the killing spectrum for a particular toxin can be quite narrow, Ganter and Quarles (1997) cross-tested all of their *P. cactophila* strains for their ability to kill one another. Although *P. cactophila* is not generally known as a killer yeast, some strains were able to kill conspecifics. When variation in this important ecological character was analysed, geographic location was once again the most important explanatory variable. The separation of Florida into separate geographic regions was also evident in the killer factor data. Some of the north Florida strains were able to kill but they killed only strains from south Florida. Thus, strains with the '*P. cactophila* phenotype' consist of both sibling species and a large number of regionally divided amictic lineages. This is not the EiE scenario. Under EiE, there should be no geographic component to genetic variation, no isolation by distance.

Genetic variation tied to collection locale is also present in both *C. sonorensis* and *M. mucilagina*. Both are asexual but *C. sonorensis* is more widely distributed, occurring in as many locales and hosts as *P. cactophila*. *Myxozyma mucilagina* is more common in North America than South America and is mostly confined to *Opuntia* and Stenocereinae rots. Variation in several traits has been measured in a set of strains from both species. For *C. sonorensis*, variation in RAPD profile, physiological profile, karyotype and locale (entered as distance from other locales) was recorded for 36 strains from a limited number of locales (Australia, Florida and Texas). All were from *Opuntia* rots. While there was little variation among

Table 8.5 Comparison of between-strain variation for several characters for two amictic cactophilic yeast with wide distributions, *C. sonorensis* (a) and *M. mucilagina* (b) (Mantel's R in the upper diagonal, probability of getting a greater R from randomised data in the lower diagonal). The *C. sonorensis* data are from Ganter et al. (2004) and the *M. mucilagina* data have not been published before. The methodology used for the *M. mucilagina* data is identical to that in Ganter et al. (2004).

(a) *Candida sonorensis*

	RAPD	**Physiological profile**	**Karyotype**	**Distance**
RAPD	-	0.26	0.27	0.26
Physiological profile	0.01	-	0.08	0.17
Karyotype	0.05	NS	-	0.99
Distance	0.05	NS	0.001	-

(b) *Myxozyma mucilagina*

	RAPD	**Physiological profile**	**Karyotype**	**Distance**
RAPD	-	0.33	0.40	0.25
Physiological profile	0.001	-	0.18	0.67
Karyotype	0.001	0.01	-	0.25
Distance	0.05	0.001	0.001	-

strains' physiological profiles, variation for both karyotype and RAPD profiles was considerable (Ganter et al., 2004). Both measures of genetic variation correlated with distance (Table 8.5).

An effect of geography is also evident in variation among *M. mucilagina* strains. Twenty-two strains from the Sonoran and Chihuahuan Deserts, Florida and Australia were studied for variation in (Ganter, unpublished data, Table 8.6). These strains represented more locales and hosts than were included in the *C. sonorensis* study. The genetic and physiological variation was a bit larger as well. Genome size varied from four to seven chromosomes and from 4.35 to 8.8 Mb. Table 8.7 shows a complex history of chromosomal variation. Once again, Mantel comparison of the similarity matrices demonstrated significant correlation between genetic variation (RAPD and karyotype), physiological variation and geographic origin (Table 8.5). *Myxozyma mucilagina* is another widespread asexual species that consists of a set of local lineages. In fact, there is no genetic evidence at this time for significant

Table 8.6 Strains of *Myxozyma mucilagina* assayed for variation in physiological profile, RAPD profile and karyotype.

#	Collection Number	Host Plant	Region	Locale
1	91–503.2	*Opuntia* sp.	Sonoran Desert	Baja California, Mexico
2	91–511.5	*Ferrocactus* sp.	Sonoran Desert	Baja California, Mexico
3	91–519.4	*Stenocereus gummosus*	Sonoran Desert	Baja California, Mexico
4	91–527.4	*Stenocereus gummosus*	Sonoran Desert	Baja California, Mexico
5	91–529.6	*Stenocereus gummosus*	Sonoran Desert	Baja California, Mexico
6	91–535.3	*Stenocereus gummosus*	Sonoran Desert	Baja California, Mexico
7	91–629.4	*Opuntia phaeacantha*	Sonoran Desert	Tucson, AZ
8	91–813.3	*Opuntia lindheimeri*	Chihuahuan Desert	Mason County, TX
9	91–843.3	*Opuntia engelmannii*	Chihuahuan Desert	Big Bend National Park, TX
10	91–890.4	*Opuntia engelmannii*	Chihuahuan Desert	Davis Mountains, TX
11	91–891.5	*Opuntia engelmannii*	Chihuahuan Desert	Davis Mountains, TX
12	93–115.3	*Opuntia stricta*	Florida	Canaveral National Seashore
13	93–124.5	*Opuntia humifisa*	Florida	Archbold Biological Station
14	93–125.5	*Opuntia humifisa*	Florida	Archbold Biological Station
15	94–153.4	*Opuntia engelmannii*	Chihuahuan Desert	Davis Mountains, TX
16	94–162.4	*Opuntia engelmannii*	Chihuahuan Desert	Davis Mountains, TX

Table 8.6 (*cont.*)

#	Collection Number	Host Plant	Region	Locale
17	96–125.3	*Opuntia stricta (inermis)*	Australia	Hemmant, QLD
18	96–125.6	*Opuntia stricta (inermis)*	Australia	Hemmant, QLD
19	96–131.5	*Opuntia tomentosa*	Australia	Marburg/Minden, QLD
20	96–162.4	*Opuntia tomentosa*	Australia	Brigalow, QLD
21	96–172.4	*Opuntia tomentosa*	Australia	Ban Ban Springs, QLD
22	96–172.5	*Opuntia tomentosa*	Australia	Ban Ban Springs, QLD

phenotypic or genotypic variation from a widely distributed cactophilic yeast that is not tied to either locale or host type.

8.4.4 Summary and alternative hypothesis

In summary, I have provided evidence that (1) the cactophilic habitat is separate from neighbouring yeast habitats, that (2) most species native to it are narrow endemics and that (3) those species with wider distributions are, in fact, asexual species consisting of a set of regionally based variant lineages (i.e. geographic races). Why are these microbes local and not ubiquitous? The answer is found in their relationship with animals: a relationship that results in the yeast's biogeography resembling that of their animal vectors.

Before leaving this section, I wish to examine an alternative to the conclusion that endemism is linked to geography. In the cactophilic system, geography and host type are partially confounded. Some locales have only one host plant present. If all locales had a unique host type, it would be impossible to separate the effect of host and geography and one could explain geographic variation as simply an example of 'but the environment selects', the phrase that qualifies 'everything is everywhere'. Finlay and Fenchel, arguing for ubiquitous dispersal of microbes, warn against mistaking local habitat availability for localised dispersal (Finlay, 2002; Finlay and Fenchel, 2002). Under this alternative hypothesis, cactophilic yeast distributions are restricted not by geography but by exclusion from cacti that fail to supply vital nutrients or contain toxic secondary chemicals. There are several reasons to reject the explanation that cactophilic narrow endemism is due to chemical differences among host plants. First, there is evidence that most cacti will support the growth of almost all cactophilic yeast species. This conclusion comes from a simple experiment testing growth by cactus type. For *Opuntia*, Stenocereinae, Pachycereinae, plus some Cereeae and Cacteae hosts, sterile Petri dishes with medium consisting of 49% wt./vol. ground cactus stem tissue, 49 wt./vol. water, and 2% agar were inoculated using replica plating with 21 cactophilic species at low density. Growth was assessed only roughly by the size of the colony. While growth was generally slower in some cactus types than others, in no case was growth of any cactophilic species prevented. This does not mean that host tissue chemistry has no influence on the community of yeast that exploit it but it does mean that the qualifying phrase in the EiE hypothesis, 'but the environment selects', is not a sufficient explanation for the absence of cactophilic species when suitable cacti are present. Second, some yeast species are found in all host types within a region but not outside of the region although suitable hosts occur there (e. g. *Starmera caribaea* (Table 8.4) and *P. insulana*). EiE predicts that these suitable hosts should have the missing yeast. The third argument against habitat variation as the explanation for endemism in cactophilic yeast expands the definition of habitat beyond host chemistry to include yeast community interactions as part of 'environment'. If everything were everywhere, but host type by yeast community interactions favoured particular yeast in particular communities, then there should be no spaces in Table 8.4. Some member of a clade should fill the niche associated with the clade in all situations where the hosts are present but this is not so and the explanation is a failure of clades to successfully disperse throughout their potential niches.

8.5 Yeast and *Drosophila*

Repeated attempts have been made to recover cactophilic yeast from desiccated rots (Ganter, unpublished data). In no case has the attempt yielded a viable yeast strain. Sterile Petri dishes containing rich medium that are exposed to the air in either temperate forest (Gilbert, 1980) or desert (Ganter, unpublished data) habitats are colonised by bacteria and filamentous fungi but not by yeast. How, then, do yeast colonise cactus rots? The answer is, of course, that yeast are brought to rot pockets by arthropods that feed on the rotting plant tissue. There is a rich literature on the many associations between animals and yeast (reviewed in Phaff and Starmer, 1987; Ganter, 2006). Necrotic rot pockets are home to many species of insect that carry yeast on their surfaces (sometimes in structures specialised for storage of microbes) and in their crops and guts (Starmer et al., 1988b). In the cactophilic system, attention has been focused on the relationship between yeast and *Drosophila*, which are often the first insects to arrive at the newly damaged host tissue (Fogleman, 1982).

The role of yeast in the diet of *Drosophila* has been recognised for many years (Baumberger, 1917, 1919; Sang, 1956, 1978; Begon, 1981). The flies and their

Table 8.7 Karyotypes of 22 *M. mucilagina* strains (from Table 8.6). Chromosome size from 650 to 2300. Number corresponds to the # column in Table 8.6. N: haploid chromosome number. Host type abbreviations: O: Opuntia, C: Ferrocactus and S: Stenocereinae. Region abbreviations: B: Baja California, S: northern Sonoran Desert, T: Texas (northern

Number	**1**	**2**	**3**	**4**	**5**	**6**	**7**	**8**	**9**	**10**
Host Type	O	C	S	S	S	S	O	O	O	O
Region	B	B	B	B	B	B	S	T	T	T
N	6	5	7	6	5	5	5	4	4	5
Genome (KB)	8800	6800	8150	6950	4850	4850	5350	4250	5300	5150
650			X		X					
700	X	X		X		X		X		X
750			X	X	X	X	X		X	
800			X	X	X			X		X
850	X	X				X	X		X	
900										
950							X			X
1000										
1050			X							
1100				X	X	X				
1150										
1200			X				X	X		X
1250										
1300	X	X								
1350										
1400										
1450				X		X				
1500			X							X
1550					X			X	X	
1600							X			

Chihuahuan Desert), F: Florida and A: Australia. Genome sizes are sums of chromosomal sizes. Chromosomes are marked with an X corresponding to the molecular weight of the band. Data from contour-clamped homogeneous electric field (CHEF) analysis (methodology used is identical to that in Ganter et al., 2004).

11	12	13	14	15	16	17	18	19	20	21	22
O	O	O	O	O	O	O	O	O	O	O	O
T	F	F	F	T	T	A	A	A	A	A	A
4	4	4	4	6	5	6	4	6	6	4	5
4350	5350	5650	5700	8250	6700	7600	5300	8450	8450	5700	6050
X	X	X			X						
			X	X		X	X				X
X	X							X	X	X	
				X							
		X	X					X	X		
					X	X					
							X				
X					X						
				X							
								X	X	X	X
				X							
							X	X	X		
										X	

Table 8.7 *(cont.)*

Number	1	2	3	4	5	6	7	8	9	10
1650										
1700	X	X								
1750										
1800										
1850										
1900										
1950										
2000	X									
2050										
2100										
2150				X					X	
2200			X							
2250	X	X								
2300										

associated yeast communities differ among habitats (Heed et al., 1976; Lachance et al., 1995; Morais et al., 1995; Rosa et al., 1995). Here, I will review what is known of the association between these two groups within the cactophilic system.

Like yeast, *Drosophila* species found in rot pockets are narrowly endemic (Fellows and Heed, 1972; Heed, 1982). The relationship between cactus and fly has been most intensively studied in the Sonoran cactophilic *Drosophila* (Heed, 1977a). Cactophilic *Drosophila* partition the environment by both cactus type (Heed, 1977a) and by geographic region for widespread cactus types like *Opuntia* (Ruiz and Heed, 1988). In some cases, differences in cactus stem chemistry are the reason for habitat partitioning. *Drosophila pachea* must breed in *Lophocereus schottii* rots because of its dietary requirement for a sterol found only in this cactus but has no congeneric competitors for *L. schottii*, its only host plant, as the other Sonoran *Drosophila* species are sensitive to the toxic effects of alkaloids found only in this cactus (Kircher, 1982). Other instances of resource partitioning by the flies are not as absolute and involve substrate chemistry, the yeast community and fly behaviour.

The relationship between cactophilic *Drosophila* and the cactophilic yeast community is a mutualism. However, the benefits provided by yeast to a fly or vice versa are general enough that, in most cases, more than one species of yeast or fly

11	12	13	14	15	16	17	18	19	20	21	22
X											
	X				X	X					X
				X				X	X		
						X					
		X	X								
							X			X	
					X						
	X	X	X	X				X	X		
						X					X

is capable of supplying some benefit. Thus, the mutualism is diffuse (not species-specific). A consequence of the diffuse nature of the mutualism is that we should not expect many one-to-one correspondences between the distribution of a species of cactophilic *Drosophila* and the distribution of particular cactophilic yeast.

Fly adults and larvae prefer yeast to bacteria as food (Vacek et al., 1985). There are numerous benefits to *Drosophila* from the presence of yeast. Some cactus tissue does not provide a complete diet for the *Drosophila* that breed in it (Starmer, 1982). This is usually because the cactus tissue is deficient in nitrogen resources. Yeast are protein-rich and complete the fly's diet. Some yeast reduce the toxicity of the host tissue for the fly by metabolising toxins. Experiments that demonstrate a 'biculture effect' have demonstrated that, except for the metabolism of specific toxins by specific yeast, the benefits supplied by yeast can usually be supplied by any of several yeast species. A 'biculture effect' is an increase in some fitness component (larval or pupal viability, reduction in development time, increase in size at eclosion) for flies reared with two yeast species compared with the midpoint performance of the flies on the relevant monocultures (Starmer and Aberdeen, 1990). The reason for the effect is not known but is probably the result of amino acid, lipid or vitamin complementarity. Ganter (2006) reviews a number of cactophilic *Drosophila*–yeast interactions ranging from conversion of larval nitrogenous waste to protein, adult and larval feeding preferences for particular yeast species, yeast production of volatile compounds that cactophilic *Drosophila* can absorb through their respiratory system and use as food, reduction in sunlight-induced fly mortality when yeast are part of the fly's diet, yeast as nuptial gifts that enhance both a male's chance of mating and the female's fecundity, and yeast lipids as possible precursors of cuticular lipids important in species recognition and mating success.

Several types of benefits supplied to yeast by cactophilic *Drosophila* have been identified so far. One benefit is the stabilisation of cactophilic yeast community dynamics. Most individual rot pockets contain between two and three yeast species. Starmer and Fogleman (1986) found that the addition of the appropriate *Drosophila* larvae changed the dynamics of pairwise interactions among yeast species and that the effect on the entire community was to stabilise it, thus preventing competitive exclusion through pairwise competition. There is some indication that animal dispersal increases the rate of outcrossing in *Saccharomyces cerevisiae* (Reuter et al., 2007) but this effect has not been demonstrated for a cactophilic yeast nor has outcrossing been shown to be universally beneficial to yeast as there are many asexual lineages. *Saccharomyces paradoxus*, a close relative of *S. cerevisiae*, apparently undergoes sex only once in 1000 generations and only 1% of sexually reproductive events are outcrossed (Tsai et al., 2008). Of course, sex might have benefits other than outcrossing as there is evidence that yeast spores survive longer in a *Drosophila* gut than do vegetative cells (Reuter et al., 2007; Coluccio et al., 2008). Increased survival in the gut is only a benefit if yeast are vectored by the flies. A dispersal benefit for yeast that undergo sex goes to the heart of the reason for cactophilic yeast endemism. Cactus rots are temporary and dispersal is a regular, unavoidable occurrence for any cactophilic yeast.

Drosophila are vital to cactophilic yeast dispersal. Fogleman and Foster (1989) found that they could induce rots in a columnar cactus, *Stenocereus gummosus*, through freeze damage. Bacterial colonisation started almost immediately after the damage occurred but yeast colonisation was delayed by two days. If they covered the newly damaged tissue with a mesh that excluded adult *Drosophila*, the bacterial community developed as though no mesh was present but there was an additional delay of two days in the arrival of yeast. *Drosophila* and larger insects were excluded by the mesh but smaller insects and mites were not. The authors concluded that *Drosophila* are normally the first insects to arrive and the first to deposit yeast in the rot but that insects are not important in bacterial colonisation, although bacterial fermentation is probably important in attracting *Drosophila* to the new rot.

Geographic (narrow) endemism is promoted by factors that limit dispersal. There are at least three consequences of vectoring by *Drosophila* that promote endemism in cactophilic yeast by limiting dispersal. The first is that the ambit of an adult fly is much smaller than the distance a desiccation-resistant spore might

Table 8.8 The overlap between the yeast found in slime fluxes and rot pockets with those found in the guts of *Drosophila* that breed in those substrates in Tucson, AZ (Ganter et al., 1986; Ganter, 1988). The proportion of strains shared with flies is the proportion of strains from a particular host plant that were members of species also found in the fly species that breed there. Similarly, the proportion of strains shared with a host plant is the proportion of strains isolated from the fly that are members of species also found in the fly's host plant.

Host plant[1]	Number of strains	Shared with flies	Flies[2]	Number of strains	Shared with host plant
Cottonwood	20	0.65	*D. b.* and *A. l.*	23	0.53
Mesquite	92	0.95	*D. carbonaria*	46	0.50
Opuntia	148	0.76	*D. l.* and *D. h.*	34	0.74
Saguaro	36	0.92	*D. n.* and *D. m.*	109	0.89
Senita	18	0.61	*D. pachea*	12	1.00
Agria	68	0.99	*D. mojavensis*	44	0.98

[1] Cottonwood: *Populus fremontii*; Mesquite: *Prosopis* sp.; *Opuntia*: *O. phaeacantha* and *O. ficus-indica*; Saguaro: *Carnegiea gigantea*; Senita: *Lophocereus schottii*; Agria: *Stenocereus gummosus*.
[2] *A. l.*: *Aulacigaster leucopeza*; *D. b.*: *Drosophila brooksae*; *D. h.*: *Drosophila hamatophila*; *D. l.*: *Drosophila longicornis*; *D. m.*: *Drosophila mettleri*; *D. n.*: *Drosophila nigrospiracula*.

be carried by wind. The second consequence comes from the endemism shown by the *Drosophila*. Cactophilic flies can have smaller ranges than the cacti they inhabit when the cactus type is widespread, such as *Opuntia*. The flies that breed in *Opuntia* in Texas are not the same flies that breed in *Opuntia* in Arizona. The third consequence is an outcome of the fly's behaviour. If the fly visits only a subset of potential yeast habitats, fly behaviour reduces the size of the yeast's realised niche compared to its fundamental niche.

Two data sets document the importance of cactophilic *Drosophila* to the distribution of cactophilic yeast (Ganter et al., 1986; Ganter, 1988). The first is a set of yeast collected from cactus rots and slime fluxes in the vicinity of Tucson, AZ (slime fluxes from several species of trees, *Opuntia* and saguaro cactus) and in Baja California, Mexico (agria and senita cactus). Each rot or flux was the breeding site of at least one *Drosophila* species (with the addition of a cosmopolitan slime flux fly, *Aulacigaster leucopeza*, in Cottonwood fluxes) and yeast were collected from flies captured at the flux or rot by allowing the fly to deposit yeast on Petri plates. Table 8.8 gives the overlap between the yeast deposited by the flies and the yeast found in the flies' breeding substrate. The overlap for the flies increases as one moves from (relatively) mesic habitat (the slime fluxes, 0.53 and 0.50 overlap) to the most xeric habitats (1.00 and 0.98 overlap). In cactus habitats, flies do not tend to harbour yeast other than those in their breeding sites.

Table 8.9 The average correlation between the yeast found in flies breeding in saguaro (a) or agria (b) cactus, the yeast in the cactus rots, and the yeast found in neighbouring cacti near Tucson, AZ (saguaro) or Punta Prieta, Baja Mexico, Mexico (agria) (Ganter, 1988). Kruskal–Wallis analysis was used to compare the average similarity among groups (ranks of the correlation coefficients calculated from the transformed proportional yeast species composition of pairs of rots, pairs of flies or a rot and a fly). n: the number of comparisons. H (corrected for ties) = 95.0 (saguaro fly analysis) or = 179.5 (agria fly analysis) and probability of a greater H < 0.0001 for either analysis.

(a) Saguaro flies

Source[1]	n	Mean correlation	Post-hoc test group
Fly – Host Rot	16	0.88	A
Saguaro rot – Saguaro-rot	91	0.66	B
Fly – Fly	120	0.58	B
Fly – Saguaro	208	0.51	BC
Opuntia – *Opuntia*	36	0.49	BC
Saguaro – *Opuntia*	126	0.46	C
Fly – *Opuntia*	144	0.24	D

(b) Agria flies

Source[1]	n	Mean correlation	Post-hoc test group
D. mojavensis – same rot	19	0.56	A
D. mojavensis – *D. mojavensis*	171	0.53	AB
Agria – Agria	210	0.40	BC
D. mojavensis – Agria	380	0.39	BC
D. mojavensis – Senita	132	0.25	CDE
Agria – Senita	55	0.19	DE
Senita – Senita	21	0.14	E

[1] Fly: either *Drosophila nigrospiracula* or *D. mettleri*; *Opuntia*:*O. phaeacantha*; Saguaro: *Carnegiea gigantea*; Senita: *Lophocereus schottii*; Agria: *Stenocereus gummosus*.

This effect is tested more explicitly in a second data set of yeast from cactus rots and the flies found on the same rots. In this set, the yeast from individual rots and individual adult flies are compared for two different species of *Drosophila* from different host cacti and different locales in the Sonoran Desert (Table 8.9). Also included are yeast from cactus rots that were at the same locale but did not host the fly. Rots in neighbouring cacti (at the same locale) that are not suitable for oviposition can still be places for the adult flies to feed and, so, yeast from neighbouring rots were included in the comparisons. For flies on saguaro, a cactus in the Pachycereinae, and agria, a cactus in the Stenocereinae, the overall pattern is the same. The yeast deposited by a fly (= the yeast dispersed by the fly) were most similar to the yeast isolated from the rot on which that fly had been captured and were least similar to the yeast found on neighbouring rots in cacti that did not host their larvae. Fly behaviour can act to reduce the likelihood of yeast being transmitted between hosts, further limiting the dispersal of the yeast.

8.6 Discussion

To summarise, the data strongly favour narrow endemism over cosmopolitanism in cactophilic yeast. Cactophilic yeast species typically occupy only a subset of their fundamental niche and that subset is often delimited by geography. The probable reason for their endemism is the diversity in cactus host chemistry and the dispersal of the yeast by animal vectors rather than by wind or water. The link between large population size and widespread dispersal is broken for these microbes. Their reliance on animal vectors allows some rare yeast species to persist at population densities far less than those attained by the dominant yeast species in the same communities. As a consequence, they do not lack biogeography as predicted by EiE (Finlay, 2002) and species richness is not unexpectedly low (Fenchel and Finlay, 2004b). Cactophilic yeast have biogeographic patterns related to their vector's biogeography and species richness at least the equal of their vectors. There is simply no fit between their biology and the predictions of EiE.

Is this the case in other yeast habitats? To many, yeast refers to a single species: *Saccharomyces cerevisiae*. The story of the taxonomy of this species is a very long one (Vaughan-Martini and Martini, 1998, list 97 synonyms for it) and involves hybridisation with congeners (Masneuf et al., 1998; de Barros Lopes et al., 2002; Sipiczki, 2008; Belloch et al., 2009). The genus has swelled and shrunk over time and new genera have been carved from it (Kurtzman, 2003). The intense interest in this species has led to the discovery of a group of phenotypically indistinguishable species that form a cluster of sibling species: the *S. cerevisiae sensu stricto* group. All of the siblings show some degree of isolation (either reproductive or genetic) but hybrids are so common that most strains examined have genes from more than one species (Sipiczki, 2008; Muller and McCusker, 2009). Even given that human activity will disperse *S. cerevisiae sensu stricto* species widely, there is evidence of some geographically based variation in *S. cerevisiae* (Naumova et al., 2003), although most studies have found a stronger association between distribution and human activity than geography (Ben-Ari et al., 2005; Aa et al., 2006). For other *S. cerevisiae sensu stricto* species, there is evidence for geographically based variation (Naumov et al., 1997b; Sweeney et al., 2004; Ayoub et al., 2006). Because human activity affects dispersal so profoundly in the *S. cerevisiae sensu stricto* group and dispersal is a key factor in endemism, the population ecology of these yeast are unique and it would be difficult to generalise from them to any other group. Carreto et al. (2008) found little geographic variation in an extensive study of wine yeast from Portugal. I agree with their conclusion that human activity influences the pattern of variation they found but I do not agree with their assertion that *S. cerevisiae* is a good model system for natural populations. Little about the population biology of *S. cerevisiae* seems natural in the sense of 'undisturbed by the activity of man'. Certainly, the finding of little geographically based genetic variation cannot be generalised to yeast that are not domesticated.

The influence of human activity is not absent from the biogeography of cactophilic yeast. *Clavispora opuntiae* is the third most common cactophilic species (if one separates the *Sporopachydermia* complex into species; Table 8.2). It is cosmopolitan in distribution but was not discussed in the section on endemism in cactophilic yeast because of human influence. It is associated not with *Drosophila* (although it has occasionally been isolated from flies) but with Lepidoptera (Lachance, 1990; Rosa et al., 1992). The larvae of some moths (e.g. species in the genera *Cactoblastis, Olycella, Laniifera* and *Sigelgaita*) are able to penetrate the cactus' cuticle and feed on stem tissue. *Cactoblastis* larvae feed on *Opuntia* and have been spread worldwide as biological control agents (Dodd, 1940; Pemberton, 1995; Perez-Sandic, 2001). They were taken from Argentina to Australia and the Caribbean and have, either through inadvertent introduction or natural dispersal, spread throughout Florida in the USA (Soberón, 2002). *Clavispora opuntiae* has accompanied the moth and patterns of genetic variation within the species reflect this (Lachance et al., 2000). Yeast genotypes from Argentina, where the moth originated, are now common in North American populations where the moth has been introduced. Human influence is not absent from the biogeography of other cactophilic yeast. There would be no Australian cactophilic yeast without it. But the evidence that human influence affects *Cl. opuntiae*'s genetic variation in the Americas, where cacti are native, makes its biogeography unique among cactophilic yeast. At this time, there is no evidence to suggest that genetic variation in North or South American populations of other widespread cactophilic yeast species is influenced by human activity.

In addition to the cactophilic habitat, there are at least three other yeast habitats where dispersal is primarily by animal vector: wood, flowers and sap (slime) fluxes. The potential diversity of yeast associated with beetles that bore in wood is high (Suh et al., 2005) and there is evidence that the beetles are the means of dispersal (Ganter, 2006). However, neither the biogeography of yeast from this habitat nor genetic variation within endemic yeast species has been well investigated so the level of endemism in this community is not understood at present. The biogeographic patterns of the yeast communities of slime fluxes and flowers are better known.

Fluxes differ from cactus rots in the availability of sugars, the persistence of some fluxes for multiple years, and the seasonality of the flow of sap. They are similar to rots in that they are the breeding sites for various insects, including *Drosophila*. A survey of teleomorphic ascomycetous yeast descriptions (Kurtzman and Fell, 1998; Morais et al., 2004; Kurtzman, 2005) reveals at least 30 sexual species known only from slime fluxes or insects that breed there (32 taxa if varieties are counted separately) and this impressive number does not include non-ascomycetous or non-sporulating yeast. Two problems make assessment of the degree of geographic endemism difficult to discern. One is that host plant is an important determinant of yeast community composition (Lachance and Starmer, 1982; Ganter et al., 1986) and geography and host plant type are confounded at a regional scale or larger. The second problem is the lack of large-scale sampling efforts in this habitat. The size of the database is such that most comparisons are anecdotal at the global scale. However, regional differences do appear. Two collections that sampled both *Quercus* and *Populus* species, one in Arizona (Ganter et al., 1986) and one in the Great Lakes region (Lachance and Starmer, 1982) are comparable in that the host plants came from the same genera. From Arizona, oaks yielded 50 isolates from 12 species of yeast. In the Great Lakes region, oaks yielded 85 yeast isolates from 24 species. The two studies had only three species in common and the isolates from those species represented 36% and 22% of the isolates in Arizona and the Great Lakes region, respectively. Although the samples are smaller, the lack of overlap was present in *Populus* as well. The counts were 20 isolates and 12 species from Arizona and 17 isolates in six species from the Great Lakes region. The overlap was only one species (5% of the Arizona sample and 20% of the Great Lakes region collection). At a larger scale, there are several species described from temperate Asian slime fluxes that have never been found in temperate North America and vice versa. The evidence points to geographic endemism in the slime-flux community, another yeast community vectored by animals.

Flowers represent a more widespread resource and a more complicated system than sap fluxes. Yeast may benefit from nectar or other rewards for pollinators (Brysch-Herzberg, 2004) and/or they may inhabit decaying, post-pollination flowers. Yeast that exploit flowers from the opening of the bud until the plant casts off the flower are in contact with the many animals that visit flowers. Yeast that use the cast-off flowers inhabit a more specialised resource and the animals involved often include beetles and *Drosophila* (Lachance et al., 2003). The flower system is only now receiving adequate study and the number of new species from the *Metschnikowia* clade, all with similar phenotypes (Lachance et al., 2001a, 2003) suggests that many new taxa will be described as more regions are sampled.

Although some yeast from this system have wide geographic ranges (Ruivo et al., 2004), geographic endemism within this system already has some supporting evidence (Lachance et al., 2001a, 2003, 2005; Rosa et al., 2007; Imanishi et al., 2008; Wardlaw et al., 2009). An interesting parallel to the cactophilic system is the high degree of geographic differentiation found in a widely distributed asexual flower yeast, *Candida ipomoeae* (Wardlaw et al., 2009).

Although studies of global distribution of other yeast are rare, there is one species of black yeast, *Aureobasidium pullulans*, for which some geographic data exist. *Aureobasidium pullulans* is a polytrophic yeast that occurs in a very wide set of habitats. A partial list includes deep seawater, mountaintops, Arctic and Antarctic sea ice, hypersaline lakes, human infections, soil and many plant surfaces (fruits, trunks and leaves). A study of genetic differentiation among strains from different habitats clusters plant-associated *A. pullulans* strains in a single clade (Zalar et al., 2008). Although this species has been found in association with insects that feed on plants (Zacchi and Vaughan-Martini, 2003; Pagnocca et al., 2008), most authors assume that strains of plant-associated *A. pullulans* are passively dispersed by wind or water droplet (splash). Loncaric et al. (2009) found no geographic component to variation among strains from fruit surfaces. Their study included mostly strains from Austria but included strains from China, South Africa, Argentina, New Zealand and Italy. Although the taxonomy and biogeography of this species is not fully understood, it may be that the plant-associated clade is wind (or water) dispersed, global in distribution, and lacking population structure linked to geography.

EiE seems to be a rather standard case of ecological controversy. By that, I mean that it is an idea that is applicable to some situations and not applicable to others. Before the limits of the idea are recognised, controversy arises between those who over-extend the concept's reach and those who feel it is generally invalid because it is invalid in their system of expertise. What is different here is that the ecological controversy involves microbes and that means microbiologists are interested. Microbiology has been accused of lacking theory (Prosser et al., 2007) and yet EiE is an example of an old, well-established theory in microbiology (Lachance, 2004; de Wit and Bouvier, 2006). The data presented here contradict EiE but do not invalidate the theory, although the data do make clear that EiE is not a universal theory of microbial distribution. The example of cactophilic yeast raises several questions about EiE. Under what circumstances does it apply? Is knowledge of the mode of dispersal sufficient to predict whether or not EiE applies?

These questions give dispersal an importance it has lacked in microbial ecology and in yeast ecology in particular. If one relies on Baas Becking's formulation, '*Everything is everywhere*, but *the environment selects*' (italics in the original; de Wit and Bouvier, 2006) then there is a risk that microbial ecology will be reduced to substrate–microbe interactions. The lack of geographic barriers, essential to EiE, means that a niche will not go unfulfilled in any region, indeed, it will be filled by the same species in all regions. The focus on selection by the environment has certainly been taken to heart by yeast biologists (Lachance, 2004; Ganter, 2006). The roles of geography and dispersal mechanism have historically been either ignored or assumed to be only weak influences in yeast ecology. One example of the consequences of the failure to consider something other than substrate is found in an investigation into the source of contaminants in cheese (Westall and Filtenborg, 1998). The search never identified the source of contamination but was confined to substrates in the factory and never examined sources outside of the factory but within the ambit of a fly.

There is another question pertaining to the cactophilic system that remains to be asked. Why are cactophilic yeast dispersed by animals instead of wind, which seems adequate for the bacteria found in the system? There is no direct answer to this question in the data presented here. The proximal cause is most probably history in that yeast have been introduced to the cactophilic system by insects and, so, were already animal-dispersed. Starmer et al. (2003) found evidence for multiple invasions of the cactophilic habitat by yeast from fruit rots and slime fluxes (presumably older habitats). Yeast in both of these habitats are animal dispersed. The origin of the association between yeast and *Drosophila* is not likely to be found in study of the cactophilic yeast.

Acknowledgements

I would like to acknowledge the generosity of W.T. Starmer, who gave me access to and permission to use his cactophilic yeast biogeography data set. I would also like to thank my colleagues Carlos Rosa, for his help collecting and identifying yeast from Brazil, Gianluigi Cardinali, for his help in identifying new species (with Livio Antonielli) and in uncovering genetic variation, and Miguel de Barros Lopes, for his help in assessing genetic variation in *P. kluyveri*. Many students contributed to the data discussed above and I thank Vanessa Williams, Jamil Scott, Eduardo Bustillo, Kevin Hillsman, Bo Li, Elvira Jaques, Bryan Quarles, Jenny Bellon, Alessandro Bolano, Monia Giammaria, Keona Washington, Jennifer Pendola, Ninette Lima and Monia Lattanzi for their contributions. Steve Benz helped with the South American collections. Finally, I would like to thank those who reviewed earlier versions of this article for their advice.

References

Aa, E., Townsend, J.P., Adams, R.I., Nielsen, K.M., Taylor, J.W. (2006). Population structure and gene evolution in *Saccharomyces cerevisiae*. *FEMS Yeast Research* **6**, 702–715.

Anderson, E.F. (2001). *The Cactus Family*. Portland, OR: Timber Press.

Ayoub, M.-J., Legras, J.-L., Saliba, R., Gaillardin, C. (2006). Application of multi locus sequence typing to the analysis of the biodiversity of indigenous *Saccharomyces cerevisiae* wine yeasts from Lebanon. *Journal of Applied Microbiology* **100**, 699–711.

Baumberger, J.P. (1917). The food of *Drosophila melanogaster* Meigen. *Proceedings of the National Academy of Science USA* **3**, 122–126.

Baumberger, J.P. (1919). A nutritional study of insects, with special reference to microorganisms and their substrata. *Journal of Experimental Zoology* **28**, 1–81.

Begon, M. (1981). Yeasts and *Drosophila*. In Ashburner, M., Carson, H.L., Thompson, J. (eds.), *The Genetics and Biology of Drosophila,* 345–384. New York, NY: Academic Press.

Belloch, C., Pérez-Torrado, R., González, S.S. et al. (2009). Chimeric genomes of natural hybrids of *Saccharomyces cerevisiae* and *Saccharomyces kudriavzevii*. *Applied and Environmental Microbiology* **75**, 2534–2544.

Ben-Ari, G., Zenvirth, D., Sherman, A. et al. (2005). Application of SNPs for assessing biodiversity and phylogeny among yeast strains. *Heredity* **95**, 493–501.

Bloom, S.A. (1981). Similarity indices in community studies: potential pitfalls. *Marine Ecology Progress Series* **5**, 125–128.

Brysch-Herzberg, M. (2004). Ecology of yeasts in plant–bumblebee mutualism in Central Europe. *FEMS Microbiology Ecology* **50**, 87–100.

Cairney, J. (2005). Basidiomycete mycelia in forest soils: dimensions, dynamics and roles in nutrient distribution. *Mycological Research* **109**, 7–20.

Carreto, L., Eiriz, E., Gomes, A. et al. (2008). Comparative genomics of wild type yeast strains unveils important genome diversity. *BMC Genomics* **9**, 524.

Coluccio, A.E., Rodriguez, R.K., Kernan, M.J., Neiman, A.M. (2008). The yeast spore wall enables spores to survive passage through the digestive tract of *Drosophila*. *PLoS ONE* **3**, e2873–e2879.

Crampin, H., Finley, K., Gerami-Nejad, M. et al. (2005). *Candida albicans* hyphae have a Spitzenkörper that is distinct from the polarisome found in yeast and pseudohyphae. *Journal of Cell Science* **118**, 2935–2947.

de Barros Lopes, M., Bellon, J.R., Shirley, N.J., Ganter, P.F. (2002). Evidence for multiple interspecific hybridization in *Saccharomyces* sensu stricto species. *FEMS Yeast Research* **1**, 323–331.

de Hoog, G.S., Smith, M.T., Guého, E. (1986). A revision of the genus *Geotrichum* and its telomorphs. *Studies in Mycology* **29**, 1–131.

de Wit, R., Bouvier, T. (2006). 'Everything is everywhere, but the environment selects'; what did Baas Becking and Beijerinck really say? *Environmental Microbiology* **8**, 755–758.

Dodd, A.P., Published under the authority of the Commonwealth Prickly Pear Board (Australia) (1940). *The Biological Campaign Against the Prickly Pear.* Brisbane: A.H. Tucker, Government Printer.

Fellows, D.P., Heed, W.B. (1972). Factors affecting host plant selection in desert-adapted cactiphilic *Drosophila*. *Ecology* **53**, 850–858.

Fenchel, T., Finlay, B.J. (2004a). Response to Lachance. *BioScience* **54**, 884–885.

Fenchel, T., Finlay, B.J. (2004b). The ubiquity of small species: patterns of local and global diversity. *BioScience* **54**, 777–784.

Ferguson, B.A., Dreisbach, T.A., Parks, C.G., Filip, G.M., Schmitt, C.L. (2003). Coarse-scale population structure of pathogenic *Armillaria* species in a mixed-conifer forest in the Blue Mountains of northeast Oregon. *Canadian Journal of Forest Management* **33**, 612–623.

Finlay, B.J. (2002). Global dispersal of free-living microbial eukaryote species. *Science* **296**, 1061–1063.

Finlay, B.J., Clarke, K.J. (1999). Ubiquitous dispersal of microbial species. *Nature* **400**, 828.

Finlay, B.J., Fenchel, T. (2002). Response to A. Coleman. *Science* **297**, 337.

Fogleman, J.C. (1982). The role of volatiles in the ecology of cactophilic *Drosophila*. In Barker, J.S.F., Starmer, W.T. (eds.), *Ecological Genetics and Evolution: the Cactus-Yeast-Drosophila Model System,* pp. 191–208. Sydney: Academic Press Inc.

Fogleman, J.C., Abril, R. (1990). Ecological and evolutionary importance of host plant chemistry. In Barker, J.S.F., MacIntyre, R.J., Starmer, W.T. (eds.), *Ecological and Evolutionary Genetics of Drosophila,* 121–143. New York, NY: Plenum Press.

Fogleman, J.C., Foster, J.L. (1989). Microbial colonization of injured cactus tissue (*Stenocereus gummosus*) and its relationship to the ecology of cactophilic *Drosophila mojavensis*. *Applied and Environmental Microbiology* **55**, 100–105.

Ganter, P.F. (1988). The vectoring of cactophilic yeasts by *Drosophila*. *Oecologia* **75**, 400–404.

Ganter, P.F. (2006). Yeast and invertebrate associations. In Rosa, C.A., Gábor, P. (eds.), *Biodiversity and Ecophysiology of Yeasts,* pp. 303–370. Berlin: Springer-Verlag.

Ganter, P.F., de Barros Lopes, M. (2000). The use of anonymous DNA markers in assessing global relatedness in the yeast species *Pichia kluyveri* Bedford and Kudrjavzev. *Canadian Journal of Microbiology* **26**, 967–980.

Ganter, P.F., Quarles, B. (1997). Analysis of population structure of cactophilic yeast from the genus *Pichia*: *Pichia cactophila* and *P. norvegensis*. *Canadian Journal of Microbiology* **43**, 35–44.

Ganter, P.F., Starmer, W.T., Lachance, M. -A., Phaff, H.J. (1986). Yeast communities from host plants and associated *Drosophila* in southern Arizona: new isolations and analysis of the relative importance of hosts and vectors on community composition. *Oecologia* **70**, 386–392.

Ganter, P.F., Cardinali, G., Giammaria, M., Quarles, B. (2004). Correlations among measures of phenotypic and genetic variation within an oligotrophic asexual yeast, *Candida sonorensis,* collected from *Opuntia*. *FEMS Yeast Research* **4**, 527–540.

Ganter, P.F., Cardinali, G., Boundy-Mills, K. (2010). *Pichia insulana* sp. nov., a novel cactophilic yeast from the Caribbean. *International Journal of Systematic and Evolutionary Microbiology* **60**, 1001–1007.

Gibson, A.C. (1982). Phylogenetic relationships of Pachycereeae. In Barker, J.S.F., Starmer, W.T. (eds.), *Ecological Genetics and Evolution: the Cactus-Yeast-Drosophila Model System,* pp. 3–16. Sydney: Academic Press.

Gilbert, D.G. (1980). Dispersal of yeasts and bacteria by *Drosophila* in a temperate forest. *Oecologia* **46**, 135–137.

Groenewald, M., Daniel, H.-M., Robert, V., Poot, G.A., Smith, M.T. (2008). Polyphasic re-examination of *Debaryomyces hansenii* strains and reinstatement of *D. hansenii, D. fabryi* and *D. subglobosus*. *Persoonia* **21**, 17–27.

Heed, W.B. (1977a). Ecology and genetics of Sonoran Desert *Drosophila*. In Brussard, P.F. (ed.), *Ecological Genetics: the Interface,* pp. 109–126. New York, NY: Springer-Verlag.

Heed, W.B. (1977b). A new cactus-feeding but soil-breeding species of Drosophila (Diptera: Drosophilidae). *Proceedings of the Entomological Society of Washington* **79**, 649–654.

Heed, W.B. (1982). The origin of *Drosophila* in the Sonoran Desert. In Barker, J.S.F., Starmer, W.T. (eds.), *Ecological Genetics and Evolution: the Cactus-Yeast-Drosophila Model System,* pp. 65–80. Sydney: Academic Press

Heed, W.B. (1989). Origin of *Drosophila* of the Sonoran Desert revisited. In search for a founder event and the description of a new species in the *eremophila* complex. In Giddings, L.V., Kaneshiro, K.Y., Anderson, W.W. (eds.), *Genetics, Speciation and the Founder Principle,* pp. 253–278. New York, NY: Oxford University Press.

Heed, W.B., Starmer, W.T., Miranda, M., Miller, M.W., Phaff, H.J. (1976). An analysis of the yeast flora associated with cactiphilic Drosophila and their host plants in the Sonoran Desert and its relation to temperate and tropical associations. *Ecology* **57**, 151–160.

Hibbett, D.S. (2004). Trends in morphological evolution in homobasidiomycetes inferred using maximum likelihood: a comparison of binary and multistate approaches. *Systematic Biology* **53**, 889–903.

Holzschu, D.L., Phaff, H.J. (1982). Taxonomy and evolution of some ascomycetous cactophilic yeasts. In Barker, J.S.F., Starmer, W.T. (eds.), *Ecological Genetics and Evolution: the Cactus-Yeast-Drosophila Model System,* pp. 127–141. Sydney: Academic Press

Holzschu, D.L., Phaff, H.J., Tredick, J., Hedgecock, D. (1983). *Pichia pseudocactophila,* a new species of yeast occurring in necrotic tissue of columnar cacti in the North American Sonoran desert. *Canadian Journal of Microbiology* **29**, 1314–1322.

Holzschu, D.L., Phaff, H.J., Tredick, J., Hedgecock, D. (1985). Resolution of the varietal relationship within the species *Pichia opuntiae* and establishment of a new species, *Pichia thermotolerans* comb. nov. *International Journal of Systematic Bacteriology* **35**, 457–461.

Imanishi, I., Jindamorakot, S., Mikata, K. et al. (2008). Two new ascomycetous anamorphic yeast species related to *Candida* friedrichii – *Candida jaroonii* sp. nov., and *Candida songkhlaensis* sp. nov. – isolated in Thailand. *Antonie van Leeuwenhoek* **94**, 267–276.

Jacques, N., Mallet, S., Casaregola, S. (2009). Delimitation of the species of the *Debaryomyces hansenii* complex by intron sequence analysis. *International Journal of Systematic and Evolutionary Microbiology* **59**, 1242–1251.

James, T.Y., Kauff, F., Schoch, C.L. et al. (2006). Reconstructing the early evolution of Fungi using a six-gene phylogeny. *Nature* **443**, 818–822.

Kircher, H. (1982). Chemical composition of cacti and its relationship to Sonoran desert *Drosophila*. In Barker J.S.F., Starmer, W.T. (eds.), *Ecological Genetics and Evolution: the Cactus-Yeast-Drosophila Model System*, pp. 143–158. Sydney: Academic Press.

Kurtzman, C.P. (1998). *Pichia* E.C. Hansen emend. Kurtzman. In Kurtzman, C.P., Fell, J.W. (eds.), *The Yeasts, A Taxonomic Study*, pp. 273–352. Amsterdam: Elsevier.

Kurtzman, C.P. (2003). Phylogenetic circumscription of *Saccharomyces, Kluyveromyces* and other members of the Saccharomycetaceae, and the proposal of the new genera *Lachancea, Nakaseomyces, Naumovia, Vanderwaltozyma* and *Zygotorulaspora*. *FEMS Yeast Research* **4**, 233–245.

Kurtzman, C.P. (2005). Description of *Komagataella phaffii* sp. nov. and the transfer of *Pichia pseudopastoris* to the methylotrophic yeast genus *Komagataella*. *International Journal of Systematic and Evolutionary Microbiology* **55**, 973–976.

Kurtzman, C.P., Fell, J.W. (1998). *The Yeasts, A Taxonomic Study*. 4th Edition. Amsterdam: Elsevier.

Kurtzman, C.P., Robnett, C.J. (1995). Molecular relationships among hyphal ascomycetous yeasts and yeastlike taxa. *Canadian Journal of Botany* **73**, S824–S830.

Kurtzman, C.P., Robnett, C. J. (2010). Systematics of methanol assimilating yeasts and neighboring taxa from multigene sequence analysis and the proposal of *Peterozyma gen.nov.*, a new memberof the Saccharomycetales. *FEMS Yeast Research* **10**, 353–361.

Kurtzman, C.P., Robnett, C.J., Basehoar-Powers, E. (2008). Phylogenetic relationships among species of *Pichia, Issatchenkia* and *Williopsis* determined from multigene sequence analysis, and the proposal of Barnettozyma gen. nov., Lindnera gen. nov. and Wickerhamomyces gen. nov. *FEMS Yeast Research* **8**, 939–954.

Lachance, M.-A. (1990). Ribosomal DNA spacer variation in the cactophilic yeast *Clavispora opuntiae*. *Molecular Biology and Evolution* **7**, 178–193.

Lachance, M.-A. (2004). Here and there or everywhere? *BioScience* **54**, 884.

Lachance, M.-A., Bowles, J.M., Kwon, S. et al. (2001a). *Metschnikowia lochheadii* and *Metschnikowia drosophilae*, two new yeast species isolated from insects associated with flowers. *Canadian Journal of Microbiology* **47**, 103–109.

Lachance, M.-A., Bowles, J.M., Starmer, W.T. (2003). *Metschnikowia santaceciliae, Candida hawaiiana*, and *Candida kipukae*, three new yeast species associated with insects of tropical morning glory. *FEMS Yeast Research* **3**, 97–103.

Lachance, M.-A., Ewing, C.P., Bowles, J.M., Starmer, W.T. (2005). *Metschnikowia hamakuensis* sp. nov., *Metschnikowia kamakouana* sp. nov. and *Metschnikowia mauinuiana* sp. nov., three endemic yeasts from Hawaiian nitidulid beetles. *International Journal of Systematic and Evolutionary Microbiology* **55**, 1369–1377.

Lachance, M.-A., Gilbert, D.G., Starmer, W.T. (1995). Yeast communities associated with *Drosophila* species and related flies in an eastern oak-pine forest: a comparison with western communities. *Journal of Industrial Microbiology* **14**, 484–494.

Lachance, M.-A., Kaden, J.E., Phaff, H.J., Starmer, W.T. (2001b). Phylogenetic structure of the *Sporopachydermia cereana* species complex. *International*

Journal of Systematic and Evolutionary Microbiology **51**, 237–247.

Lachance, M.-A., Starmer, W.T. (1982). Evolutionary significance of physiological relationships among yeast communities associated with trees. *Canadian Journal of Botany* **60**, 285–293.

Lachance, M.-A., Starmer, W.T. (2008). *Kurtzmaniella* gen. nov. and description of the heterothallic, haplontic yeast species *Kurtzmaniella cleridarum* sp. nov., the teleomorph of *Candida cleridarum*. *International Journal of Systematic and Evolutionary Microbiology* **58**, 520–524.

Lachance, M.-A., Nair, P., Lo, P. (1994). Mating in the heterothallic haploid yeast *Clavispora opuntiae*, with special reference to mating type imbalances in local populations. *Yeast* **10**, 895–906.

Lachance, M.-A., Starmer, W.T., Bowles, J.M., Phaff, H.J., Rosa, C.A. (2000). Ribosomal DNA, species structure, and biogeography of the cactophilic yeast *Clavispora opuntiae*. *Canadian Journal of Microbiology* **46**, 195–210.

Leask, B.G.S., Yarrow, D. (1976). *Pichia norvegensis* sp. nov. *Sabouraudia* **14**, 61–63.

Loncaric, I., Oberlerchner, J.T., Heissenberger, B., Moosbeckhofer, R. (2009). Phenotypic and genotypic diversity among strains of *Aureobasidium pullulans* in comparison with related species. *Antonie van Leeuwenhoek* **95**, 165–178.

Lutzoni, F., Kauff, F., Cox, C.J. et al. (2004). Assembling the fungal tree of life: progress, classification, and evolution of subcellular traits. *American Journal of Botany* **91**, 1446–1480.

Magliani, W., Conti, S., Gerloni, M., Bertolotti, D., Poloneli, L. (1997). Yeast killer systems. *Clinical Microbiology Reviews* **10**, 369–400.

Marcet-Houben, M., Gabaldón, T. (2009). The tree versus the forest: the fungal tree of life and the topological diversity within the yeast phylome. *PLoS ONE* **4**, e4357–e4364.

Masneuf, I., Hansen, J., Groth, C., Piskur, P., Dubourdieu, D. (1998). New hybrids between *Saccharomyces* sensu stricto yeast species found among wine and cider production strains. *Applied and Environmental Microbiology* **64**, 3887–3892.

Morais, P.B., Rosa, C.A., Hagler, A.N., Mendonça-Hagler, L.C. (1995). Yeast communities as descriptors of habitat use by the *Drosophila fasciola* subgroup (repleta group) in Atlantic rain forests. *Oecologia* **104**, 45–51.

Morais, P.B., Teixeira, L.C.R.S., Bowles, J.M., Lachance, M.-A., Rosa, C.A. (2004). *Ogataea falcaomoraisii* sp. nov., a sporogenous methylotrophic yeast from tree exudates. *FEMS Yeast Research* **5**, 81–85.

Muller, L.A.H., McCusker, J.H. (2009). A multispecies-based taxonomic microarray reveals interspecies hybridization and introgression in *Saccharomyces cerevisiae*. *FEMS Yeast Research* **9**, 143–152.

Murray, N.D. (1982). Ecology and evolution of the *Opuntia–Cactoblastis* ecosystem in Australia. In Barker, J.S.F., Starmer, W.T. (eds.), *Ecological Genetics and Evolution: the Cactus-Yeast-Drosophila Model System*, pp. 17–30. Sydney: Academic Press.

Naumov, G.I., Naumova, E.S., Kondrativea, V.I. et al. (1997a). Genetic and molecular delineation of three sibling species in the *Hansenula polymorpha* complex. *Systematic and Applied Microbiology* **20**, 50–56.

Naumov, G.I., Naumova, E.S., Sniegowski, P.D. (1997b). Differentiation of European and Far East Asian populations of *Saccharomyces paradoxus* by allozyme analysis. *International Journal of Systematic Bacteriology* **47**, 341–344.

Naumova, E.S., Smith, M.T., Boekhout, T., de Hoog, G.S., Naumov, G.I. (2001). Molecular differentiation of sibling species in the *Galactomyces geotrichum* complex. *Antonie van Leeuwenhoek* **80**, 263–273.

Naumova, E.S., Korshunova, I.V., Jespersen, L., Naumov, G.I. (2003). Molecular genetic identification of *Saccharomyces sensu stricto* strains from African sorghum beer. *FEMS Yeast Research* **3**, 177–184.

Nguyen, H.-V., Gaillardin, C., Neuvéglise, C. (2009). Differentiation of *Debaryomyces hansenii* and *Candida famata* by rRNA gene intergenic spacer fingerprinting and reassessment of phylogenetic relationships among *D. hansenii, C. famata, D. fabryi, C. flareri* (=*D. subglobosus*) and *D. prosopidis*: description of *D. vietnamensis* sp. nov. closely related to *D. nepalensis*. *FEMS Yeast Research* **9**, 641–662.

Pagnocca, F.C., Rodrigues, A., Nagamoto, N.S., Bacci, M. (2008). Yeasts and filamentous fungi carried by the gynes of leaf-cutting ants. *Antonie van Leeuwenhoek* **94**, 517–526.

Pemberton, R.W. (1995). *Cactoblastis cactorum* (Lepidoptera: Pyralidae) in the United States: An immigrant biological control agent or an introduction of the nursery industry? *American Entomologist* **41**, 230–232.

Perez-Sandic, M. (2001). Addressing the threat of *Cactoblastis cactorum* (Lepidoptera: Pyralidae) to *Opuntia* in Mexico. *Florida Entomologist* **84**, 499–502.

Phaff, H.J., Starmer, W.T. (1987). Yeasts associated with plants, insects and soil. In Rose, A.H., Harrison, S.J. (eds.), *The Yeasts*, pp. 123–180. Orlando, FL: Academic Press.

Phaff, H.J., Starmer, W.T., Tredick-Klinc, J. (1987a). *Pichia kluyveri sensu lato.* A proposal for two new varieties and a new anamorph. In de Hoog, G.S., Smith, M.T., Weijman, A.C.M. (eds.), *The Expanding Realm of Yeast-like Fungi: Proceedings of an International Symposium on the Perspectives of Taxonomy, Ecology and Phylogeny of Yeasts and Yeast-like Fungi,* pp. 403–414. Amsterdam: Elsevier Science.

Phaff, H.J., Starmer, W.T., Tredick-Kline, J., Miranda, M., Aberdeen, V. (1987b). *Pichia barkeri*, a new species of yeast occurring in necrotic tissue of *Opuntia stricta*. *International Journal of Systematic Bacteriology* **37**, 783–796.

Phaff, H.J., Starmer, W.T., Lachance, M.-A., Aberdeen, V., Tredick-Kline, J. (1992). *Pichia caribaea*, a new species of yeast occurring in necrotic tissue of cacti in the Caribbean area. *International Journal of Systematic Bacteriology* **42**, 459–462.

Phaff, H.J., Blue, J., Hagler, A.N., Kurtzman, C.P. (1997). *Dipodascus starmeri* sp. nov., a new species of yeast occurring in cactus necroses. *International Journal of Systematic Bacteriology* **47**, 307–312.

Phaff, H.J., Vaughan-Martini, A., Starmer, W.T. (1998). *Debaryomyces prosopidis* sp. nov., a yeast from exudates of mesquite trees. *International Journal of Systematic Bacteriology* **48**, 1419–1424.

Pohl, C.H., Kock, J.L.F., van Wyk, P.W.J., Albertyn, A. (2006). *Cryptococcus anemochoreius* sp. nov., a novel anamorphic basidiomycetous yeast isolated from the atmosphere in central South Africa. *International Journal of Systematic and Evolutionary Microbiology* **56**, 2703–2706.

Prosser, J.I., Bohannan, B.J.M., Curtis, T.P. et al. (2007). The role of ecological theory in microbial ecology. *Nature Reviews Microbiology* **5**, 384–392.

Reuter, M., Bell, G., Greig, D. (2007). Increased outbreeding in yeast in response to dispersal by an insect vector. *Current Biology* **17**, R81–R83.

Rosa, C.A., Hagler, A.N., Mendonça-Hagler, L.C. et al. (1992). *Clavispora opuntiae* and other yeasts associated with the moth *Sigelgaita* sp. in the cactus *Pilosocereus arrabidae* of Rio de Janeiro, Brazil. *Antonie van Leeuwenhoek (Historical Archive)* **62**, 267–272.

Rosa, C.A., Morais, P.B., Santos, S.R. et al. (1995). Yeast communities associated with different plant resources in sandy coastal plains of southeastern Brazil. *Mycological Research* **99**, 1047–1054.

Rosa, C.A., Lachance, M.-A., Teixeira, L.C.R.S., Pimenta, R.S., Morais, P.B. (2007). *Metschnikowia cerradonensis* sp. nov., a yeast species isolated from ephemeral flowers and their nitidulid beetles in Brazil. *International Journal of Systematic and Evolutionary Microbiology* **57**, 161–165.

Ruivo, C.C.C., Lachance, M.-A., Bacci, M. et al. (2004). *Candida leandrae* sp. nov., an asexual ascomycetous yeast species isolated from tropical plants. *International Journal of Systematic and Evolutionary Microbiology* **54**, 2405–2408.

Ruiz, A., Heed, W.B. (1988). Host-plant specificity in the cactophilic *Drosophila mulleri* species complex. *Journal of Animal Ecology* **57**, 237–249.

Sang, J.H. (1956). The quantitative nutritional requirements of *Drosophila melanogaster. Journal of Experimental Biology* **33**, 45–72.

Sang, J.H. (1978). The nutritional requirements of *Drosophila*. In Ashburner, M., Wright, T.R.F. (eds.), *The Genetics and Biology of Drosophila*, pp. 159–192. New York, NY: Academic Press.

Sipiczki, M. (2008). Interspecies hybridization and recombination in *Saccharomyces* wine yeasts. *FEMS Yeast Research* **8**, 996–1007.

Soberón, J. (2002). The routes of invasion of *Cactoblastis cactorum*. pdf of a powerpoint presentation, Comisión Nacional para el Conocimiento y uso de la Biodiversidad (CONABIO), Puerto Vallarta, Mexico.

Spatafora, J.W., Sung, G.H., Johnson, D. et al. (2006). A five-gene phylogeny of Pezizomycotina. *Mycologia* **98**, 1018–1028.

Starmer, W.T. (1982). Associations and interactions among yeasts, *Drosophila*, and their habitats. In Barker, J.S.F., Starmer, W.T. (eds.), *Ecological Genetics and Evolution: The Cactus-Yeast-Drosophila Model System, 159–174.* Sydney: Academic Press.

Starmer, W.T., Aberdeen, V. (1990). The nutritional importance of pure and mixed cultures of yeasts in the development of *Drosophila mulleri* larvae in *Opuntia* tissues and its relationship to host plant shifts. In Barker, J.S.F., Starmer, W.T., MacIntyre, R.J. (eds.), *Ecological and Evolutionary Genetics of Drosophila,* 145–160. New York, NY: Plenum Press.

Starmer, W.T., Fogleman, J.C. (1986). Coadaptation of *Drosophila* and yeasts in their natural habitat. *Journal of Chemical Ecology* **12**, 1035–1053.

Starmer, W.T., Phaff, H.J., Miranda, M., Miller, M.W. (1978). *Pichia amethionina*, a new heterothallic yeast associated with the decaying stems of cereoid cacti. *International Journal of Systematic Bacteriology* **28**, 433–441.

Starmer, W.T., Phaff, H.J., Miranda, M., Miller, M.W., Barker, J.S.F. (1979). *Pichia opuntiae*, a new heterothallic yeast found in decaying cladodes of *Opuntiae inermis* and in necrotic tissue of cereoid cacti. *International Journal of Systematic Bacteriology* **29**, 159–167.

Starmer, W.T., Kircher, H.W., Phaff, H.J. (1980). Evolution and speciation of host plant specific yeasts. *Evolution* **34**, 137–146.

Starmer, W.T., Phaff, H.J., Tredick, J., Miranda, M., Aberdeen, V. (1984). *Pichia antillensis*, a new species of yeast associated with necrotic stems of cactus in the Lesser Antilles. *International Journal of Systematic Bacteriology* **34**, 350–354.

Starmer, W.T., Ganter, P.F., Phaff, H.J. (1986). Quantum and continuous evolution of the DNA base composition in the yeast genus *Pichia*. *Evolution* **40**, 1263–1274.

Starmer, W.T., Aberdeen, V., Lachance, M.-A. (1988a). The yeast community associated with decaying *Opuntia stricta* (Haworth) in Florida with regard to the moth *Cactoblastis cactorum* (Berg). *Florida Scientist* **51**, 7–11.

Starmer, W.T., Phaff, H.J., Bowles, J.M., Lachance, M.-A. (1988b). Yeasts vectored by insects feeding on decaying saguaro cactus. *Southwestern Naturalist* **33**, 362–363.

Starmer, W.T., Lachance, M.-A., Phaff, H.J., Heed, W.B. (1990). The biogeography of yeasts associated with decaying cactus tissue in North America, the Caribbean, and Northern Venezuela. In Hecht, M.K., Wallace, B., MacIntyre, R.J. (eds.), *Evolutionary Biology*, pp. 253–296. New York, NY: Plenum Press.

Starmer, W.T., Ganter, P.F., Aberdeen, V. (1992). Geographic distribution and genetics of killer phenotypes for the yeast *Pichia kluyveri* across the United States. *Applied and Environmental Microbiology* **58**, 990–997.

Starmer, W.T., Phaff, H.J., Ganter, P.F., Lachance, M.-A. (2001). *Candida orba* sp. nov., a new cactus-specific yeast species from Queensland, Australia. *International Journal of Systematic and Evolutionary Microbiology* **51**, 699–705.

Starmer, W.T., Schmedicke, R.A., Lachance, M.-A. (2003). The origin of the cactus-yeast community. *FEMS Yeast Research* **3**, 441–448.

Sudbery, P., Gow, N., Berman, J. (2004). The distinct morphogenic states of *Candida albicans*. *Trends in Microbiology* **12**, 317–324.

Suh, S.O., McHugh, J.V., Pollock, D.D., Blackwell, M. (2005). The beetle gut: a hyperdiverse source of novel yeasts. *Mycological Research* **109**, 261–265.

Sweeney, J.Y., Kuehne, H.A., Sniegowski, P.D. (2004). Sympatric natural *Saccharomyces cerevisiae* and *S. paradoxus* populations have different thermal growth profiles. *FEMS Yeast Research* **4**, 521–525.

Takashima, M., Sugita, T., Shinoda, T., Nakase, T. (2003). Three new combinations from the *Cryptococcus laurentii* complex: *Cryptococcus aureus*, *Cryptococcus carnescens* and *Cryptococcus peneaus*. *International Journal of Systematic and Evolutionary Microbiology* **53**, 1187–1194.

Tsai, I.J., Bensasson, D., Burt, A., Koufopanou, V. (2008). Population genomics of the wild yeast *Saccharomyces paradoxus*: quantifying the life cycle. *Proceedings of the National Academy of Sciences USA* **105**, 4957–4962.

Vacek, D.C., East, P.D., Barker, J.S.F., Soliman, M.H. (1985). Feeding and oviposition preferences of *Drosophila buzzatii* for microbial species isolated from its natural environment. *Biological Journal of the Linnean Society* **24**, 175–187.

Vaughan-Martini, A., Martini, A. (1998). *Saccharomyces* Meyen ex Reess. In Kurtzman, C.P., Fell, J.W. (eds.), *The Yeasts, A Taxonomic Study*, pp. 358–371. Amsterdam: Elsevier.

Vaughan-Martini, A.E., Kurtzman, C.P., Meyer, S.A., O ' Neill, E.B. (2005). Two new species in the *Pichia guilliermondii* clade: *Pichia caribbica* sp. nov., the ascosporic state of *Candida fermentati*, and *Candida carpophila* comb. nov. *FEMS Yeast Research* **5**, 463–469.

Vegelius, J., Janson, S., Johansson, F. (1986). Measures of similarity between distributions. *Quality and Quantity* **20**, 437–441.

von Arx, J.A. (1977). Notes on *Dipodascus*, *Endomyces*, and *Geotrichum* with the description of two new species. *Antonie van Leeuwenhoek* **43**, 333–340.

Wardlaw, A.M., Berkers, T.E., Man, K.C., Lachance, M.-A. (2009). Population structure of two beetle-associated yeasts: comparison of a New World asexual and an endemic Nearctic sexual species in the Metschnikowia clade. *Antonie van Leeuwenhoek* **96**, 1–15.

Westall, S., Filtenborg, O. (1998). Spoilage yeasts of decorated soft cheese packed in modified atmosphere. *Food Microbiology* **15**, 243–249.

Yamada, Y., Higashi, T., Ando, S., Mikata, K. (1997). The phylogeny of strains of species of the genus *Pichia* Hansen (Saccharomycetacea) based on the partial sequences of the 18S ribosomal RNA: the proposals of *Phaffomyces* and *Starmera*, the new genera. *Bulletin of the Faculty of Agriculture, Shizuoka University* **47**, 23–35.

Yamada, Y., Kawasaki, H., Nagatsuka, Y., Mikata, K., Seki-T. (1999). The phylogeny of the cactophilic yeasts based on the 18S ribosomal RNA gene sequences: The proposals of *Phaffomyces antillensis* and *Starmera caribaea*, new combinations. *Bioscience Biotechnology and Biochemistry* **63**, 827–832.

Zacchi, L., Vaughan-Martini, A. (2003). Distribution of three yeast and yeast-like species within a population of soft scale insects (*Saissetia oleae*) as a function of developmental age. *Annals of Microbiology* **53**, 43–46.

Zalar, P., Gostinčar, C., de Hoog, G.S. et al. (2008). Redefinition of *Aureobasidium pullulans* and its varieties. *Studies in Mycology* **61**, 21–38.

Part IV

Pluricellular eukaryotes

9

Coalescent analyses reveal contrasting patterns of intercontinental gene flow in arctic-alpine and boreal-temperate fungi

JÓZSEF GEML

National Herbarium of the Netherlands, Netherlands Centre for Biodiversity Naturalis, Leiden University, Leiden, the Netherlands

9.1 Introduction

Many microbial prokaryote and eukaryote morphospecies have been observed to have essentially global distributions (Finlay, 2002). Based on this observation, it has been suggested that populations are so large and dispersal is so effective in these organisms that any tendencies toward geographic isolation and speciation are swamped by gene flow and, hence, that microbes lack detectable biogeography (Fenchel and Finlay, 2004). Where geographic patterning of genetic structure is seen in microbes, the question arises as to whether it is due to selection by the habitat or historical limitations on dispersal (Martiny et al., 2006). Given the tremendous uncharted diversity of fungi, their critical roles in ecosystems, their potential for extensive dispersal by humans (e.g. with agricultural and forestry products), their importance to public health (e.g. mycoses), their valued roles in diverse human societies (e.g. in diverse foods), and the threats posed by rapidly changing climates, it is imperative to determine to what degree fungi have biogeographies.

Biogeography of Microscopic Organisms: Is Everything Small Everywhere?, ed. Diego Fontaneto. Published by Cambridge University Press.

Fungi share some features with prokaryotic microbes: many species are unicellular and nearly all fungi disperse via single-celled mitotic and/or meiotic spores. They also share many features with other eukaryotes, including discrete evolutionary lineages that are not interconnected by lateral gene transfer. A number of recent molecular phylogenetic studies summarised in Taylor et al. (2006) have demonstrated intercontinental genetic breaks in cosmopolitan fungal morphospecies. Thus, the emerging pattern is that many fungi do have biogeographies. However, at present, our understanding of these patterns is very coarse. Furthermore, although ecological attributes likely play a role in the dispersal capacities of fungal taxa, it is virtually unknown to what extent fungi inhabiting the same biome share phylogeographic patterns and whether or not there is any general trend among them.

Comparative phylogeographic analyses, such as those detailed in this paper, can contribute to broader studies of ecology and evolution in at least two principal ways: (1) phylogeography provides an evolutionary and geographic context for the species comprising ecological communities, permitting determination of historical and spatial influences on patterns of species richness (e.g. Ricklefs and Schluter, 1993); and (2) an understanding of historical responses to changes in the landscape and the identification of evolutionarily isolated areas can inform conservation strategies (Moritz and Faith, 1998). In this paper, my focus is on comparing population structures and long-distance gene flow in arctic and boreal fungi.

While phylogeography of arctic plants and animals have been extensively studied (e.g. Reiss et al., 1999; Tremblay and Schoen, 1999; Abbott and Comes, 2003; Brunhoff et al., 2003; Fedorov et al., 2003; Flagstad and Røed, 2003; Wickström et al., 2003; Alsos et al., 2005, 2007; Dalén et al., 2005; Parmesan, 2006; Eidesen et al., 2007; Schönswetter et al., 2007; Marthinsen et al., 2008), the systematics, ecology and phylogeographic studies of fungi in arctic regions remain scarce, despite their critical roles in the functioning of these nutrient-poor ecosystems (Callaghan et al., 2004; Printzen, 2008). Studying migration in high-latitude fungi, i.e. the degree to which they are able to colonise newly exposed, suitable habitats (e.g. following receding glaciers) and to exchange genes with populations inhabiting different geographic regions, is especially relevant for climate change studies. Climate warming is expected to cause a northward shift in the distribution of many arctic species, and the long-distance dispersal capability of individual species will greatly influence the composition of future arctic communities (Alsos et al., 2007).

I examined intraspecific genetic diversity, genetic structure and estimated long-distance gene flow in selected arctic-alpine and boreal-temperate fungi with circumpolar or disjunct distributions to attempt to answer the following questions: (1) What is the extent of intercontinental gene flow in the selected species? (2) Do species from a certain biome share similar migration patterns? (3) If some differences exist among species and/or biomes, what may be the underlying ecological, historical and geographic reasons?

Species that are widely distributed in the northern hemisphere and represent different taxonomical and ecological groups were chosen for the analyses: the arctic-alpine lichens *Flavocetraria cucullata* (Bellardi) Kärnefelt and A. Thell, *Flavocetraria nivalis* (L.) Kärnefelt and A. Thell and *Dactylina arctica* (Hook. f.) Nyl., the arctic-alpine ectomycorrhizal agaric *Cortinarius favrei* D.M. Hend, the arctic alpine lichenised agaric *Lichenomphalia umbellifera* (L.) Redhead, Lutzoni, Moncalvo and Vilgalys, the temperate-boreal ectomycorrhizal agarics *Amanita muscaria* (L.) Hooker, *Amanita pantherina* (DC.) Krombh., and *Lactarius deliciosus* (L.) Gray, the boreal-temperate wood-rotting *Grifola frondosa* (Dicks.) Gray, and the temperate lichen *Trapeliopsis glaucolepidea* (Nyl.) Gotth. Schneid.

Applying non-parametric permutation tests, migration estimates and genealogies generated using coalescent methods, I found a moderate to high amount of intercontinental gene flow in arctic fungi and very little or no intercontinental gene flow in boreal and temperate species, regardless of their systematic positions. To my knowledge, these results provide the most complete characterisation of intraspecific genetic diversity in arctic fungi and suggest that many arctic species have strong potential to adapt to the changing Arctic by tracking their ecological niche and to maintain high genetic diversity through long-distance dispersal and gene flow among distant populations.

9.2 Materials and methods

The DNA sequence data analysed in this paper were compiled from new data and sequences available from GenBank. I concentrated my efforts on the internal transcribed spacer (ITS) region of the nuclear ribosomal DNA repeat, because this locus has been useful in earlier phylogeographic studies in a variety of fungi (e.g. Shen et al., 2002; Oda et al., 2004; Palice and Printzen, 2004; Geml et al., 2008, 2010; Bergemann et al., 2009) and because it is the most frequently sequenced fungal locus and, thus, it usually provides the greatest sample size and geographic coverage for any species. Because my main purpose was to estimate transoceanic gene flow, only species with several sequences from both Eurasian and North (or Central) American samples were included. The final data sets for the individual species contained the following sequences (GenBank accession numbers): *Flavocetraria cucullata* (FJ914765-FJ914812), *Flavocetraria nivalis* (GU067685-GU067729), *Dactylina arctica* (GU981748-GU981760), *Lichenomphalia umbellifera* (AY293955-AY293961, GU810926–810969), *Cortinarius favrei* (DQ295071-DQ295085, AF182798, AF325575, GU234036, GU234040, GU234070, GU234087, GU234096, GU234128, GU981746-GU981747), *Amanita muscaria*

(AB080777-AB080795, AB080980-AB080984, AB081294-AB081296, AB096048-AB096052, EU071889, EU071893, EU071896-EU071936), *Amanita pantherina* (AB080774-AB080776, AB080784-AB080786, AB080973-AB080978, AB096043-AB096047, AB103329, EF493269, EU525997, EU909452, GQ401354), *Grifola frondosa* (AY049091-AY049141), *Lactarius deliciosus* (AF230892, AF249283-AF249284, AY332557, DQ116886-DQ116904, EF685050-EF685059, EU423914-EU423923) and *Trapeliopsis glaucolepidea* (AY600064-AY600082).

Multiple sequence alignments were made using Clustal W (Thompson et al., 1997) and subsequently were corrected manually. Identical ITS sequences were collapsed into haplotypes using SNAP Map (Aylor et al., 2006) after excluding insertion or deletions (indels) and infinite-sites violations. The analyses presented here assume an infinite sites model, under which a polymorphic site is caused by exactly one mutation and there can be no more than two segregating bases. Base substitutions were categorised as phylogenetically uninformative or informative, and as transitions or transversions. Site compatibility matrices were generated from each haplotype data set using SNAP Clade and Matrix (Markwordt et al., 2003; Bowden et al., 2008) to examine compatibility/incompatibility among all variable sites, with any resultant incompatible sites removed from the data set. Genetic differentiation among geographic populations was analysed using SNAP Map, Seqtomatrix and Permtest (Hudson et al., 1992) implemented in SNAP Workbench (Price and Carbone, 2005). Permtest is a non-parametric permutation method based on Monte Carlo simulations that estimates Hudson's test statistics (K_{ST}, K_S and K_T) under the null hypothesis of no genetic differentiation. K_{ST} is equal to $1 - K_S/K_T$, where K_S is a weighted mean of $K1$ and $K2$ (mean number of differences between sequences in subpopulations 1 and 2, respectively) and K_T represents the mean number of differences between two sequences regardless of the subpopulation to which they belong. The null hypothesis of no genetic differentiation is rejected ($P < 0.05$) when K_S is small and K_{ST} is close to 1. For this test, specimens were assigned to geographic groups based on continents (North America or Eurasia).

Two independent methods were used to determine whether there was any evidence of transoceanic migration between pairs of populations inhabiting different continents. First, I used MDIV (Nielsen and Wakeley, 2001), implemented in SNAP Workbench (Price and Carbone, 2005), employing both likelihood and Bayesian methods using Markov chain Monte Carlo (MCMC) coalescent simulations to estimate the migration (M), population mean mutation rate (*Theta*), and divergence time (T). Here, M equals 2 × the net effective population size (N_e) multiplied by m (migration rate), while *Theta* is $4 \times N_e$ multiplied by μ (mutation rate) (Watterson, 1975). Ages were measured in coalescent units of $2N$, where N is the population size. This approach assumes that all populations descended from one panmictic population that may or may not have been followed by migration. For each data set, the data were simulated assuming an infinite sites model with uniform prior. I used 2 000 000 steps in the chain for estimating the posterior probability distribution and an initial 500 000 steps to ensure that enough genealogies were simulated before approximating the posterior distribution. Second, if MDIV showed evidence of migration, MIGRATE was used to estimate migration rates assuming equilibrium migration rates (symmetrical or asymmetrical) in the history of the populations (Beerli and Felsenstein, 2001). I applied the following specifications for the MIGRATE maximum likelihood analyses: M^* (migration rate m divided by mutation rate μ) and *Theta* generated from the F_{ST} calculation, migration model with variable *Theta*, and constant mutation rate. The numbers of immigrants per generation ($4N_em$) were calculated by multiplying *Theta* of the receiving population with the population migration rate M^*. Subsequently, I reconstructed the genealogy with the highest root probability and the ages of mutations in the sample using coalescent simulations in Genetree v. 9.0 (Griffiths and Tavaré, 1994). Ages were measured in coalescent units of $2N$, where N is the population size.

9.3 Results

Sample size, number of haplotypes and polymorphic sites, and estimates of Hudson's test statistics (K_{ST}, K_S and K_T) on population subdivision for each species are shown in Table 9.1. For boreal-temperate species, non-parametric permutation tests always indicated strong geographic structure corresponding to continents. The opposite was true for arctic species, where no or weak (marginally significant) genetic differentiations were found among North American and Eurasian populations.

In the combined approach detailed above, I utilised the complementary strengths of MDIV and MIGRATE to estimate the extent of transoceanic gene flow. For example, MIGRATE was used to estimate the direction of migration, but could not distinguish between shared ancestral polymorphism and recurrent gene flow; while MDIV was used to determine if the diversity patterns in North American and Eurasian populations were the result of retention of ancestral polymorphism or recent gene flow. Although values for *Theta* were comparable among all species (data not shown), estimates for long-distance gene flow were widely different in boreal-temperate vs. arctic-alpine species (Fig 9.1). In all boreal species, MDIV showed evidence for no intercontinental gene flow ($M = 0$) and statistically significant, non-zero population divergence time (T, data not shown). On the other hand, in all arctic species, MDIV estimated moderate to high gene flow between North American and Eurasian populations and estimated no population divergence (T not significantly different from 0). MIGRATE detected bidirectional gene flow in most arctic population pairs. Based on these results, simulations in Genetree were

Table 9.1 Sample size, number of haplotypes and polymorphic sites, and genetic differentiation between Eurasian vs. American populations according to Hudson's test statistics (K_{ST}, K_S and K_T). Significance was evaluated by performing 1000 permutations for each species.

Species	Sample size	Number of haplotypes	Polymorphic sites*	Eurasian vs. American populations			
				K_{ST}	K_S	K_T	P
Arctic/Alpine							
Cortinarius favrei	29	14	20	0.032	2.627	2.714	0.092
Dactylina arctica	13	7	18	–0.075	5.348	4.974	0.876
Flavocetraria cucullata	49	21	40	0.043	7.411	7.744	0.022
Flavocetraria nivalis	41	14	21	0.043	2.182	2.281	0.019
Lichenomphalia umbellifera	51	24	46	–0.008	3.689	3.658	0.891
Boreal/Temperate							
Amanita muscaria	61	29	49	0.520	4.126	8.598	<0.001
Amanita pantherina	22	16	42	0.461	5.979	11.104	<0.001
Grifola frondosa	46	8	16	0.829	1.028	6.045	<0.001
Lactarius deliciosus	41	15	47	0.694	2.464	8.056	<0.001
Trapeliopsis glaucolepidea	58	13	22	0.252	2.465	3.297	<0.001

* After recoding indels and removing infinite-sites violations.

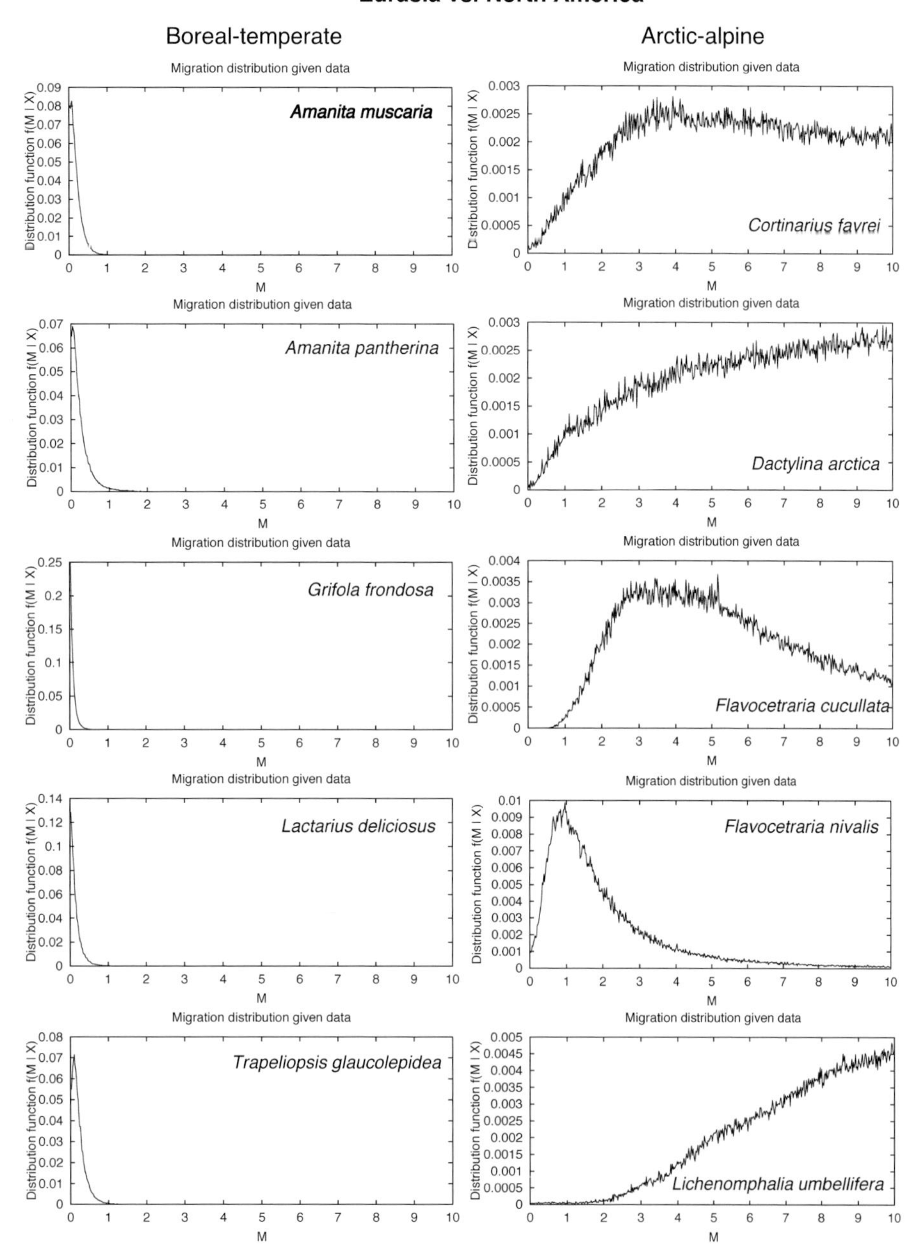

Fig 9.1 Posterior probability distributions of migration ($M = 2N_em$) estimated between transoceanic population pairs of arctic/alpine and boreal/temperate species using Markov chain Monte Carlo coalescent simulations in MDIV. For each data set, the data were simulated assuming an infinite sites model, using 2 000 000 steps in the chain, and an initial 500 000 steps to ensure that enough genealogies were simulated before approximating the posterior distribution.

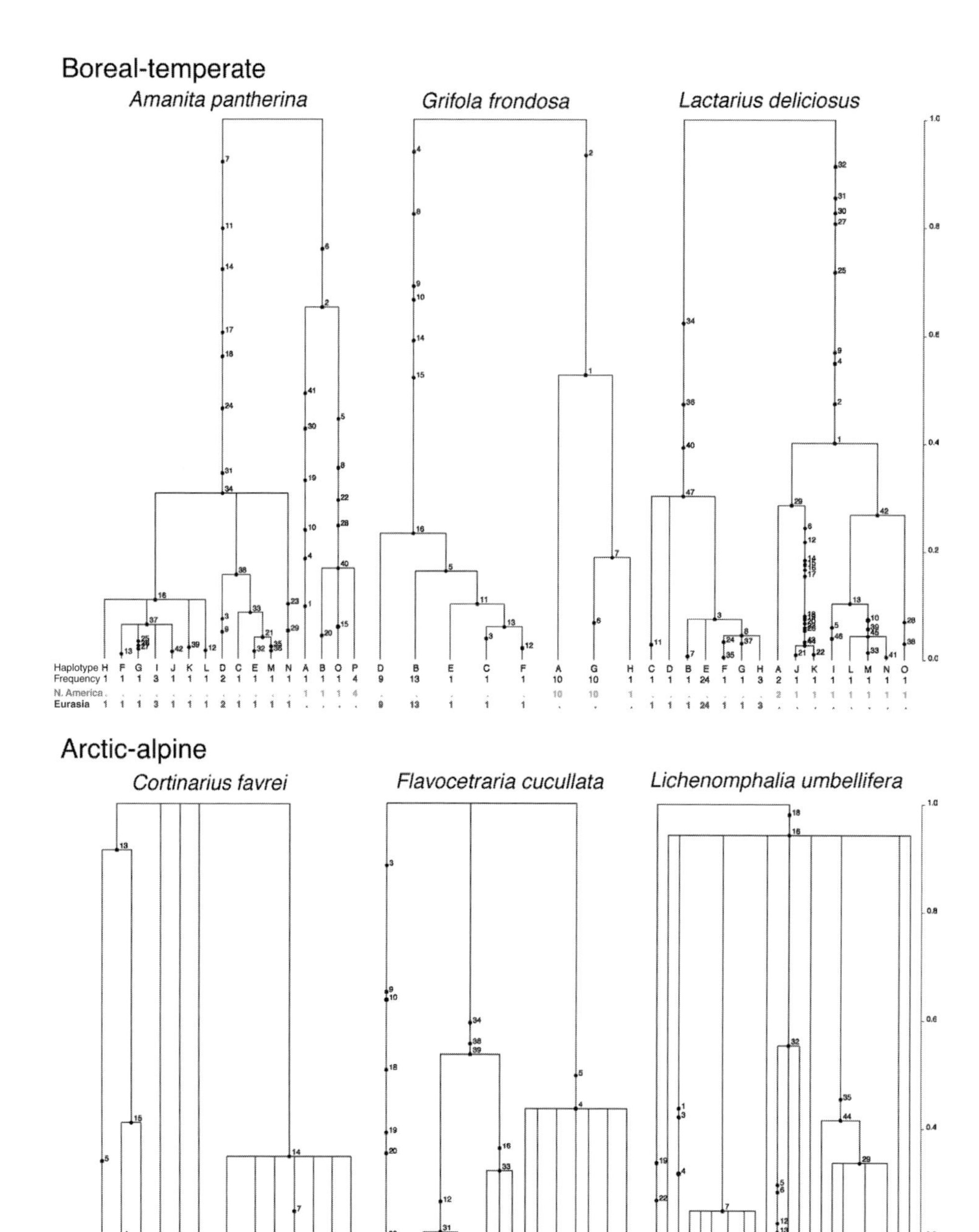

Fig 9.2 Examples of coalescent-based genealogies with the highest root probabilities showing the distribution of mutations for the ITS region. The inferred genealogies are based on 2 000 000 simulations of the coalescent. The time scale is in coalescent units of $2N$, where N is the population size. Mutations and bifurcations are time ordered from the top (past) to the bottom (present). The letters and numbers below the trees designate the distinct haplotypes, their observed frequencies in total and in the different geographic regions.

conducted assuming moderately high level of migration ($M = 0.1$) in arctic species, and very low level of migration ($M = 0.001$) in boreal-temperate taxa. As expected, the coalescent-based genealogies showed strong historical population divisions in the boreal group, but not in the arctic taxa, and were informative with respect to inferring the mutational history and variation between and within geographic regions. Examples of coalescent-based genealogies are shown in Fig 9.2 for three arctic-alpine and three boreal-temperate species.

9.4 Discussion

The main goal of this chapter was to estimate intercontinental migration between populations of arctic-alpine and boreal-temperate fungi in the northern hemisphere with the purpose of gaining some insights in the possible mechanisms that play roles in the dispersal capacities of fungi. Beside the possible theoretical advancement in our knowledge regarding long-distance dispersal, the varying capacities of fungi to migrate over vast areas have practical implications in shaping the composition of past, present and future communities during shifts in species distributions due to climatic changes.

Many boreal, temperate or tropical fungi show strong intercontinental, sometimes even intracontinental, phylogeographic patterns and limited dispersal, and there is an increasing amount of geographic endemism being discovered (e.g. Taylor et al., 2006 and references therein; Geml et al., 2008; Bergemann et al., 2009). In recent years, molecular tools have revealed several examples of distinct phylogeographic groups within complexes that were previously treated as morphological species. The existence of multiple phylogenetic lineages within a morphological species, mostly with non-overlapping geographic distributions, in itself is a powerful argument against the 'everything is everywhere' theory that is a commonly held view regarding the distribution of microbial taxa (Finlay, 2002). For example, morphological species complexes of fungi from the northern hemisphere have generally been shown to include two major lineages, a Eurasian and a North American (e.g. Shen et al., 2002; Taylor et al., 2006 and references therein; Geml et al., 2008). Because in most studied fungi, the allopatric phylogenetic clades inhabit similar environments in different continents, this implies a phylogenetic structure that has arisen as a result of the lack of intercontinental dispersal. All of the boreal-temperate species analysed here share this pattern and had intraspecific phylogenetic groups corresponding to continents.

Arctic fungi seem very different from these examples, as patterns of genetic diversity described above were not observed in any of the arctic-alpine species discussed in this paper. Instead, clades inferred within each species contained specimens from distant geographic regions, and geographic population pairs exhibited moderate to high transoceanic gene flow. This suggests that, in response to climatic fluctuations, these species have been able to migrate over large distances due to efficient long-distance dispersal capability. In addition, large and diverse populations have served as sources for such migrants, as suggested by the number of haplotypes (Table 9.1). The high observed genetic diversity in the Arctic indicates long-term survival at northern high latitudes, while the estimated migration rates and the no or weak geographic population structure suggest continuing long-distance gene flow between continents that has prevented pronounced genetic differentiation. Similar patterns of circumpolar genetic diversity have been detected in some other arctic organisms, for example in highly mobile animals, such as the arctic fox, *Alopex lagopus* (Dalén et al., 2005) and the snowy owl, *Bubo scandiacus* (Marthinsen et al., 2008), as well as in the arctic-alpine lineage of the bog blueberry *Vaccinium uliginosum* (Alsos et al., 2005).

One cannot avoid posing the question: Why do arctic-alpine fungi differ in phylogeographic trends from more southern species? The most likely explanation is that climatic changes during the Quaternary dramatically influenced the distribution of flora and fauna, particularly at higher latitudes. During glacial maxima, plants, fungi and animals were restricted to unglaciated refugia, from which they recolonised newly exposed areas in warmer interglacial periods (see Abbott and Brochmann, 2003). Although climatic changes have caused some shifts in species distributions in most terrestrial habitats around the globe, changes have been the greatest at high latitudes, and so were the geographic distances the species had to cover to track their ecological niche and to colonise newly available habitats following glacial retreats. It is, therefore, very likely, that many arctic fungi, particularly the keystone taxa with circumpolar distribution, have been selected for mobility during the glacial cycles, as suggested for plants (Brochmann and Brysting, 2008).

Although the mode of dispersal is not known for arctic fungi, it is likely that wind dispersal is important, as expected for most fungi. Wind dispersal should be particularly effective in the Arctic as a result of the open landscape, strong winds and extensive snow and ice cover, as has also been shown for arctic plants (Alsos et al., 2007). In this regard, the sea ice may be of particular importance for intercontinental dispersal, as it provides a dry surface bridging the continents and archipelagos. Obviously, this is an important difference to boreal-temperate species, which would need to cross greater distances of open water to reach another continent. Although such transoceanic dispersals are often documented in fungi in mid-latitudes in the southern hemisphere (Moyersoen et al., 2003; Moncalvo and Buchanan, 2008), facilitated by extremely strong and mostly unidirectional wind currents, long-distance dispersal in mid-latitudes in the northern hemisphere seem to be rarer. Besides wind, other possible means of dispersal include spores being carried by migratory animals, driftwood and drifting sea ice, which may also favour arctic fungi due to ocean currents and animal migrations (particularly birds) linking continents over shorter distances.

The implications of my results may not be restricted to species discussed in this paper, but can be important for studies on the biodiversity, ecology and conservation of arctic fungi in general. Reconstruction of phylogeographic patterns of arctic organisms is of paramount importance, because knowledge of both past migrational history and present-day genetic diversity are essential to improve our predictions on how arctic species and communities will respond to global change. The high genetic diversity and the efficient long-distance dispersal capability of the arctic taxa analysed here suggest that these species, and perhaps other arctic fungi as well, will probably be able to track their potential niche in the changing Arctic.

Acknowledgements

This research was sponsored by the University of Alaska International Polar Year Office, the National Centre for Biosystematics (University of Oslo) and the Humboldt Foundation. I thank the following persons for providing specimens: Katriina Bendiksen, Ronald Daanen, Dmitry Dobrynin, Arve Elvebakk, Guido Grosse, Gro Gulden, Alexander Kovalenko, Patrick Kuss, François Lutzoni, David McGuire, Molly McMullen, Olga Morozova, Claude Roy, Siri Rui, Nina Sazanova, Lutz Schirrmeister, Einar Timdal, Ina Timling, Irina Urbnavichene and Mikhail Zhurbenko. Special thanks go to D. Lee Taylor (University of Alaska Fairbanks) for the attentive mentoring he provided during my 5-year stay in Alaska. Also, I am grateful to Christian Brochmann (University of Oslo) and Frank Kauff (University of Kaiserslautern) for their financial and professional support during research stays at their institutes, and to Mikhail Zhurbenko for his hospitality at the Komarov Botanical Institute during my sampling visit.

References

Abbott, R.J., Brochmann, C. (2003). History and evolution of the arctic flora: in the footsteps of Eric Hultén. *Molecular Ecology* **12**, 299–213.

Abbott, R.J., Comes, H.P. (2003). Evolution in the Arctic: a phylogeographic analysis of the circumarctic plant, *Saxifraga oppositifolia* (Purple saxifrage). *New Phytologist* **161**, 211–224.

Alsos, I.G., Engelskjøn, T., Gielly, L., Taberlet, P., Brochmann, C. (2005). Impact of ice ages on circumpolar molecular diversity: insights from an ecological key species. *Molecular Ecology* **14**, 2739–2753.

Alsos, I.G., Eidesen, P.B., Ehrich, D. et al. (2007) Frequent long-distance plant colonization in the changing Arctic. *Science* **316**, 1606–1609.

Aylor, D.L., Price, E.W., Carbone, I. (2006). SNAP: Combine and Map modules for multilocus population genetic analysis. *Bioinformatics* **22**, 1399–1401.

Beerli, P., Felsenstein, J. (2001). Maximum-likelihood estimation of a migration matrix and effective population sizes in n subpopulations by using a coalescent approach. *Proceedings of the National Academy of Sciences USA* **98**, 4563–4568.

Bergemann, S.E., Smith, M.A., Parrent, J.L., Gilbert, G.S., Garbelotto, M. (2009). Genetic population structure and distribution of a fungal polypore, *Datronia caperata* (Polyporaceae), in mangrove forests of Central America. *Journal of Biogeography* **36**, 266–279.

Bowden, L.C., Price, E.W., Carbone, I. (2008). *SNAP Clade and Matrix*, Version 2. Department of Plant Pathology, North Carolina State University, Raleigh, NC. Available at: http://www.cals.ncsu.edu/plantpath/faculty/carbone/home.html (accessed on 25 February 2009).

Brochmann, C., Brysting, A.K. (2008). The Arctic – an evolutionary freezer? *Plant Ecology and Diversity* **1**, 181–195.

Brunhoff, C., Galbreath, K.E., Fedorov, V.B., Cook, J.A., Aarola, M. (2003). Holarctic phylogeography of the root vole (*Microtus oeconomus*): implications for late Quaternary biogeography of high latitudes. *Molecular Ecology* **12**, 957–968.

Callaghan, T.V., Björn, L.O., Chernov, Y. et al . (2004). Biodiversity, distributions and adaptations of Arctic species in the context of environmental change. *Ambio* **33**, 404–417.

Dalén, L., Fuglei, E., Hersteinsson, P. et al. (2005). Population history and genetic structure of a circumpolar species: the arctic fox. *Biological Journal of the Linnean Society* **84**, 79–89.

Eidesen, P.B., Carlsen, T., Molau, U., Brochmann, C. (2007). Repeatedly out of Beringia: *Cassiope tetragona* embraces the Arctic. *Journal of Biogeography* **34**, 1559–1574.

Fedorov, V.B., Goropashnaya, A.V., Jaarola, M., Cook, J.A. (2003). Phylogeography of lemmings (*Lemmus*): no evidence for postglacial colonization of Arctic from the Beringian refugium. *Molecular Ecology* **12**, 725–731.

Fenchel, T., Finlay, B. J. (2004). Here and there or everywhere? Response from Fenchel and Finlay. *Bioscience* **54**, 777–784.

Finlay, B. J. (2002). Global dispersal of free-living microbial eukaryote species. *Science* **296**, 1061–1063.

Flagstad, O., Røed, K.H. (2003). Refugial origins of reindeer (*Rangifer tarandus* L.) inferred from mitochondrial DNA sequences. *Evolution* **57**, 658–670.

Geml, J., Tulloss, R.E., Laursen, G.A., Sazanova, N.A., Taylor, D.L. (2008). Evidence for strong inter- and intracontinental phylogeographic structure in *Amanita muscaria*, a wind-dispersed ectomycorrhizal basidiomycete. *Molecular Phylogenetics and Evolution* **48**, 694–701.

Geml, J., Kauff, F., Brochmann, C., Taylor, D.L. (2010). Surviving climate changes: high genetic diversity and transoceanic gene flow in two arctic-alpine lichens, *Flavocetraria cucullata* and *F. nivalis* (Parmeliaceae, Ascomycota). *Journal of Biogeography* **37**, 1529–1542.

Griffiths, R.C., Tavaré, S. (1994). Ancestral inference in population genetics. *Statistical Sciences* **9**, 307–319.

Hudson, R.R., Boos, D.D., Kaplan, N.L. (1992). A statistical test for detecting geographic subdivision. *Molecular Biology and Evolution* **9**, 138–151.

Markwordt, J., Doshi, R., Carbone, I. (2003). *SNAP Clade and Matrix*. Department of Plant Pathology, North Carolina State University, Raleigh, NC. Available at: http://www.cals.ncsu.edu/plantpath/faculty/carbone/home.html (accessed on 25 February 2009).

Marthinsen, G., Wennerbeg, L., Solheim, R., Lifjeld, J.T. (2008). No phylogeographic structure in the circumpolar snowy owl (*Bubo scandiacus*). *Conservation Genetics* **10**, 923–933.

Martiny, J.B.H., Bohannan, B.J.M., Brown, J.H et al. (2006). Microbial biogeography: putting microorganisms on the map. *Nature Reviews Microbiology* **4**, 102–112.

Moncalvo, J.M., Buchanan, P.K. (2008). Molecular evidence for long distance dispersal across the southern hemisphere in the *Ganoderma applanatum-australe* species complex (*Basidiomycota*). *Mycological Research* **112**, 425–436.

Moritz, C., Faith, D.P. (1998). Comparative phylogeography and the identification of genetically divergent areas for conservation. *Molecular Ecology* **7**, 419–429.

Moyersoen, B., Beever, R.E., Martin, F. (2003). Genetic diversity of *Pisolithus* in New Zealand indicates multiple long-distance dispersal from Australia. *New Phytologist* **160**, 569–579.

Nielsen, R., Wakeley, J. (2001). Distinguishing migration from isolation: a Markov chain Monte Carlo approach. *Genetics* **158**, 885–896.

Oda, T., Tanaka, C., Tsuda, M. (2004). Molecular phylogeny and biogeography of the widely distributed *Amanita* species, *A. muscaria* and *A. pantherina*. *Mycological Research* **108**, 885–896.

Palice, Z., Printzen, C. (2004). Genetic variability in tropical and temperate populations of *Trapeliopsis glaucolepidea*: evidence against long range dispersal in a lichen with disjunct distribution. *Mycotaxon* **90**, 43–54.

Parmesan. C. (2006). Ecological and evolutionary responses to recent climate change. *Annual Review of Ecology, Evolution and Systematics* **37**, 637–669.

Price, E.W., Carbone, I. (2005). SNAP: workbench management tool for evolutionary population genetic analysis. *Bioinformatics* **21**, 402–404.

Printzen, C. (2008). Uncharted terrain: the phylogeography of arctic and boreal lichens. *Plant Ecology and Diversity* **1**, 265–271.

Reiss, A.R., Ashworth, A.C., Schwert, D.P. (1999). Molecular genetic evidence for the post-Pleistocene divergence of populations of the arctic-alpine ground beetle *Amara alpina* (Paykull) (Coleoptera: Carabidae). *Journal of Biogeography* **26**, 785–794.

Ricklefs, R.E., Schluter, D (1993). *Species Diversity in Ecological Communities: Historical and Geographical Perspectives*. Chicago, IL: University of Chicago Press.

Schönswetter, P., Suda, J., Popp, M., Weiss-Schneeweiss, H., Brochmann, C. (2007). Circumpolar phylogeography of *Juncus biglumis* (Juncaceae) inferred from AFLP fingerprints, cpDNA sequences, nuclear DNA content and chromosome numbers. *Molecular Phylogenetics and Evolution* **42**, 92–103.

Shen, Q., Geiser, D. M., Royse, D. J. (2002). Molecular phylogenetic analysis of *Grifola frondosa* (maitake) reveals a species partition separating eastern North American and Asian isolates. *Mycologia* **94**, 472–482.

Taylor, J.W., Turner, E., Townsend, J.P., Dettman, J.R., Jacobson, D. (2006). Eukaryotic microbes, species recognition and the geographic limits of species: examples from the kingdom Fungi. *Philosophical Transactions of the Royal Society B: Biological Science* **361**, 1947–1963.

Thompson, J.D., Gibson, T.J., Plewniak, F., Jeanmougin, F., Higgins, D.G. (1997). The CLUSTAL X windows interface: flexible strategies for multiple sequence alignment aided by quality analysis tools. *Nucleic Acids Research* **25**, 4876–4882.

Tremblay, N.O., Schoen, D.J. (1999). Molecular phylogeography of *Dryas integrifolia*: glacial refugia and postglacial recolonization. *Molecular Ecology* **8**, 1187–1198.

Watterson, G.A. (1975). On the number of segregating sites in genetic models without recombination. *Theoretical Population Biology* **10**, 256–276.

Wickström, L.M., Haukisalmi, V., Varis, S. et al. (2003). Phylogeography of the circumpolar *Paranoplocephala arctica* species complex (Cestoda: Anoplocephalidae) parasitizing collared lemmings (*Dicrostomyx* spp.). *Molecular Ecology* **12**, 3359–3371.

10

Biogeography and phylogeography of lichen fungi and their photobionts

SILKE WERTH

Biodiversity and Conservation Biology, WSL Swiss Federal Research Institute, Birmensdorf, Switzerland

10.1 Introduction

Lichens are the classical example of mutualistic symbiosis (de Bary, 1879). The first researcher to recognise the dual nature of lichens was the Swiss botanist and lichenologist Simon Schwendener (1868), whose theory on the algal–fungal association of lichens was rejected by the leading lichenologists at the time, to become widely accepted only in the twentieth century when lichens had been resynthesised from aposymbiotic cultures for the first time (for review, see Honegger, 2000). The lichen symbiosis is shaped by a fungus ('mycobiont') which forms an intimate association with a photosynthetic partner ('photobiont'). In the lichen symbiosis, a single fungal species associates either with one or several species of green algae or cyanobacteria, or sometimes with both taxa ('tripartite' lichen).

More than 15 000 species of lichen fungi are known to date, belonging to about 1000 genera (Kirk et al., 2008); 98% of these species are ascomycetes (Honegger, 2008), the remaining are basidiomycetes or fungi of unclear systematic position (Tehler and Wedin, 2008; Printzen, 2010). In contrast, only about 100 photobiont species (Lücking et al., 2009) belonging to about 40 genera have been reported

Biogeography of Microscopic Organisms: Is Everything Small Everywhere?, ed. Diego Fontaneto. Published by Cambridge University Press.

(Ahmadjian, 1967; Tschermak-Woess, 1988; Büdel, 1992; Nyati et al., 2007). This vast discrepancy in numbers implies that a large number of lichen fungi share a common pool of photobionts. The photobionts of lichens include eukaryotic green algae (phylum Chlorophyta), other eukaryotic algae (phylum Heterokontophyta), as well as cyanobacteria. Most photobionts belong to four genera: the green algae *Trebouxia* and *Trentepohlia* and the cyanobacteria *Nostoc* and *Scytonema* (Ahmadjian, 1967; Tschermak-Woess, 1988; Büdel, 1992; Friedl and Büdel, 2008).

As poikilohydric organisms, lichens have an extraordinarily high stress tolerance, similar to that of the poikilohydric bryophytes (Chapter 12). When dry (< 5% water content), the physiological activity of lichens approximates zero, and they can survive extreme conditions – for example, a remarkable range of temperatures (–196 °C to +60 °C) (Kappen, 1973; Beckett et al., 2008). In an experiment performed by the European Space Agency, two lichen species, *Rhizocarpon geographicum* and *Xanthoria elegans*, were launched into space, and exposed to space conditions for 16 days mounted on an Earth-orbiting satellite. Upon return to earth, the same specimens regained full metabolic activity within 24 hours, after having survived exposure to high-vacuum, extreme temperatures, as well as vast levels of UV and cosmic radiation – conditions lethal for bacteria and most other microorganisms (Sancho et al., 2007). Another study compared the performance of the symbiotic phenotype of the widespread foliose lichen *X. elegans* to that of axenic cultures of its photobiont and mycobiont under simulated space conditions. In the symbiotic state, tolerance to environmental stresses such as exposure to vacuum and different levels of UV radiation was considerably enhanced (de Vera et al., 2008). Both studies highlight the remarkable ability of lichens to tolerate extreme environmental conditions.

Their ability to tolerate extreme physiological stresses distinguishes lichens as successful colonists of habitats that few higher plants or animals are able to survive in – the surfaces of bare rocks, leaves, wood or tree trunks. Lichens are common in most terrestrial ecosystems (Nash, 1996), and they reach especially high abundance in extreme environments such as the Maritime Antarctic or coastal deserts, where they may be more pervasive than vascular plants (Rundel, 1978; Øvstedal and Lewis Smith, 2001; Seymour et al., 2005; Knudsen and Werth, 2008). Temperate and tropical rain forests also host a high abundance and as well as an amazing diversity of lichens (Kantvilas et al., 1985; Lücking, 2003, 2008; Goward and Spribille, 2005; Caceres et al., 2007).

Lichen fungi are able to reproduce sexually, as well as asexually. The relative frequency of sexual and vegetative propagation is species specific; however, examples of species are known which reproduce sexually on one continent and asexually on another ('species pairs', Poelt, 1970). Few genera of lichen fungi are exclusively asexual, most are either exclusively sexual or possess a mixed reproductive strategy (Bowler and Rundel, 1975). Sexual lichen fungi reproduce with microscopic ascospores (ascomycetes) or with basidiospores (basidiomycetes), both of which contain only the fungal partner. The size of these spores ranges from a few mm, e.g. 2–5 mm in *Polysporina* spp. (Smith et al., 2009) to 200–400 mm, for instance in *Varicellaria rhodocarpa* (Poelt, 1969), which is a similar size range as that of pollen grains. Conidia are fungal cells which are produced by mitotic cell division. Macroconidia have been suggested to operate as vegetative propagules in lichen fungi. For instance, *Vezdaea aestivalis* has been suggested to be able to reproduce vegetatively with macroconidia, 7–19 mm in size (Scheidegger, 1995). Other species with macroconidea are *Micarea* spp. (Coppins, 1983) and *Candelariella biatorina* (Westberg, 2007). The bacteria-sized microconidia of *Xanthoria parietina* do not contain mitochondria and are thus not able to function as vegetative propagules, which might be a common phenomenon in lichen-forming ascomycetes (Honegger et al., 2004). Instead, microconidia have been hypothesised to be spermatia that fertilise trichogynes to form dicaryotic, ascogenic hyphae and hence apothecia, the fruiting bodies of ascomycetes. Asexual lichen propagules such as soredia, isidia or thallus fragments are usually larger than spores, and contain both fungal hyphae and photobiont cells, thus codispersing both partners of the symbiosis.

Cyanobacterial photobionts develop hormogonia, specific reproductive structures (Geitler, 1934). When in symbiotic state, green-algal lichen photobionts are not known to form specific propagules, but algal 'escape' from lichen thalli under humid conditions has been noticed (Wornik and Grube, 2010). The formation of sporangia has been noticed from lichenised photobionts; these sporangia may develop either into aplanospores or zoospores (Scheidegger, 1985). As they are able to survive the gut passage of lichenivorous mites, lichen photobionts could be dispersed by these invertebrates (Meier et al., 2002) and form free-living colonies derived from faeces. There is evidence for the existence of free-living *Trebouxia* colonies, but the free-living populations appear to be small and not persistent (Bubrick et al., 1984; Mukhtar et al., 1994; Schroeter and Sancho, 1996; Sanders, 2005; Hedenås et al., 2007; Wornik and Grube, 2010).

Thanks to the microscopic size of their propagules, lichens should be rather good dispersers, and dispersal distances of at least 200 m have been measured in the field for soredia of epiphytic lichens (Werth et al., 2006a). Another argument in favour of high dispersal capability is that newly exposed substrates such as rocks in glacier forefields or newly emerging volcanic islands have been found to be colonised by several species within about a decade (Kristinsson, 1972; Winchester and Harrison, 2000). The island of Surtsey, situated *c.* 30 km from the mainland of Iceland and *c.* 15 km from the nearest island, was formed by volcanic eruptions between 1963 and 1967. Since its emergence, Surtsey has been colonised by 87 lichen species in a time period of 39 years and the number of species has increased more or less continuously since the first three species were observed in 1970 (Kristinsson and Heiðmarsson, 2009).

In summary, due to their microscopic propagules and large stress tolerance, lichen fungi as well as their photobionts should have the potential for substantial long-distance dispersal. Here, I present evidence from biogeography and phylogeography relating to the hypotheses that (1) current ranges of lichen fungi and their photobionts are a result of extensive long-distance dispersal ('Everything is everywhere' hypothesis) and (2) that current ranges are the result of long-term isolation and divergence of population ('Vicariance' hypothesis).

10.2 Biogeography of lichen fungi

Lichen fungi show distinct distribution patterns resembling those of higher plants, and many species are very widespread (Lücking, 2003; Feuerer and Hawksworth, 2007). Wide distributions are thought to be favoured by high dispersal ability and generalist ecological features. Moreover, older species of lichen fungi have had more time to disperse and reach extensive distributions, relative to younger species. Indeed, some species of lichen fungi appear to be very old, as indicated by long net diversification intervals relative to those of insects, other animals and higher plants (Lücking et al., 2008).

Four studies have treated the similarities of lichen biota across various regions of the world applying a statistical framework. Martinez et al. (2003) compared the world distributions of 66 species of *Peltigera* across 230 biogeographic provinces. They identified boreal and arctic parts of North America and Europe as well as temperate and subtropical areas in eastern Asia as the regions of highest species richness, and with the highest frequency of endemics.

Feuerer and Hawksworth (2007) analysed the known occurrences of lichen species based on species checklists from 132 geographic areas, grouped into 35 floristic regions. They used a clustering approach to compare the species composition among floristic regions and to group them into units relevant for lichens. The four main geographic regions that were identified to host distinct lichen floras were a Holarctic, a Pantropical, an Oceanian and a Subantarctic–Australian region.

Lücking (2003) analysed worldwide biogeographic relationships in foliicolous lichens using multidimensional scaling, cluster analysis and cladistic approaches, and found six major lichenogeographic regions – three tropical and three extratropical lichen biota. He interpreted the similarities among Neotropic and African Paleotropic foliicolous lichen biota as an indication of a shared Western Gondwanan element – sets of vicariant species that occupied a common ancestral Gondwanan region, and were separated by continental drift ('biotic ferry hypothesis'). In contrast, the similarities among the African Paleotropic and Australian/Caledonian Paleotropic lichen biota were suggested to have arisen by more recent long-distance dispersal among continents.

Using a phylogenetic approach, Lücking et al. (2008) investigated the historical biogeography of the widespread tropical foliicolous genus *Chroodiscus*. They proposed an evolutionary scenario suggesting that the present distribution of *Chroodiscus* species is the result of historic dispersal events over intermediate distances facilitated by proximity among continental shelves during the Mid-Cretaceous, rather than recent transoceanic dispersal.

One emerging biogeographic pattern from the various studies of lichen fungi is that at the species level, distributions are often far more widespread than those of phanerogams, as indicated by the few biogeographic regions that could be defined based on floristic similarities (Lücking, 2003; Feuerer and Hawksworth, 2007). Moreover, the classical biogeographic studies have pointed out ecological similarities among the distribution patterns of lichen fungi.

One common problem of the biogeographic studies of lichens that rely on distribution patterns of morphospecies (i.e. morphologically defined species) is that at least a part of the widely distributed fungal morphospecies could be genetically incompatible, cryptic species with more restricted distributions (e.g. Kroken and Taylor, 2001; Lücking, 2003). Another major problem is that current distribution patterns do not allow conclusions about the historical processes that caused them: similar distributions observed today may reflect completely different underlying histories (Otte et al., 2005). It is thus to be expected that species with the same distribution type will exhibit a variety of different phylogeographic patterns and that the role of long-distance dispersal vs. vicariance in creating wide ranges and disjunctions may vary among species. Probably the most convincing evidence to investigate the phenomenon of wide and disjunct ranges and to clarify the role of long-distance dispersal in creating them comes from molecular studies – in particular those that test specific hypotheses about the divergence of populations in different geographic regions, or on different continents.

10.2.1 Wide distributions

If dispersal were no factor constraining the geographic distribution of lichen fungi, cosmopolitan or wide distributions should be relatively common. Indeed, many lichen fungi have astoundingly wide geographic distributions, across multiple climatic zones or continents; such widespread distributions are rare in higher plants (Otte et al., 2002). *Lobaria pulmonaria* ranges across most of the northern hemisphere including subtropical, Mediterranean, temperate and boreal woodlands and has a few occurrences on the southern hemisphere (Yoshimura, 1971), a distribution which has been described as incomplete circum-bipolar and tropical montane/subalpine. Other examples of species following this distribution type are *Hypogymnia physodes* and *Platismatia glauca* (Litterski, 1999). Other lichen fungi are circumpolar, with widespread distributions in the northern hemisphere, e.g. *Cladonia coniocraea* (Litterski, 1999), some with circumpolar arctic-alpine distributions such as *Melanelia tominii* (Otte et al., 2005). Another type of wide distribution occurs in tropical and subtropical areas and those species may also reach temperate or boreal oceanic areas (Litterski, 1999). *Cetraria aculeata* and *C. muricata* are widespread in the northern hemisphere including the Mediterranean and Arctic zone, with additional occurrences in the tropics and in South America, South Africa and Antarctica (Kärnefelt, 1986).

Wide distributions that are ecologically structured are indeed very common among lichen fungi, but only few taxa are truly cosmopolitan (Litterski, 1999; Otte et al., 2002, 2005). Even the most widespread species do not occur everywhere – most show some ecological tendencies, often related to macroclimatic gradients (Werth et al., 2005). For instance, they may grow predominantly in oceanic areas, reflected in western distributions in North America and Eurasia (Otte et al., 2002, 2005). Moreover, in warm climates, some widespread arctic or boreal circumpolar lichens tend to occur at high altitudes (Codogno and Sancho, 1991). Therefore, the various geographic distribution patterns of widespread lichens are partly driven by the species-specific ecology.

10.2.2 Disjunct distributions

Disjunct distributions are known for numerous lichens. For example, several disjunct lichen fungi are known from the boreal zone of western Fennoscandia, Newfoundland and the Pacific Northwest region of North America; notable examples of this distribution pattern are *Cavernularia hultenii*, *Erioderma pedicellatum* and *Lobaria hallii* (Ahlner, 1948; Bjerke, 2003; Otte et al., 2005).

Other lichens exhibit impressive disjunctions between the southern and northern hemisphere ('bipolar' distribution, Galloway and Aptroot, 1995). Some species showing circum-bipolar disjunctions are completely absent in the tropics; examples include *Parmeliopsis ambigua* and *Verrucaria maura* (Litterski, 1999). Others are additionally found in tropical montane and subalpine areas (Litterski, 1999), and are then often widespread in the cool and temperate zones of the northern and southern hemisphere. In subtropical and tropical climates, these species only occur in higher altitudes, corresponding to montane and subalpine zones; e.g. *Cladonia rangiferina* (Litterski, 1999).

10.2.3 Endemic distributions

Possible dispersal restrictions in lichen fungi should be reflected in endemic distribution patterns. Indeed, several areas of the world are known to host endemic lichen fungi; examples include the European (Otte et al., 2002) and North American Mediterranean region (Moberg and Nash, 1999; Nash et al., 2002), Macaronesia (Krog and Østhagen, 1978, 1980), or the Galapagos Islands (Aptroot and Bungartz, 2007) – largely regions which also harbour endemic vascular plant floras. However, lichen fungi may be restricted in distribution for several reasons other than dispersal limitation. First, many rare and threatened species have a very specialised ecology – e.g. specific microclimatic or habitat-related requirements – which may only be met in a small geographic area. In these cases, even if dispersal were extensive, it would not lead to successful establishment of new individuals unless the specialised ecological conditions are met. Second, some endemics may have arisen through recent speciation, and may not have had enough time to increase their distribution ranges. Third, it cannot be excluded that some sexual lichen fungi are highly specialised with respect to their photobionts and that restricted distributions might result from lack of suitable algal strains. Fourth, the ranges of some endemic species may be poorly known, and they may turn out to be more widespread than anticipated, when under scrutiny.

For all these reasons, it is to be expected that endemic lichen fungi are a rather heterogeneous group and that there is no common underlying process which has led to their current distribution.

10.3 Phylogeography of lichen fungi

Are wide intercontinental and disjunct distributions the remaining fragments of historically continuous distributions? Or, have populations on different continents been founded by long-distance dispersal? If the latter is true, do the disjunctions simply reflect the ecology of widely dispersing taxa, excluding them from occurrences in intermediate sites (Otte et al., 2005)? Some observations have been made that favour the dispersal hypothesis. The similarity among lichen biota of the Antarctic islands is apparently highly correlated with south polar wind patterns, rather than with geographic distance (Muñoz et al., 2004). Similarly, the sharing of species of *Ramalina* spp. among different volcanic islands in the New Zealand geographic area largely resembles inter-island connectivity by wind (Bannister and Blanchon, 2003). If long-distance dispersal was indeed a frequent phenomenon, this should be reflected in the sharing of identical genotypes around the poles in regions of high wind connectivity, and phylogeographic studies of DNA sequence variation may shed light on this phenomenon.

However, current ranges may also reflect palaeodistributions across formerly connected land masses which got separated by continental drift (Lücking, 2003). In these cases, no long-distance dispersal would be necessary to explain current intercontinental distributions. Given the long time frame, one would expect speciation; various congeneric phylogenetic species might in this case occupy different continents.

Wide intercontinental and disjunct ranges could also have been created either by many long-distance (e.g. intercontinental) events, or few long-distance but a lot of dispersal over medium or short distances founding intermediate populations acting as 'stepping stones', which may or may not have survived to the present time. If intercontinental long-distance dispersal was a common phenomenon in lichens, one should expect to see genetic homogeneity of populations distributed on different continents. If dispersal were rare, and it occurred mainly over shorter distances and via intermediate stepping stone populations, one would predict substantial genetic differentiation among lichen populations from different continents, and from different geographic regions.

Another important aspect regarding long-distance dispersal of lichen fungi is its timing throughout the evolutionary history of species. Even in cases where long-distance dispersal may seem an unlikely possibility at present, it could be that at some point in the past, specific environmental conditions facilitated dispersal. Accordingly, the relative role of dispersal (and vicariance) may vary throughout the evolutionary history of a species. Moreover, whether one can successfully detect intercontinental long-distance dispersal, for example using demographic genetic analyses, depends on how many generations have passed since; such historic genetic signals tend to fade over time, eventually to become undetectable (Excoffier et al., 2009).

10.3.1 Molecular studies supporting extensive gene flow

While intercontinental dispersal with bird vectors or through wind currents has been proposed to be important in creating disjunct bipolar distribution patterns (Galloway and Aptroot, 1995), few studies so far have shown intercontinental long-distance dispersal in lichen fungi. Buschbom (2007) studied migration and divergence in *Porpidia flavicunda*, a circumpolar arctic lichen fungus. There was substantial genetic differentiation of a Canadian population from populations in Europe and Greenland. However, haplotypes were widely shared among geographic regions, and substantial transoceanic migration was inferred for *P. flavicunda*.

Similarly, substantial gene flow was suggested for the bipolar disjunct lichen fungi *Cladonia arbuscula* and *C. mitis*, directed towards the southern hemisphere – either via stepping stone populations in the Andes, or through intercontinental long-distance dispersal across the equator (Myllys et al., 2003). The *C. arbuscula* and *C. mitis* consisted of one clade including all samples of *C. mitis*, and three clades with specimens of *C. arbuscula*. The clades showed no geographic structure – for both fungal species, samples from South America occurred intermingled with samples from northern Europe.

Both examples above are drawn from taxa with a widespread distribution in arctic parts of the northern hemisphere. Other arctic lichen fungi also show little population differentiation and high migration rates, supporting the dispersal hypothesis (Chapter 9). The high gene flow inferred from taxa located at high latitudes corresponds well with the observation of high wind connectivity among sites, which may facilitate efficient long-distance dispersal (Muñoz et al., 2004). Moreover, the genetic similarities among populations of arctic lichen fungi may reflect their biogeographic history, as arctic taxa may not have experienced severe population bottlenecks during the Pleistocene glaciations, when boreal and temperate species were limited to refugia (Hewitt, 1999).

10.3.2 Molecular studies supporting restricted gene flow

The majority of phylogeographic studies suggest that gene flow is restricted among populations of lichen fungi, and intercontinental long-distance dispersal is rare (Printzen et al., 2003; Palice and Printzen, 2004; Walser et al., 2005; Altermann, 2009; Otálora et al., 2010). Even along continuous ranges, some lichen fungi exhibit striking geographic genetic structure (Walser et al., 2005; Werth et al., 2006b, 2007; Cassie and Piercey-Normore, 2008; Altermann, 2009; Widmer, 2009).

Walser et al. (2005) compared populations of *Lobaria pulmonaria* from Europe and western North America, and the allele sizes of two fungal and two algal microsatellite markers *sensu* Widmer et al. (2010) differed markedly among populations from the two continents, suggesting that intercontinental dispersal of both symbionts must be rare.

Printzen et al. (2003) studied phylogeographic patterns in the disjunct lichen fungus *Cavernularia hultenii*, a species distributed in northern Europe, Newfoundland and western North America and tightly associated with oceanic boreal coniferous forests. The shallow genealogy of *C. hultenii* DNA sequence data with the occurrence of region-specific tip haplotypes suggested the fragmentation of a continuous ancestral distribution. Indeed, during the Mid-Pliocene warming period, coniferous forests were widespread in humid northern areas, promoting the occurrence of a more or less continuous distribution of *C. hultenii*, which became fragmented when the boreal forest belt shifted southwards into drier continental areas during the Pleistocene. Moreover, while no evidence for recent intercontinental long-distance dispersal was found for *C. hultenii*, there was support for recent population expansion in connection with a postglacial range expansion of the lichen fungus into previously glaciated regions of western North America.

The lichen fungus *Trapeliopsis glaucolepidea* exhibits a disjunct range in Europe, East Africa, New Guinea and tropical America. Using DNA sequences from central America and Europe, Palice and Printzen (2004) found no overlap of haplotypes among continents, but the populations on both continents did not represent monophyletic lineages – one haplotype found in Europe was closely related to haplotypes from central America, whereas the remaining European haplotypes represented two distinct European lineages. The results suggested the absence of recent intercontinental long-distance dispersal, and the long-term isolation of populations on the two continents.

Phylogenetic analyses showed that the species pair *Letharia vulpina* and *L. columbiana* contained six evolutionary lineages which may represent cryptic species; some of these had non-overlapping geographic distributions (Kroken and Taylor, 2001). A second example of a widespread lichen fungus with geographically restricted evolutionary lineages is *Parmelia saxatilis*. This species exhibited two monophyletic clades, one of which was restricted to either oceanic or cold climates, while the other occurred in the Mediterranean region (Crespo et al., 2002). Third, the species complex *Leptogium furfuraceum/L. pseudofurfuraceum* contained four monophyletic evolutionary lineages, which may represent cryptic species (Otálora et al., 2010). Europe, North and South America and East Africa each hosted a different evolutionary lineage. Interestingly, South American specimens were more closely related to individuals from East Africa, rather than to those from North America. Rare episodes of transoceanic dispersal, followed by genetic divergence in isolation, characterised the evolution of the species complex and the long-distance dispersal events predated the Pleistocene (Otálora et al., 2010).

Taking into account the previous results from the above fungal species which exhibited geographically restricted evolutionary lineages, it is to be expected that numerous other widespread lichen fungi may include different evolutionary lineages in contrasting parts of their ranges. The distinct geographic structure of the fungal side of the symbiosis may facilitate adaptation to local environmental conditions. Future studies should focus on investigating the adaptive component of genetic variability in populations of lichen fungi.

10.3.3 Conclusions regarding the 'Everything is everywhere' hypothesis

Is the 'Everything is everywhere' hypothesis true for lichen fungi, taking into account the results of distribution-based and molecular studies, or does the 'Vicariance' hypothesis receive support? It is important to emphasise that our current knowledge of the biogeography of lichen fungi is very limited, which makes it hard to draw general conclusions.

Nevertheless, both hypotheses have received some support. Among the few biogeographic studies of lichen fungi, the 'Everything is everywhere' hypothesis has been empirically supported by two molecular studies so far (section 10.3.1). The remaining studies supported the 'Vicariance' hypothesis, for instance by inferring substantial genetic divergence among populations from different geographic regions or continents, or by revealing the existence of cryptic species (section 10.3.2). Thus, the importance of long-distance dispersal vs. vicariance seems to depend largely on the species under consideration, and its species-specific biogeographic history. Given the large diversity of lichen fungi and the complexity of their biological interactions, ecological relationships and biogeographic histories, it does not come as a surprise that one hypothesis does not fit all.

10.4 Geographic patterns of photobionts

The biogeographic patterns of lichen photobionts at the species level have received little attention, which may partly reflect taxonomic uncertainties. Photobionts are traditionally identified based on their morphology, and for many genera, phylogenies that amalgamate morphology with molecular phylogenies are still lacking. Also, for several groups of lichen photobionts, species concepts still require clarification, and algal morphospecies may contain multiple phylogenetic species (Kroken and Taylor, 2000; Blaha et al., 2006; Skaloud and Peksa, 2010). Novel species and evolutionary lineages of photobionts are still being discovered routinely in molecular investigations (Helms et al., 2001; Blaha et al., 2006; Lücking et al., 2009; Skaloud and Peksa, 2010).

The probably most well-studied photobiont genus in terms of molecular phylogeny is *Trebouxia* (Friedl and Rokitta, 1997; Dahlkild et al., 2001; Helms et al., 2001). However, even for this genus, no detailed biogeographic studies have as of yet been published. Some species of this genus are widely distributed (Blaha et al., 2006). For instance, the morphospecies *T. impressa* is known from Antarctica and Europe and *T. jamesii* s.l. has additionally been reported from North America (Romeike et al., 2002; Blaha et al., 2006). Also some genotypes of *Trebouxia* spp. are widespread across Europe (Wornik and Grube, 2010).

Besides unclear species concepts, a second complication for elucidating biogeographic patterns of the photobionts is that photobiont species are frequently associated with a large set of fungal species, and across the distributional range of a particular lichen-forming fungal species, the photobiont partner may change (Kroken and Taylor, 2000). Thus, photobiont distribution patterns may not necessarily mirror the distribution of fungal species. Third, some photobiont species may occur in free-living populations or in aerial algal assemblages (Tschermak-Woess, 1978; Bubrick et al., 1984; Mukhtar et al., 1994; Rindi and Guiry, 2003; Sanders, 2005; Handa et al., 2007; Hedenås et al., 2007), and biogeographic studies of a particular photobiont species would need to sample from these populations as well. For these reasons, the biogeography of photobionts at the species level is rather tricky to study, and still poorly known. Nevertheless, some interesting biogeographic patterns have been revealed.

One pattern that has long been recognised, is that lichens with *Trebouxia* photobionts are more or less ubiquitous whereas lichens with *Trentepohlia* photobionts (e.g. Roccellaceae) are most common in the Mediterranean, subtropics and tropics (Rundel, 1978). The algal group *Trentepohliales* is most frequent and diverse in low latitudes (Rindi and Lopez-Bautista, 2008), and the increasing frequency of association with *Trentepohlia* spp. towards the equator may reflect the overall abundance of this algal group. In an ecophysiological study, Nash et al. (1987) determined the cold resistance of lichens associated with *Trentepohlia* vs. *Trebouxia*. They found that lichens that associated with *Trentepohlia* spp. had a much lower tolerance of cold temperatures, which suggested that sensitivity to freezing might be one of the reasons why *Trentepohlia*-associated lichens are generally found in warmer areas than those associated with *Trebouxia*. Moreover, the drought tolerance of *Trebouxia* spp. appears to be higher than that of *Trentepohlia* spp. (Scheidegger C., personal communication).

Photobiont species are often associated with particular environmental conditions. For instance, several species seem to be associated with specific habitat types (Yahr et al., 2006; Werth and Sork, 2010). Some species of *Trebouxia* appear to occur predominantly in cold climates, e.g. high altitudes or latitudes, whereas others are rare under such conditions (Blaha et al., 2006; Ohmura et al., 2006; Nelsen and Gargas, 2009), or occur in tropical climates (Cordeiro et al., 2005), pointing to different ecological amplitudes.

In some lichens, considerable photobiont variation has been observed within populations (Wornik and Grube, 2010). Also, the same fungal species was found to be associated with multiple photobiont species in the same site (Guzow-Krzeminska, 2006). In contrast, other lichen fungi are associated with different photobiont species in contrasting parts of their ranges (Blaha et al., 2006; Nelsen and Gargas, 2009). Moreover, the photobiont populations associated with several lichen fungi have been shown to exhibit substantial genetic structure related to geography, habitat or both – more structure than their mycobionts (Yahr et al., 2006; Altermann, 2009; Nelsen and Gargas, 2009; Werth and Sork, 2010). The association with a spatially structured, locally adapted photobiont pool enables lichen fungi to survive in different habitats, and to be capable of surviving under various selection pressures (Blaha et al., 2006; Nelsen and Gargas, 2009; Werth and Sork, 2010). These examples highlight that despite the wide distributions of many lichen fungi and presumably many photobiont species, the lichen symbiosis is not an association among random partners, but is fine-tuned in space, along environmental gradients.

Acknowledgements

I cordially thank Diego Fontaneto for inviting me to make this contribution, and Christian Printzen for suggesting me as an author. Ariel Bergamini, Christoph Scheidegger and Martin Westberg provided highly appreciated comments on a previous draft. This work received financial support from the Swiss National

Foundation (grants 3100AO-105830, 31003A_1276346/1, PBBEA-111207) and Bundesamt für Umwelt BAFU.

References

Ahlner, S. (1948). Utbredningstyper bland nordiska barrträdslavar. *Acta Phytogeographica Suecica* **22**, 1–257.

Ahmadjian, V. (1967). A guide to the algae occurring as lichen symbionts: isolation, culture, cultural physiology, and identification. *Phycologia* **6**, 127–160.

Altermann, S. (2009). *Geographic Structure in a Symbiotic Mutualism*. PhD thesis, University of California at Santa Cruz.

Aptroot, A., Bungartz, F. (2007). The lichen genus *Ramalina* on the Galapagos. *Lichenologist* **36**, 519–542.

Bannister, J.M., Blanchon, D.J. (2003). The lichen genus *Ramalina* Ach. (Ramalinaceae) on the outlying islands of the New Zealand geographic area. *Lichenologist* **35**, 137–146.

Beckett, R.P., Kranner, I., Minibayeva, F.V. (2008). Stress physiology and the symbiosis. In Nash, T.H. (ed.), *Lichen Biology*, pp. 134–151. Cambridge: Cambridge University Press.

Bjerke, J.W. (2003). The northern distribution range of *Lobaria hallii* in Europe and Greenland. *Graphis Scripta* **14**, 27–31.

Blaha, J., Baloch, E., Grube, M. (2006). High photobiont diversity associated with the euryoecious lichen-forming ascomycete *Lecanora rupicola* (Lecanoraceae, Ascomycota). *Biological Journal of the Linnean Society* **88**, 283–293.

Bowler, P.A., Rundel, P.W. (1975). Reproductive strategies in lichens. *Botanical Journal of the Linnean Society* **70**, 325–340.

Bubrick, P., Galun, M., Frensdorff, A. (1984). Observations on free-living *Trebouxia* Depuymaly and *Pseudotrebouxia* Archibald, and evidence that both symbionts from *Xanthoria parientina* (L.) Th. Fr. can be found free-living in nature. *New Phytologist* **97**, 455–462.

Büdel, B. (1992). Taxonomy of lichenized procaryotic blue-green algae. In Reisser, W. (ed.), *Algae and symbioses*, pp. 301–324. Bristol: Biopress.

Buschbom, J. (2007). Migration between continents: geographical structure and long-distance gene flow in *Porpidia flavicunda* (lichen-forming Ascomycota). *Molecular Ecology* **16**, 1835–1846.

Caceres, M.E.S., Lücking, R., Rambold, G. (2007). Phorophyte specificity and environmental parameters versus stochasticity as determinants for species composition of corticolous crustose lichen communities in the Atlantic rain forest of northeastern Brazil. *Mycological Progress* **6**, 117–136.

Cassie, D.M., Piercey-Normore, M.D. (2008). Dispersal in a sterile lichen-forming fungus, *Thamnolia subuliformis* (Ascomycotina: Icmadophilaceae). *Botany-Botanique* **86**, 751–762.

Codogno, M., Sancho, L.G. (1991). Distribution patterns of the lichen family Umbilicariaceae in the W Mediterranean Basin (Iberian

Peninsula, S France and Italy). *Botanika Chronika* **10**, 901–910.

Coppins, B.J. (1983). A taxonomic study of the lichen genus *Micarea* in Europe. *Bulletin of the British Museum (Natural History), Botany Series* **11**, 17–214.

Cordeiro, L.M.C., Reis, R.A., Cruz, L.M. et al. (2005). Molecular studies of photobionts of selected lichens from the coastal vegetation of Brazil. *FEMS Microbiology Ecology* **54**, 381–390.

Crespo, A., Molina, M.C., Blanco, O. et al. (2002). rDNA ITS and beta-tubulin gene sequence analyses reveal two monophyletic groups within the cosmopolitan lichen *Parmelia saxatilis*. *Mycological Research* **106**, 788–795.

Dahlkild, A., Kallersjo, M., Lohtander, K., Tehler, A. (2001). Photobiont diversity in the Physciaceae (Lecanorales). *The Bryologist* **104**, 527–536.

de Bary, H.A. (1879). *The Phenomenon of Symbiosis*. Privately printed in Strasburg.

de Vera, J.P., Rettberg, P., Ott, S. (2008). Life at the limits: Capacities of isolated and cultured lichen symbionts to resist extreme environmental stresses. *Origins of Life and Evolution of Biospheres* **38**, 457–468.

Excoffier, L., Foll, M., Petit, R.J. (2009). Genetic consequences of range expansions. *Annual Review of Ecology Evolution and Systematics* **40**, 481–501.

Feuerer, T., Hawksworth, D.L. (2007). Biodiversity of lichens, including a world-wide analysis of checklist data based on Takhtajan's floristic regions. *Biodiversity and Conservation* **16**, 85–98.

Friedl, T., Büdel, B. (2008). Photobionts. In Nash, T.H. (ed.), *Lichen Biology*, pp. 9–26. Cambridge: Cambridge University Press.

Friedl, T., Rokitta, C. (1997). Species relationships in the lichen alga *Trebouxia* (Chlorophyta, Trebouxiophyceae): Molecular phylogenetic analyses of nuclear-encoded large subunit rRNA gene sequences. *Symbiosis* **23**, 125–148.

Galloway, D.J., Aptroot, A. (1995). Bipolar lichens: A review. *Cryptogamic Botany* **5**, 184–191.

Geitler, L. (1934). Beiträge zur Kenntnis der Flechtensymbiose. *Journal für Protistenkunde* **82**, 51–85.

Goward, T., Spribille, T. (2005). Lichenological evidence for the recognition of inland rain forests in western North America. *Journal of Biogeography* **32**, 1209–1219.

Guzow-Krzeminska, B. (2006). Photobiont flexibility in the lichen *Protoparmeliopsis muralis* as revealed by ITS rDNA analyses. *Lichenologist* **38**, 469–476.

Handa, S., Ohmura, Y., Nakano, T., Nakahara-Tsubota, M. (2007). Airborne green microalgae (Chlorophyta) in snowfall. *Hikobia* **15**, 109–120.

Hedenås, H., Blomberg, P., Ericson, L. (2007). Significance of old aspen (*Populus tremula*) trees for the occurrence of lichen photobionts. *Biological Conservation* **135**, 380–387.

Helms, G., Friedl, T., Rambold, G., Mayrhofer, H. (2001). Identification of photobionts from the lichen family Physciaceae using algal-specific ITS rDNA sequencing. *Lichenologist* **33**, 73–86.

Hewitt, G.M. (1999). Post-glacial re-colonization of European biota. *Biological Journal of the Linnean Society* **68**, 87–112.

Honegger, R. (2000). Simon Schwendener (1829–1919) and the dual hypothesis of lichens. *Bryologist* **103**, 307–313.

Honegger, R. (2008). Mycobionts. In Nash, T.H. (ed.), *Lichen Biology*, pp. 27–39. Cambridge: Cambridge University Press.

Honegger, R., Zippler, U., Gansner, H., Scherrer, S. (2004). Mating systems in the genus *Xanthoria* (lichen-forming ascomycetes). *Mycological Research* **108**, 480–488.

Kantvilas, G., James, P.W., Jarman, S.J. (1985). Macrolichens in Tasmanian rainforests. *Lichenologist* **17**, 67–84.

Kappen, L. (1973). Response to extreme environments. In Ahmadjian, V., Hale, M.E. (eds.), *The Lichens*, pp. 311–380. New York, NY: Academic Press.

Kärnefelt, I. (1986). The genera *Bryocaulon*, *Coelocaulon* and *Cornicularia* and formerly associated taxa. *Opera Botanica* **86**, 1–90.

Kirk, P.M., Cannon, P.F., Minter, D.W., Stalpers, J.A. (2008). *Dictionary of the Fungi*. Wallingford: CAB International.

Knudsen, K., Werth, S. (2008). Lichens of the Granite Mountains, Sweeney Granite Mountain desert research center, southwestern Mojave Desert, San Bernardino county, California. *Evansia* **25**, 15–20.

Kristinsson, H. (1972). Studies on lichen colonization in Surtsey 1970. *Surtsey Research Programme Report* **5**, 77.

Kristinsson, H., Heiðmarsson, S. (2009). Colonization of lichens on Surtsey 1970–2006. *Surtsey Research Programme Report* **12**, 81–104.

Krog, H., Østhagen, H. (1978). Three new *Ramalina* species from Macaronesia. *Norwegian Journal of Botany* **25**, 55–59.

Krog, H., Østhagen, H. (1980). The genus *Ramalina* in the Canary Islands. *Norwegian Journal of Botany* **27**, 255–296.

Kroken, S., Taylor, J.W. (2000). Phylogenetic species, reproductive mode, and specificity of the green alga *Trebouxia* forming lichens with the fungal genus *Letharia*. *The Bryologist* **103**, 645–660.

Kroken, S., Taylor, J.W. (2001). A gene genealogical approach to recognize phylogenetic species boundaries in the lichenized fungus *Letharia*. *Mycologia* **93**, 38–53.

Litterski, B. (1999). Arealkundliche Studien - ein Beitrag zur Bewertung der Flechtenvielfalt. *Courier Forschungsinstitut Senckenberg* **215**, 137–142.

Lücking, R. (2003). Takhtajan's floristic regions and foliicolous lichen biogeography: a compatibility analysis. *Lichenologist* **35**, 33–54.

Lücking, R. (2008). *Foliicolous Lichenized Fungi*. Organization for Flora Neotropica and The New York Botanical Garden Press, Bronx, New York.

Lücking, R., Papong, K., Thammathaworn, A., Boonpragob, K. (2008). Historical biogeography and phenotype-phylogeny of *Chroodiscus* (lichenized Ascomycota: Ostropales: Graphidaceae). *Journal of Biogeography* **35**, 2311–2327.

Lücking, R., Lawrey, J.D., Sikaroodi, M. et al. (2009). Do lichens domesticate photobionts like farmers domesticate crops? Evidence from a previously unrecognized lineage of filamentous cyanobacteria. *American Journal of Botany* **96**, 1409–1418.

Martinez, I., Burgaz, A.R., Vitikainen, O., Escudero, A. (2003). Distribution patterns in the genus *Peltigera* Willd. *Lichenologist* **35**, 301–323.

Meier, F.A., Scherrer, S., Honegger, R. (2002). Faecal pellets of lichenivorous mites contain viable cells of the lichen-forming ascomycete *Xanthoria parietina* and its green algal photobiont, *Trebouxia arboricola*. *Biological Journal of the Linnean Society* **76**, 259–268.

Moberg, R., Nash, T.H. (1999). The genus *Heterodermia* in the Sonoran desert area. *Bryologist* **102**, 1–14.

Mukhtar, A., Garty, J., Galun, M. (1994). Does the lichen alga *Trebouxia* occur free-living in nature – further immunological evidence. *Symbiosis* **17**, 247–253.

Muñoz, J., Felicisimo, A.M., Cabezas, F., Burgaz, A.R., Martinez, I. (2004). Wind as a long-distance dispersal vehicle in the southern hemisphere. *Science* **304**, 1144–1147.

Myllys, L., Stenroos, S., Thell, A., Ahti, T. (2003). Phylogeny of bipolar *Cladonia arbuscula* and *Cladonia mitis* (Lecanorales, Euascomycetes). *Molecular Phylogenetics and Evolution* **27**, 58–69.

Nash, T.H. (1996). *Lichen Biology*. Cambridge: Cambridge University Press.

Nash, T.H., Kappen, L., Lösch, R., Larson, D.W., Matthes-Sears, U. (1987). Cold resistance of lichens with *Trentepohlia* or *Trebouxia* photobionts from the North American West coast. *FLORA* **179**, 241–251.

Nash, T.H., Ryan, B.D., Gries, C., Bungartz, F. (2002). *Lichen Flora of the Greater Sonoran Desert region*. Tempe, AZ: Arizona State University.

Nelsen, M.P., Gargas, A. (2009). Symbiont flexibility in *Thamnolia vermicularis* (Pertusariales: Icmadophilaceae). *Bryologist* **112**, 404–417.

Nyati, S., Beck, A., Honegger, R. (2007). Fine structure and phylogeny of green algal photobionts in the microfilamentous genus *Psoroglaena* (Verrucariaceae, lichen-forming ascomycetes). *Plant Biology* **9**, 390–399.

Ohmura, Y., Kawachi, M., Kasai, F., Watanabe, M.M., Takeshita, S. (2006). Genetic combinations of symbionts in a vegetatively reproducing lichen, *Parmotrema tinctorum*, based on ITS rDNA sequences. *Bryologist* **109**, 43–59.

Otálora, M.A.G., Martínez, I., Aragón, G., Molina, M.C. (2010). Phylogeography and divergence date estimates of a lichen species complex with a disjunct distribution pattern. *American Journal of Botany* **97**, 216–223.

Otte, V., Esslinger, T.L., Litterski, B. (2002). Biogeographical research on European species of the lichen genus *Physconia*. *Journal of Biogeography* **29**, 1125–1141.

Otte, V., Esslinger, T.L., Litterski, B. (2005). Global distribution of the European species of the lichen genus *Melanelia* Essl. *Journal of Biogeography* **32**, 1221–1241.

Øvstedal, D.O., Lewis Smith, R.I. (2001). *Lichens of Antarctica and South Georgia. A Guide to their Identification and Ecology*. Cambridge: Cambridge University Press.

Palice, Z., Printzen, C. (2004). Genetic variability in tropical and temperate populations of *Trapeliopsis glaucolepidea*: Evidence against long-range dispersal in a lichen with disjunct distribution. *Mycotaxon* **90**, 43–54.

Poelt, J. (1969). *Bestimmungsschlüssel europäischer Flechten*. Cramer, Germany.

Poelt, J. (1970). Das Konzept der Artenpaare bei den Flechten. *Cladistics* **125**, 77–81.

Printzen, C. (2010). Lichen systematics: the role of morphological and molecular data to reconstruct phylogenetic relationships. *Progress in Botany* **71**, 233–275.

Printzen, C., Ekman, S., Tønsberg, T. (2003). Phylogeography of *Cavernularia hultenii*: evidence of slow genetic drift in a widely disjunct lichen. *Molecular Ecology* **12**, 1473–1486.

Rindi, F., Guiry, M.D. (2003). Composition and distribution of subaerial algal assemblages in Galway City, western Ireland. *Cryptogamie Algologie* **24**, 245–267.

Rindi, F., Lopez-Bautista, J.M. (2008). Diversity and ecology of Trentepohliales (Ulvophyceae, Chlorophyta) in French Guiana. *Cryptogamie Algologie* **29**, 13–43.

Romeike, J., Friedl, T., Helms, G., Ott, S. (2002). Genetic diversity of algal and fungal partners in four species of *Umbilicaria* (lichenized ascomycetes) along a transect of the Antarctic peninsula. *Molecular Biology and Evolution* **19**, 1209–1217.

Rundel, P.W. (1978). Ecological relationships of desert fog zone lichens. *The Bryologist* **81**, 277–293.

Sancho, L.G., de la Torre, R., Horneck, G. et al. (2007). Lichens survive in space: results from the 2005 LICHENS experiment. *Astrobiology* **7**, 443–454.

Sanders, W.B. (2005). Observing microscopic phases of lichen life cycles on transparent substrata placed *in situ*. *Lichenologist* **37**, 373–382.

Scheidegger, C. (1985). Systematische Studien zur Krustenflechte *Anzina carneonivea* (Trapeliaceae, Lecanorales). *Nova Hedwigia* **41**, 191–218.

Scheidegger, C. (1995). Reproductive strategies in *Vezdaea* (Lecanorales, lichenized Ascomycetes): a low-temperature scanning electron microscopy study of a ruderal species. *Cryptogamic Botany* **5**, 163–171.

Schroeter, B., Sancho, L.G. (1996). Lichens growing on glass in Antarctica. *Lichenologist* **28**, 385–390.

Schwendener, S. (1868). Ueber die Beziehungen zwischen Algen und Flechtengonidien. *Botanische Zeitung* **26**, 289–292.

Seymour, F.A., Crittenden, P.D., Dyer, P.S. (2005). Sex in the extremes: lichen-forming fungi. *Mycologist* **19**, 51–58.

Skaloud, P., Peksa, O. (2010). Evolutionary inferences based on ITS rDNA and actin sequences reveal extensive diversity of the common lichen alga *Asterochloris* (Trebouxiophyceae, Chlorophyta). *Molecular Phylogenetics and Evolution* **54**, 36–46.

Smith, C.W., Aptroot, A., Coppins, B. et al. (2009). *The Lichens of Great Britain and Ireland*. London: British Lichen Society.

Tehler, A., Wedin, M. (2008). Systematics of lichenized fungi. In Nash, T.H. (ed.), *Lichen Biology*, pp. 336–352. Cambridge: Cambridge University Press.

Tschermak-Woess, E. (1978). *Myrmecia reticulata* as a phycobiont and free-living – free-living Treouxia – problem of *Stenocybe septata*. *Lichenologist* **10**, 69–79.

Tschermak-Woess, E. (1988). The algal partner. In Galun, M. (ed.), *CRC Handbook of Lichenology*, pp. 39–92. Boca Raton, FL: CRC Press.

Walser, J.C., Holderegger, R., Gugerli, F., Hoebee, S.E., Scheidegger, C. (2005). Microsatellites reveal regional population differentiation and isolation in *Lobaria pulmonaria*, an epiphytic lichen. *Molecular Ecology* **14**, 457–467.

Werth, S., Sork, V.L. (2010). Identity and genetic structure of the photobiont of the epiphytic lichen *Ramalina menziesii* on three oak species in southern California. *American Journal of Botany* **97**, 821–830.

Werth, S., Tømmervik, H., Elvebakk, A. (2005). Epiphytic macrolichen communities along regional gradients in northern Norway. *Journal of Vegetation Science* **16**, 199–208.

Werth, S., Wagner, H.H., Gugerli, F. et al. (2006a). Quantifying dispersal and establishment limitation in a population of an epiphytic lichen. *Ecology* **87**, 2037–2046.

Werth, S., Wagner, H.H., Holderegger, R., Kalwij, J.M., Scheidegger, C. (2006b). Effect of disturbances on the genetic diversity of an old-forest associated lichen. *Molecular Ecology* **15**, 911–921.

Werth, S., Gugerli, F., Holderegger, R. et al. (2007). Landscape-level gene flow in *Lobaria pulmonaria*, an epiphytic lichen. *Molecular Ecology* **16**, 2807–2815.

Westberg, M. (2007). *Candelariella* (Candelariaceae) in western United States and northern Mexico: the species with biatorine apothecia. *Bryologist* **110**, 365–374.

Widmer, I. (2009). *Evolutionary History and Phylogeography of a Lichen Symbiosis*. Ph.D. thesis, University of Berne, Berne.

Widmer, I., Dal Grande, F., Cornejo, C., Scheidegger, C. (2010). Highly variable microsatellite markers for the fungal and algal symbionts of the lichen *Lobaria pulmonaria* and challenges in developing biont-specific molecular markers for fungal associations. *Fungal Biology* **114**, 538–544.

Winchester, V., Harrison, S. (2000). Dendrochronology and lichenometry: colonization, growth rates and dating of geomorphological events on the east side of the North Patagonian Icefield, Chile. *Geomorphology* **34**, 181–194.

Wornik, S., Grube, M. (2010). Joint dispersal does not imply maintenance of partnerships in lichen symbioses. *Microbial Ecology* **59**, 150–157.

Yahr, R., Vilgalys, R., DePriest, P.T. (2006). Geographic variation in algal partners of *Cladonia subtenuis* (Cladoniaceae) highlights the dynamic nature of a lichen symbiosis. *New Phytologist* **171**, 847–860.

Yoshimura, I. (1971). The genus *Lobaria* of Eastern Asia. *Journal of the Hattori Botanical Laboratory* **34**, 231–364.

11

Biogeography of mosses and allies: does size matter?

NAGORE G. MEDINA, ISABEL DRAPER AND FRANCISCO LARA

Departamento de Biología (Botánica), Facultad de Ciencias, Universidad Autónoma de Madrid, Madrid, Spain

11.1 Introduction

Bryophytes are the second largest group of embryophytes, or green land plants, after the very diverse Angiosperms. They comprise three main lineages (Frey, 2009; Goffinet and Shaw, 2009): mosses (Division or Phylum Bryophyta), that are currently estimated to include 12 500–13 000 species; liverworts (Marchantiophyta), that are thought to number 5000 or a few more; and hornworts (Anthocerotophyta), with only 100–150 species. This adds up to around 18 000 species, although estimates range from 14 000 to 25 000.

Bryophytes in general, and especially mosses and liverworts, are highly successful plants. They display a high level of diversity, are almost universally present in land environments, and play a significant role in many terrestrial and freshwater ecosystems (Vanderpoorten and Goffinet, 2009). Although often inconspicuous, mosses and liverworts can be found even in the world's toughest environments, such as freezing and hot deserts. Moreover, in some harsh environments, such as

Biogeography of Microscopic Organisms: Is Everything Small Everywhere?, ed. Diego Fontaneto. Published by Cambridge University Press.

the epiphytic stratum of temperate woodlands or the terrestrial ecosystems of the tundra, they are the chief group of organisms, together with lichens. Peatlands, which cover *c.* 3% of the Earth's land surface (Limpens et al., 2008), are a particular and outstanding case of habitat where bryophytes generally prevail, commonly with species of *Sphagnum* as the dominant vegetation element. Even if bryophytes are particularly diverse and luxuriant in tropical montane cloud forests and in humid temperate woodlands, they can be found more or less abundantly in all environments where land plants can survive (Gignac, 2001). Furthermore, in global terms, bryophytes contribute to a significant proportion of the production and biomass in a variety of ecosystems (Vanderpoorten and Goffinet, 2009).

Mosses and their allies are the smallest green land plants. They can be minute: some leafy liverworts are almost invisible to the naked eye and the tiniest mosses are less than 2 mm tall. At the opposite end, some hepatics reach 30 cm and several mosses can exceed 70 cm in length. The majority of bryophytes, however, measure between 0.5 and 10 cm. They are reputed to be, both structurally and physiologically, the 'simplest' land plants. Actually, bryophytes exhibit a wide range of structural complexity, although they cannot develop complex supporting or conducting tissues because, unlike vascular plants, bryophytes always lack lignin. The bryophyte life cycle is distinguished from that of all other embryophytes in the predominance of the haploid generation, the gametophyte being the photosynthetic phase. Compared with tracheophytes (vascular plants, including ferns and allies, conifers and angiosperms), the gametophyte of bryophytes is therefore very complex. On the other hand, the diploid phase of bryophytes consists of a single sporangium on an unbranched leafless stalk, attached to the gametophyte and nutritionally depending on it. The sporophyte generation of bryophytes is thus the least complicated among the land plants.

Bryophytes lack mechanisms and structural systems that allow an effective control of water relations. They are therefore poikilohydrous, a trait shared with algae and lichens, but uncommon among pteridophytes and angiosperms (Pugnaire and Valladares, 2007). This characteristic could be interpreted in terms of physiological simplicity, but is actually an alternative and successful life strategy since bryophytes combine poikilohydry with another essential feature: the capacity of maintaining latent life (quiescence) after desiccation, with a high faculty of reviviscence without damage after rehydration (Oliver et al., 2005). In fact, their ability to survive cold and dry conditions is unparalleled in other principal plant groups (Glime, 2007). In the absence of major barriers for gas and water exchange, the hydration state in most bryophytes is dependent on ambient humidity. However, many species have evolved morphological structures or architectural characteristics that modify water uptake and storage rates, and that limit water loss from shoot surfaces. As a consequence, species of bryophytes differ greatly in their evaporative exchange properties (Rice et al., 2001). Since desiccation tolerance and other ecophysiological traits also vary among bryophytes (Proctor, 2009), different species exhibit a broad range of physiological optima and ecological amplitudes.

Bryophytes, like pteridophytes but unlike flowering plants, propagate by means of sexual spores (meiospores). These are produced in the capsule (sporangium) of the sporophyte and germinate into a protonema that subsequently produces the gametophores or green plants. Gametophytes can be monoecious and then the processes of fertilisation and consequent generation of new sporophytes have no special difficulties. However, among mosses dioecious species are just as common as monoecious ones, and more so among liverworts; in some cases, sporophytes have never been found in certain populations or even species. Bryophyte spores are quite diverse in size, although in general they are small, usually between 10–20 μm in diameter (Frahm, 2008). Hence, they are potentially adequate for wind dispersal, which is indeed the most common dispersal mechanism, although animal and water dispersal also occur (e.g. Porley and Hodgetts, 2005; Marino et al., 2009). There is a tremendous variation in spore production among bryophytes, the number of spores per capsule being in the interval of 10^4 to 10^6 for most species (Vanderpoorten and Goffinet, 2009). In addition, mosses and allies propagate through vegetative structures. Asexual propagules can be undifferentiated (fragments of gametophyte structures) or more or less specialised reproductive bodies (brood bodies or gemmae), and have various sizes and shapes. It is suspected that vegetative propagation plays an essential role in the dispersal of bryophytes and it has a chief importance for colony expansion when the plants have already initiated its establishment (Glime, 2007).

Many of the characteristics of bryophytes enumerated above (size, way of life, dispersal mechanisms, etc.) suggest that these organisms are excellent candidates for corroborating Baas Becking's hypothesis of 'Everything is everywhere' (Baas Becking, 1934). In fact, traditional thoughts revolve around the assumptions that (1) bryophytes display no major dispersal restrictions because of their minute wind-transported diaspores (spores or vegetative structures for dispersal) and (2) since they are small their distributions are largely depending on the microenvironment rather than on macroclimatic characteristics (cf. Schuster, 1983). In the following, we intend to show to what extent this idea is valid nowadays. We use both a classical approach, based on morphological concepts of taxa, and a phylogeographic approach, based on molecular data.

11.2 The classical approach

In classical biogeography the analysis of distribution ranges has been a keystone in the attempts to disentangle the relative importance of the major factors affecting bryophyte distributions.

11.2.1 Wide distribution ranges

Bryophytes tend to show wider distribution ranges than flowering plants. Many species have geographic ranges that include more than one continent, and cosmopolitan distributions are said to be relatively frequent (Shaw and Goffinet, 2000; Frahm, 2008; Vanderpoorten and Goffinet, 2009). At the family and generic level, worldwide distributions are more or less the rule. Thus, more than 75% of the families of bryophytes are widespread in both hemispheres (Tan and Pócs, 2000). However, at the species level, extremely wide distributions are not as common as could be expected. There are no accurate data on how many cosmopolitan bryophyte species there are in total, but based on regional Holarctic bryofloras we can get some insights. Among European liverworts and hornworts the level of cosmopolitan species ranges from 3.1% (Frey et al., 2006) to 5.1% (Dierssen, 2001), for a flora of 453 species (Grolle and Long, 2000). For mosses, following the broad criterion of Dierssen (2001), 93 European species are considered to be cosmopolitan, which represents 7.2% of the 1292 species known from this continent (Hill et al., 2006). Finally, in eastern North America, Crum and Anderson (1981) considered only five mosses out of 765 to be cosmopolitan. Although the number of cosmopolitan species depends on the definition of the concept (the data on European mosses include several species regarded as cosmopolitan that are not considered so in the flora of eastern North America), the rates obtained are not very high in any case. Considering that the provided data were gathered from two well-known continents, it can be assumed that most of the world's cosmopolitan bryophytes are included. If this is true, the global percentage of cosmopolitan bryophytes might be far below 1%.

The idea that very wide distributions are rare is supported by the fact that most bryophyte species are limited to certain regions, even if their ranges are usually larger than the ones found among other terrestrial plants. The question is whether these distributions are restricted because of environmental factors or if there are other meaningful biogeographic constraints. To illustrate what we mean, let us look at the European species with affinities for a Mediterranean climate. If we assume that bryophytes are wide-ranging plants in which transcontinental dispersal is common, it should be expected that species with marked Mediterranean affinity in Europe occur in other parts of the world with a similar climate type. In an unpublished study of the European species with a Mediterranean affinity we found that 62 out of 117 (61.5%) species are restricted to the Palaearctic. Of the remaining species, 23 (19.7%) are also present in California, 11 (9.4%) are recorded in South Africa, seven (6.0%) occur in Chile, four (3.4%) are present in Australia, and only one species is present in all Mediterranean climate zones. Interestingly, many of the species from the Mediterranean basin have not established successful populations in other continents, even if they produce high quantities of small spores (down to less than 18 μm), as is the case for *Orthotrichum philibertii* or *Anomobryum lusitanicum*. Despite their apparently having the means for spreading over long distances, a high percentage of the bryophytes show distributions with strong geographic constraints.

Wide distribution patterns can arise from different mechanisms, such as stepping-stone or long-distance dispersal. Stepping-stone dispersal is a result of numerous effective short-distance dispersal events and therefore requires connections (present or past) between landmasses for range expansion and possibly implies long periods of time to attain wide ranges. Consequently, some of the ranges discussed above, including the Mediterranean–South American disjunctions are difficult to explain solely by the stepping-stone mechanism. If long-distance dispersal is responsible for wide species ranges, dispersal could have occurred in recent times, after the separation of landmasses. This requires that species produce small spores and have access to adequate dispersal agents (e.g. air currents). Whereas the transport of spores to the new locality is necessary this is not sufficient for colonisation: spore survival, establishment and persistence are also essential. Thus, for the Mediterranean species it appears that several biological or ecological features can prevent species from effective dispersal (including both transport and establishment) across long distances.

11.2.2 Endemic ranges

As we have seen, many bryophytes have geographically limited but relatively large distribution areas. There are also endemic bryophytes with very restricted distributions (narrow endemisms); among many others, good examples are *Vandiemenia ratkowskiana*, a liverwort endemic to Tasmania and only known from two localities that are separated by *c.* 70 km (Furuki and Dalton, 2008), or *Renauldia lycopodioides*, a moss known from a few localities in Tanzania and Kenya (O'Shea, 2006). Nonetheless, bryophytes show low levels of endemism in most regions. Low incidence of endemism is usually understood as an indicator of the relevance of long-distance dispersal. On the other hand, a high percentage of endemism is interpreted as indicative of a high degree of isolation of a given flora, both in time and space, and can be related to the prevalence of short-distance dispersal in most of the included species. Due to the high dispersal capacity of many species, a large number of bryophytes could potentially maintain the genetic connectivity between populations, even if they grow in localities that are several thousand kilometres apart. Under such a scenario, allopatric speciation should be relatively rare among bryophytes. In agreement with this idea, bryophytes show significantly lower rates of endemism than flowering plants. Some islands in the Mediterranean Sea that are known to have nearly 10% endemic vascular plants, such as Corsica and Sardinia, have just one endemic moss (Sotiaux et al., 2009). Other good examples are the Canary Islands, where

Table 11.1. Percentage of endemic bryophytes in selected areas of high endemicity level.

Territories	Moss	Liverwort	Source of the data
Hawai	29	48	Staples et al., 2004; Staples and Imada, 2006
New Caledonia	50	48	Morat, 1993
New Zealand	23	50	Fife, 1995; Engel and Glenny, 2008
Madagascar	39	?	O'Shea, 2006
Andean mountain range	31	?	Churchill, 2009

more than 21% of the vascular plants are endemic (Machado, 2002) compared with only 1.4% of the mosses (González-Mancebo et al., 2008). In the Iberian Peninsula the rate of vascular plants endemism exceeds 25% (Sáinz Ollero and Moreno Saiz, 2002), but less than 1% of the bryophyte species are endemic. In the Galapagos Islands only 15 liverworts and six mosses are 'proven' endemics (Porley and Hodgetts, 2005). Even if the rate of endemism among bryophytes is lower than that found in larger organisms, there are however many cases around the world where the rates of endemism indicate isolation of the floras. Typical examples are those areas that exhibit the world's highest endemism rates for vascular plants (exceeding 70%) and that also present corresponding rates of liverwort and moss endemism, ranging from 23% to 50% (Table 11.1). From such data we can infer that biogeographic barriers can have different significance for bryophytes and flowering plants. Short sea distances, such as the ones that separate the Mediterranean islands from the continent, or mountain ranges comparable to the Pyrenees, appear not to represent true barriers for bryophyte dispersal, whereas they are obviously limiting the distributions of a number of flowering plants. However, this does not mean that there are no obstacles at all to bryophyte dissemination and there are many examples that indicate the existence of barriers (not always physical). Thus, the ecologically isolated Andean region has a significant level of bryophyte endemism, while New Zealand, New Caledonia and Hawaii are equally unique in this respect as a consequence of their strong geographic isolation.

Several species are potentially capable of maintaining transcontinental connectivity between populations through long-distance dispersal, which could prevent population differentiation even in remote islands, but it is clear that there are a number of cases in which isolation and allopatric speciation occur. How important then is long-distance dispersal in relation to endemism? Surprisingly, only one of the eight endemic mosses of the British Isles produces

Fig 11.1 The epiphytic moss *Orthotrichum handiense* growing on the branches of *Asteriscus sericeus.*

sporophytes (Smith, 2004; Porley and Hodgetts, 2005), which indicates the relevance of the lack of sexual reproduction and spore production for maintaining isolation. Although seemingly remarkable, the case above is relatively untypical and many endemic species do indeed produce sporophytes and high quantities of small spores and could potentially connect even distant populations. For example, *Orthotrichum handiense* is a moss restricted to a small area on Fuerteventura (Canary Islands) and represents an outstanding example of local endemism (Figs 11.1, 11.2) because it has a population that is relatively rich in individuals and a high production of sporophytes (Lara et al., 2003). This shows that when evaluating the relevance of dispersal capacity and connectivity among populations, not only lack of spore production, but also many other factors can hamper the effectiveness of dispersal, such as accessibility to transport means, survival and establishment success. In a set of experiments on spore viability in New Zealand mosses, spores of endemic species were found to have low survival rates after desiccation, freezing and UV exposure (van

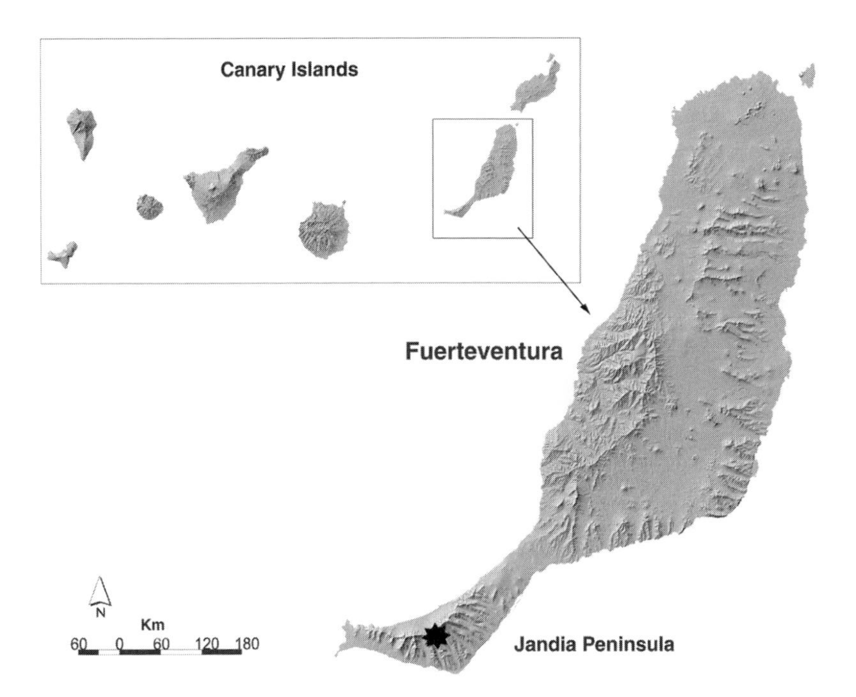

Fig 11.2 Location (star) of the known population of *O. handiense*.

Zanten and Pócs, 1981; van Zanten and Gradstein, 1988), pointing out the paramount importance of spore endurance to survive the harsh conditions during long-distance dispersal. Equally critical but less explored is the establishment phase. It is thought that establishment of new shoots from spores germinating under natural conditions depends on a complex set of factors and may occur only rarely (e.g. Boatman and Lark, 1971; Miles and Longton, 1990; but see discussion in Sundberg and Rydin, 2002).

Another unknown factor is how easily diaspores can access the agents of dispersal, which might be especially relevant for the isolation in small organisms that are found in specific microenvironments. Some bryophyte species occupy sheltered microhabitats where wind transport is very unlikely. A good example of this is the so-called 'rockhouses', in the Southern Appalachian Mountains (USA). These are deep narrow gorges that harbour a remarkably isolated bryoflora with several endemic mosses and a relatively high number of species that have their closest relatives in the tropics (Billings and Anderson, 1966; Farrar, 1998). The greatest concentration of endemism occurs near waterfalls and in sheltered microenvironments where conditions are stable and spore dissemination by wind may be prevented by the surface tension of water in a permanently wet environment. The result is a humid and mild refugium for certain bryophytes, isolated from major air currents.

11.2.3 Disjunct ranges

Up to here, we have looked at insights provided by analyses of continuous ranges. However, perhaps more than wide ranges and rates of endemism, the paradigmatic examples that summarise the controversy on the relative importance of the main processes determining species geographic ranges are the disjunctions. Discontinuous distribution ranges can be interpreted either as the result of fragmentation of ancient continuous distributions or as consequences of long-range dispersals. Supporting the hypothesis of relict ranges is the fact that most bryophyte disjunct distributions are highly congruent with the continental drift hypothesis (Schofield and Crum, 1972; Schuster, 1983). Indeed, the bryophyte disjunctions parallel those found in spermatophytes, which suggests that the historic events that shaped species distribution for flowering plants are also relevant for bryophytes. Schofield (1988) analysed the disjunctions between Europe and North America and concluded that the western North American–western European bryophyte disjunctions have a relict origin. He based his conclusion on inferences made from the biological and ecological characteristics of the disjunct species. This kind of analysis gives indirect but consistent results regarding the importance of past geologic and climatic events in causing the observed disjunct patterns.

If we accept the relict origin of the disjunctions we have to assume that some species have remained unchanged (at least morphologically) for very long periods of time. At the population level there is evidence for the ability of bryophytes for long-term survival. For example, the perennial species *Anastrophyllum saxicola* is known for producing extensive populations of clones that likely survived a minimum of 2000 years (Longton and Schuster, 1983). This could indicate that some species are capable of persisting, even under suboptimal conditions, as living fossils (Hallingbäck, 2002). It is unknown how common these rates of survival are and, in any case, to assume species stability at a geologic time scale is a very different matter. Although records of old fossils are scarce within bryophytes, there are a few remarkable examples of morphological stability, and at least some species have remained unchanged for more than 45 million years (cf. Taylor et al., 2009). However, to provide evidence for the African–Tropical American or Pangaean distributions, species need to remain without apparent changes for more than 100 and 180 millions of years respectively.

Certain intriguing disjunct distributions that are difficult to interpret under a historic and geologic perspective provide support for the long-distance dispersal hypothesis. Perhaps the best examples are the bipolar disjunctions, distributions that include boreal or temperate regions in both hemispheres but with an absence from the landmasses in between. There are 18 bryophytes species in Antarctica that are known to have this type of distribution (Ochyra et al., 2008). In most cases, species with bipolar disjunctions have a predominantly Holarctic distribution and a few isolated populations in the southern hemisphere and therefore they probably originated by long-distance dispersal from source populations in the northern hemisphere (Ochyra and Buck, 2003).

In spite of their support for the long-distance dispersal hypothesis, erratic and unconnected distributions such as bipolar disjunctions are infrequent among bryophytes. Most species show geographic ranges that are congruent with general floristic patterns and therefore different biogeographic units (kingdoms, subkingdoms, regions etc.) can be recognised (e.g. Schofield, 1992). The general patterns of vascular plants and bryophytes are analogous, suggesting that the historical and biological mechanisms shaping the species distributions are similar in both groups (Schofield, 1992). Because of the low likelihood of a repetition of a unidirectional stochastic event, it has been argued that the concordant bryofloras are a proof of the lack of importance of neutral long-dispersal events in shaping species distributions (Schuster, 1983). However, it has been increasingly understood that wind dispersal is not just a stochastic neutral process. On the contrary, intercontinental dispersal events by wind are considered part of a directional, congruent and consistent process which could give rise to patterns of concordant floras (McDowall, 2004; Muñoz et al., 2004; Cook and Crisp, 2005), especially in groups of high dispersal potential such as bryophytes. Despite the latter, the similarity across floristic realms presents a picture highly concordant with historic connectivity and appears difficult to explain solely by wind connectivity. For example, Pócs (1998) analysed the phytogeographic affinities of the bryophyte flora of the Arc Mountains in eastern Africa showing how the old crystalline mountains of this cordillera host a significantly higher number of Lemurian (Madagascan) elements, than the neighbouring younger mountains do. Furthermore, even if nowadays the Eastern Arch is not the closest area to Madagascar, it contains the highest number of Lemurian bryophytes in continental Africa. This strongly reflects ancient links between Eastern Africa and Madagascar, two dissected parts of Gondwanaland.

11.2.4 Diversity patterns

As we have seen, bryophytes show a wide variety of distribution patterns, from very extensive to very narrow ranges, including both continuous and disjunct distributions. Bryophyte distributions frequently parallel those of flowering plants, which suggests that both groups were influenced by the same factors. However, a detailed analysis of their distribution patterns shows important discrepancies which could have resulted from the fact that bryophytes are smaller than vascular plants, that they have a more ancient origin, and/or that they have different modes of dispersal. In some respects these differences make bryophytes more similar to other small-sized organisms with passive wind dispersal. We will now focus on bryophyte diversity patterns to explore whether these are shared with vascular plants or with microorganisms.

The latitudinal gradient of species richness has traditionally been thought to be one of the few general rules in biogeography (Hawkins et al., 2003). The decrease in species richness towards the poles is consistent across a wide variety of organisms with high levels of organisation. This latitudinal gradient is lacking in some micro-organisms, which has raised doubts about the generality of the pattern. For bryophytes it has repeatedly been stated that the tropics harbour the richest bryofloras of the world (e.g. Argent, 1979; Frahm et al., 2003). However, recently published global maps of moss (Mutke and Barthlott, 2005) and liverwort (von Konrat et al., 2008) species richness do not show unambiguous evidence of such a latitudinal gradient. Tropical regions in South America are consistently richer in moss species than regions at higher latitudes. This pattern is not so clear in liverworts, and some countries of tropical Africa show remarkably low species richness for both mosses and liverworts. Furthermore, in a study of moss species richness in the tropical Andes, Churchill (2009) suggested that the global pattern of richness in bryophytes is probably the opposite of that observed in other groups. In his study he argues that the richest areas of the world are the temperate and boreal forests of the northern hemisphere. Although the compilation effort made is huge in both the studies of Mutke and Barthlott (2005) and von Konrat et al. (2008), it is important to note that the information on bryophytes species richness is far from complete, and some of the observed patterns may be flawed by knowledge gaps and uneven sampling efforts, as already noted by the authors. A good example of this is that many tropical African countries, such as Guinea Bissau and The Gambia, still have just one species recorded (O'Shea, 2006). In addition, both Mutke and Barthlott (2005) and von Konrat et al. (2008) based their overviews on geopolitical units (countries and states), which clearly oversimplifies the species richness patterns and can produce misleading results (Mutke and Barthlott, 2005).

This problem of too coarse units, as well as the bias caused by gaps in knowledge, could also have influenced the conclusions by Shaw et al. (2005). These authors published one of the few attempts of statistically quantifying the latitudinal gradient in mosses, and failed to show a strong relationship between species richness and latitude.

Despite the lack of a general unambiguous pattern in liverworts, the Lejeuneaceae, one of the largest liverwort families, is clearly most speciose in the tropical regions of the world (von Konrat et al., 2008). An analysis of beta diversity of pleurocarpous mosses showed a higher species turnover in the tropics than in temperate and boreal regions, indicating the existence of a latitudinal gradient of moss diversity (Hedenäs, 2007). In addition, the hot spots proposed by Tan and Pócs (2000) are generally found in the same areas that are traditionally considered to have a high diversity for vascular plants. At present we must therefore conclude that it is not yet possible to infer whether bryophytes really lack a latitudinal diversity gradient or if such suggestions are due to incomplete data at the global scale.

The differences in diversity patterns among taxonomic groups can also be addressed in terms of species–area relationships. Species–area curves are one of the most frequently cited correlations in geographic ecology. With some exceptions (e.g. Kimmerer and Driscoll, 2000) bryophytes show a strong relationship between area size and species richness (Tangney et al., 1990; Ingerpuu et al., 2001; Nakanishi, 2001; Virtanen and Oksanen, 2007). However, the slope of the curve is supposedly less steep in bryophytes than in vascular plants and other large organisms (Ingerpuu et al., 2001; Peintinger et al., 2003). Bryophytes therefore seem to occupy an intermediate position between microorganisms, for which a species–area relationship is rarely found (Fenchel and Finlay, 2004), and large organisms, where the slope is steeper. These results support the hypothesis that less steep species–area slopes are found in small organisms (Drakare et al., 2006) and suggest that the greater dispersal abilities of the latter (Mouquet and Loreau, 2002; Hovestadt and Poethke, 2005) lead to lower species turnover and thus to more similar communities across regions. In agreement with this, Hillebrand et al. (2001) found that the decay in community similarity with distance is slower for small than large organisms, which corroborates the importance of dispersal ability in small organisms in shaping communities. Bryophytes do not completely fulfil the expectations regarding this tendency. For example, a meta-community analysis by Löbel et al. (2006) showed a strong spatial aggregation in bryophytes and, more significantly, a study of spatial structure in communities of mosses and other taxonomic groups demonstrated that the distance decay in similarity for bryophytes was comparable to that in wind-dispersed vascular plants (Nekola and White, 1999). Although these results are not conclusive, other works that do not directly address the structure of communities have also highlighted the aggregated distribution of bryophytes (Hedenäs et al., 1989; Söderström and Jonsson, 1989; Kuusinen and Penttinen, 1999). This kind of pattern suggests the existence of similar processes shaping communities of vascular plants (where spatial aggregation is a well-known pattern) and bryophytes.

11.3 The phylogeographic approach

In bryology, as in other fields of biology, the use of molecular techniques has become increasingly common during the last 20 years. These techniques have especially been applied in phylogenetic studies, but also in taxonomy and biogeography (cf. Shaw et al., 2002; Vanderpoorten and Goffinet, 2009; Heinrichs et al., 2009a). In many cases molecular analyses have confirmed earlier hypotheses based on morphological data. However, there are also numerous examples where taxonomical units defined by molecular similarity are incongruent with morphologically defined taxa. This has led both to the recognition of cryptic species and to the synonymisation of taxa, and has re-opened the species concept debate (see e.g. Mishler, 2009 vs. Zander, 2007). In some cases, molecular-based taxonomic units show clear biogeographic patterns that allow more or less uncontroversial interpretations, but there are also studies in which the biogeographic history of clades is difficult to assess. There are even cases where contradictory conclusions can be drawn from similar molecular data sets.

Molecular studies have revealed variation not only between, but also within morphologically defined bryophyte species (morphospecies). In some cases the molecular variation within morphologically stable species is even larger than among morphologically clearly different taxa. This has led to the recognition of cryptic bryophyte species (see revision in Shaw, 2001, and Heinrichs et al., 2009a). When morphological studies were carried out after the molecular variation was detected, this has sometimes led to the description of new species (e.g. Szweykowski et al., 2005; Hedenäs et al., 2009) or to the re-establishment of previously synonymised taxa (e.g. Rycroft et al., 2004; Cano et al., 2005; Hedenäs, 2008; Oguri et al., 2008; Draper and Hedenäs, 2008, 2009). On the contrary, there are also examples where taxonomically problematic bryophyte morphotypes were impossible to distinguish on the basis of the studied markers. Therefore some taxa, both at the infraspecific and specific levels, have been synonymised based on molecular evidence. This has especially happened for endemic taxa that were described mostly on the basis of their isolated occurrences (e.g. Heinrichs et al., 2004a, 2004b; Werner et al., 2009). In spite of all the mentioned taxonomic changes, the number of described bryophyte species remains approximately constant. This sharply contrasts with the tendency observed in other groups of small organisms, where the introduction of molecular techniques has meant a cut-across classic taxonomy and has multiplied the estimated number of species up to ten times (e.g. Foissner, 1999). The difference between the two categories probably lies in that the number of morphological characters available for species delimitation is much higher in bryophytes than in many other small organisms. Among microorganisms, the DNA information has revealed a vast molecular diversity within relatively few morphologically recognised species that are globally distributed (Spratt et al., 2006).

If molecular techniques are useful, although not revolutionary for our understanding of bryophyte taxonomy at the species level, they are more widely used for understanding bryophyte phylogeny and they have, in many cases, revealed unsuspected relationships (cf. Renzaglia et al., 2007). Since phylogenetic relationships reflect the evolutionary history, these can also be used to infer dispersal and colonisation histories at various taxonomic levels. As an example, the geographic distributions of haplotypes within morphospecies allow us to infer species origins and/or diversification areas, as well as their dispersal routes.

11.3.1 Phylogeography of endemics

Molecular techniques have a great potential to elucidate the geographic history for both widely and narrowly distributed species, and have often confirmed endemics as distinct taxonomic entities. A nice example of the latter among bryophytes is found in the genus *Echinodium* (Stech et al., 2008). This genus was earlier thought to comprise six species, four restricted to the Macaronesian archipelago and two to the Australasian and Pacific regions. It has now been demonstrated that this is an artificial group consisting of an endemic Macaronesian genus (*Echinodium* s.str., three species), one Macaronesian *Isothecium* species and two Australasian members possibly in connection with *Thamnobryum* that show convergent morphological evolution. Stech et al. (2008) not only elucidate the taxonomy and phylogeny of this group of species, but they also clarify the biogeography of a vicariance that is otherwise difficult to explain. Studies of molecular variation involving endemic taxa in a phylogenetic framework can also lead to conclusions regarding speciation processes. Thus, Vanderpoorten and Long (2006) interpreted the nested position of the Azorean endemic liverwort *Leptoscyphus azoricus* within populations of the Neotropical *L. porphyrius* as an example of recent speciation caused by a long-distance dispersal. Similarly, Hedderson and Zander (2007) postulated that the South African endemic moss *Triquetrella mxinwana* originated during the Pliocene–Pleistocene as a result of long-distance dispersal, on the basis of low molecular divergence levels and a chronology consistent with the existence of the niche where it grows.

11.3.2 Phylogeography of widely distributed taxa

Haplotype diversity and molecular variation in species distributed throughout large and more or less continuous areas have been used by several authors to infer possible areas of origin, glacial refugia and migration routes. Hedenäs and Eldenäs (2007) hypothesised that the two cryptic species of *Hamatocaulis vernicosus* complex occurring in Europe diverged before the last periods of glaciations, based on their occurrence also in America. The species occurring in southern Europe could have survived somewhere in the northern Mediterranean region, from where it re-colonised earlier glaciated or periglacial areas in central and northern Europe. Hedderson and Nowell (2006) deduced that glacial refugia for *Homalothecium sericeum* occurred both in southern Europe (Balkan and Italian peninsulas) and in the British islands and adjacent mainland, on the basis of the greater haplotype diversity and occurrence of unique haplotypes in these areas. As a final example, Hedenäs (2009a) used molecular data and fossil evidence to postulate a late Tertiary origin of ancestral haplotypes of *Scorpidium cossonii* in cold pockets in a then partly sparsely forest-covered Arctic. Subsequently, haplotypes evolved adaptations to warmer climates and that allowed colonisation of temperate wetlands and also gave rise to the morphospecies *S. scorpioides*.

Analyses of molecular variation among populations have also been used to address the origin of disjunct distributions. If disjunctions originated as a consequence of recent or repeated and continuing long-distance dispersal events, we should expect that molecular sequences from separate populations are rather similar. On the other hand, if an original distribution area was fragmented and populations remained isolated for a long period thereafter different mutations should have accumulated in the respective areas. In the latter case molecular differentiation between populations should be relatively higher (e.g. Shaw et al., 2002). In bryophyte studies both fragmentation and long-distance dispersal have been proposed to explain extant species disjunctions. McDaniel and Shaw (2003) analysed the distribution of different chloroplast haplotypes of the subantarctic moss *Pyrrhobryum mnioides*. They found evidence for recent or ongoing migration across the Tasman Sea but not for intercontinental dispersal between Australasia and South America or along the Andes between Patagonia and the Neotropics. From the degree of molecular differentiation, they concluded that Australasian and South American populations have been isolated for approximately 80 million years, after the Gondwanan fragmentation. Likewise, Stech and Dohrmann (2004) found a strong geographic structure in haplotype distribution in the widespread moss *Campylopus pilifer*, which they interpreted as a probable result of divergent evolution of the populations, after the segregation of Gondwana. However, they also assumed long-range dispersal or introduction events to explain anomalous positions of some haplotypes that deviate from the general pattern. As a third example, Heinrichs et al. (2006) suggested that the current distribution of the liverwort family Plagiochilaceae is the result of the breakup of Gondwana, in combination with short-distance and rare long-distance dispersal events.

Other studies suggest long-distance dispersal to be the main process to explain disjunct distributions, such as Vanderpoorten et al. (2008), Shaw et al. (2003) and Werner et al. (2003) for several moss species occurring both in North America and Europe, and Shaw et al. (2008) for several taxa in Australasia and South America. For liverworts, long-distance dispersal was proposed to explain especially tropical American–African disjunctions (e.g. Feldberg et al., 2007; Heinrichs et al., 2009b). Most of these cases exhibit a low degree of molecular differentiation among populations. As mentioned above, this probably reflects either recent divergence or repeated and continuing dispersal events. In addition, the molecular dating estimates for some liverworts, such as *Marchesinia brachiata*, support an Oligocene (that is, post-Gondwanan) divergence (Heinrichs et al., 2009b). Also the biogeography of *Herbertus* cannot be solely explained by Gondwanan divergence (Feldberg et al., 2007), since extremely low mutation rates would need to be assumed. All these examples except one (cf. Shaw et al., 2003) show some geographic structure in their haplotype distributions, since European, African and American samples are more closely related within than between continents. Thus, barriers to gene flow must exist at the continental scale and this probably indicates that intercontinental gene flow is not necessarily a continuous, repeated and/or currently ongoing process.

It can now be concluded that disjunct distributions in bryophytes probably originated from different processes, and that it is not possible to infer a general pattern that is valid for all species. In line with the idea that each species has an individual history, it is especially significant that the distributions of congeneric species are sometimes explained by different processes. Hentschel et al. (2007) found that the extant distribution patterns of the *Porella* species *P. swartziana*, *P. cordaeana* and *P. platyphylla*, are probably a result of long-distance dispersal events. On the other hand, Freitas and Brehm (2001) concluded that the present distribution of *P. canariensis* is the result of a fragmentation of an originally larger continuous area, and they interpreted the geographic structure of the studied haplotypes as indicating a lack of present-day gene flow among regions.

In some cases, similar data on a single species can be interpreted in conflicting ways. Based on the demonstrated low capacity for long-distance dispersal of spores in the moss *Lopidium concinnum* (van Zanten, 1978), Frey et al. (1999) interpreted the molecular similarity among populations from New Zealand, Brazil and Chile as indicating a Gondwanan origin and slight molecular divergence and speciation in geological times (stenoevolution). McDaniel and Shaw (2003) and Shaw et al. (2008) considered that a Gondwanan origin for this moss would imply unacceptable rates for chloroplast sequences evolution, and that recent or ongoing dispersal is a more likely explanation for the species' present distribution. Whether one or the other interpretation is more plausible could be approached by dating the molecular phylogenies. Although this method, if used properly, can be extremely useful, it is sometimes difficult to assess divergence times precisely (Cook and Crisp, 2005; Heads, 2005). For bryophytes the fossil record is relatively incomplete compared with the situation for some other organisms and only a few such studies therefore exist for this group. The problems can be illustrated by the survey on *Plagiochila* by Heinrichs et al. (2006), who used a single fossil specimen to date their phylogeny. Since the specimen could not be unequivocally assigned to a single node, they explored several possible scenarios to minimise the effect of incorrect assignments and concluded that resolving the diversification time-frame is critical to understand the historical biogeography of their study group. Another example that shows how the dating of phylogenies is decisive for unravelling biogeographic patterns is that of Huttunen et al. (2008). Their phylogeny recovered for the moss genus *Homalothecium* shows a strong phylogeographic signal that suggested two main lineages, one including only American species and the other one with Western-Palaearctic species. Such a strong geographic structure is usually interpreted as indicating lack of long-distance dispersal events. The authors estimated the divergence time for the two lineages by using absolute rates of molecular evolution from the literature and factoring uncertainties around those estimates using probabilistic calibration priors. The different scenarios that could be reconstructed from their dating suggested that the a priori probable Laurasian origin would involve unsustainable nucleotide substitution rates and they therefore suggested that the present distribution is instead a result of transoceanic long-distance dispersal. A similar conclusion was reached by Devos and Vanderpoorten (2009) to explain the present distribution of the liverwort genus *Leptoscyphus*, also based on different calibration points depending on several assumptions. These dated phylogenies suggest that long-distance dispersal did probably play an important role in shaping the distribution patterns of some bryophyte species or lineages. However, one should not forget that when such species or haplotypes show clear geographic structures, this only indicates that dispersal between the areas occurred at a certain more or less remote moment in time.

Finally, it should also be considered that in bryophytes the study of different gene regions has sometimes led to incongruent phylogenetic topologies. As was already discussed by Hedenäs (2009b) this should be taken into account, since molecular phylogenies based on too few specimens per taxon may not reflect the actual complexity and can lead to misleading conclusions. Moreover, the biogeographic patterns inferred by different analysis techniques or different molecular markers can be completely divergent. Even when molecular techniques provide much interesting and valuable information on bryophyte biogeography, care and further work are needed before more definitive conclusions are possible.

11.4 Concluding remarks

As a summary of the numerous studies published, some main conclusions arise: several distribution patterns are found among the bryophytes, and different processes can explain each of them. Long-distance dispersal by wind has apparently played a chief role in at least some cases, whereas most of the known distribution patterns are better explained if other mechanisms, such as continental drift, stepping-stone migration and dispersal by anthropogenic or other agents, are also taken into account.

Bryophytes are an ancient group of land plants and have had time enough to reach a very high level of diversification. They are remarkably heterogeneous from many points of view, such as structurally, physiologically and ecologically. Allorge (1947) stated that among mosses and allies, as well as in most plants, species greatly differ in their ecological amplitude; the same can be said about their dispersal capacity and their evolutionary potential.

Given the complexity of the matter in hand, it seems that not a single hypothesis will be enough to explain the intricacy of the observed patterns. Biogeography of bryophytes more likely depends on a complex set of interacting phenomena, including both long-distance dispersal and effective stepping-stone propagation acting across long periods of time.

Finally, bryophytes are small plants and their distribution areas are in general larger than those of flowering plants. Still, the distribution patterns of species belonging to both groups are to a high degree the result of similar processes.

References

Allorge, P. (1947). Essai de bryogéographie de la péninsule Ibérique. *Enciclopédie Biogéographique et Écologique*. Paris: P. Lechevalier.

Argent, G.C.G. (1979). Systematics of tropical mosses. In Clarke, G., Duckett, J. (eds.), *Bryophyte Systematics*, pp. 185–194. London: Academic Press.

Baas Becking, L.G.M. (1934). *Geobiologie of inleiding tot de milieukunde*. The Hague: Van Stockum and Zoon.

Billings, W.D., Anderson, L.E. (1966). Some microclimatic characteristics of habitats of endemic and disjunct bryophytes in the southern Blue Ridge. *The Bryologist* **69**, 76–95.

Boatman, D.J., Lark, P.M. (1971). Inorganic nutrition of the protonemata of *Sphagnum papillosum* Lindb., *S. magellanicum* Brid. and *S. cuspidatum* Ehrh. *New Phytologist* **70**, 1053–1059.

Cano, M.J., Werner, O., Guerra, J. (2005). A morphometric and molecular study in *Tortula subulata* complex (Pottiaceae, Bryophyta). *Botanical Journal of the Linnean Society* **149**, 333–350.

Churchill, S.P. (2009). Moss diversity and endemism of the tropical Andes 1. *Annals of the Missouri Botanical Garden* **96**, 434–449.

Cook, L.G., Crisp, M.D. (2005). Directional asymmetry of long-distance dispersal and colonization could mislead reconstructions of biogeography. *Journal of Biogeography* **32**, 741–754.

Crum, H.A., Anderson, L.E. (1981). *Mosses of Eastern North America*, Vol. 1. New York, NY: Columbia University Press.

Devos, N., Vanderpoorten, A. (2009). Range disjunctions, speciation, and morphological transformation rates in the liverwort genus *Leptoscyphus*. *Evolution* **63**, 779–792.

Dierssen, K. (2001). Distribution, ecological amplitude and phytosociological characterization of European bryophytes. *Bryophytorum Bibliotheca* **56**, 1–289.

Drakare, S., Lennon, J.J., Hillebrand, H. (2006). The imprint of the geographical, evolutionary and ecological context on species-area relationships. *Ecology Letters* **9**, 215–227.

Draper, I., Hedenäs, L. (2008). *Sciuro-hypnum tromsoeense* (Kaurin & Arnell) Draper & Hedenäs, a distinct species from the European mountains. *Journal of Bryology* **30**, 271–278.

Draper, I., Hedenäs, L. (2009). *Sciuro-hypnum dovrense* (Limpr.) Draper & Hedenäs comb. nov., a distinct Eurasian alpine species. *Cryptogamie, Bryologie* **30**, 289–299.

Engel, J.J., Glenny, D. (2008). *A Flora of the Liverworts and Hornworts of*

New Zealand, Vol. 1. Monographs in Systematic Botany from the Missouri Botanical Garden. St Louis, MI: Missouri Botanical Garden.

Farrar, D.R. (1998). The tropical flora of rockhouse cliff formations in the eastern United States. *Journal of the Torrey Botanical Society* **125**, 91–108.

Feldberg, K., Hentschel, J., Wilson, R. et al. (2007). Phylogenetic biogeography of the leafy liverwort *Herbertus* (Jungermanniales, Herbertaceae) based on nuclear and chloroplast DNA sequence data: correlation between genetic variation and geographical distribution. *Journal of Biogeography* **34**, 688–698.

Fenchel, T., Finlay, B.J. (2004). The ubiquity of small species: patterns of local and global diversity. *Bioscience* **54**, 777–784.

Fife, A.J. (1995). Checklist of the mosses of New Zealand. *The Bryologist* **98**, 313–337.

Foissner, W. (1999). Protist diversity: estimates of the near-imponderable. *Protist* **150**, 363–368.

Frahm, J.-P. (2008). Diversity, dispersal and biogeography of bryophytes (mosses). *Biodiversity and Conservation* **17**, 277–284.

Frahm, J.-P., O'Shea, B., Pócs, T. et al. (2003). Manual of tropical bryology. *Tropical Bryology* **23**, 5–194.

Freitas, H., Brehm, A. (2001). Genetic diversity of the Macaronesian leafy liverwort *Porella canariensis* inferred from RAPD markers. *The American Genetic Association* **92**, 339–345.

Frey, W. (ed.). (2009). *Syllabus of Plant Families,* Vol. 3. *Bryophytes and Seedless Vascular Plants.* Berlin: Borntraeger.

Frey, W., Stech, M., Meissner, K. (1999). Chloroplast DNA-relationship in palaeoaustral *Lopidium concinnum* (Hypopterygiaceae, Musci). An example of stenoevolution in mosses. Studies in austral temperate rain forest bryophytes 2. *Plant Systematics and Evolution* **218**, 67–75.

Frey, W., Frahm, J.-P., Fischer, E., Lobin, W. (2006). *The Liverworts, Mosses and Ferns of Europe.* Colchester: Harley Books.

Furuki, T., Dalton, P.J. (2008). *Vandiemenia ratkowskiana* Hewson (Marchantiophyta): a revised description and reassessment of its taxonomic status. *Journal of Bryology* **30**, 48–54.

Gignac, L.D. (2001). Bryophytes as indicators of climate change. *The Bryologist* **104**, 410–420.

Glime, J.M. (2007). *Bryophyte Ecology,* Vol. 1. *Physiological Ecology.* EBook sponsored by Michigan Technological University and the International Association of Bryologists. Accessed October 2009 at www.bryoecol.mtu.edu.

Goffinet, B., Shaw, A.J. (eds.). (2009). *Bryophyte Biology.* Cambridge: Cambridge University Press.

González-Mancebo, J.M., Romaguera, F., Ros, R.M., Patiño, J., Werner, O. (2008). Bryophyte flora of the Canary Islands: an updated compilation of the species list with an analysis of distribution patterns in the context of the Macaronesian Region. *Cryptogamie, Bryologie* **29**, 315–357.

Grolle, R., Long, D.G. (2000). An annotated check-list of the Hepaticae and Anthocerotae of Europe and Macaronesia. *Journal of Bryology* **22**, 103–140.

Hallingbäck, T. (2002). Globally widespread bryophytes, but rare in Europe. *Portugaliae Acta Biologica* **20**, 11–24.

Hawkins, B.A., Porter, E.E., Felizola Diniz-Filho, J.A. (2003). Productivity and history as predictors of the latitudinal diversity gradient of terrestrial birds. *Ecology* **84**, 1608–1623.

Heads, M. (2005). Dating nodes on molecular phylogenies: a critique of molecular biogeography. *Cladistics* **21**, 62–78.

Hedderson, T.A., Nowell, T.L. (2006). Phylogeography of *Homalothecium sericeum* (Hedw.) Br. Eur.; toward a reconstruction of glacial survival and postglacial migration. *Journal of Bryology* **28**, 283–292.

Hedderson, T.A., Zander, R.H. (2007). *Triquetrella mxinwana*, a new moss species from South Africa, with a phylogenetic and biogeographic hypothesis for the genus. *Journal of Bryology* **29**, 151–160.

Hedenäs, L. (2007). Global diversity patterns among pleurocarpous mosses. *The Bryologist* **110**, 319–331.

Hedenäs, L. (2008). Molecular variation and speciation in *Antitrichia curtipendula* s.l. (Leucodontaceae, Bryophyta). *Botanical Journal of the Linnean Society* **156**, 341–354.

Hedenäs, L. (2009a). Relationships among arctic and non-arctic haplotypes of the moss species *Scorpidium cossonii* and *Scorpidium scorpioides* (Calliergonaceae). *Plant Systematics and Evolution* **277**, 217–231.

Hedenäs, L. (2009b). Haplotype variation of relevance to global and European phylogeography in *Sarmentypnum exannulatum* (Bryophyta: Calliergonaceae). *Journal of Bryology* **31**, 145–158.

Hedenäs, L., Herben, T., Rydin, H., Söderström, L. (1989). Ecology of the invading moss species *Orthodontium lineare* in Sweden: Spatial distribution and population structure. *Holarctic Ecology* **12**, 163–172.

Hedenäs, L., Eldenäs, P. (2007). Cryptic speciation, habitat differentiation, and geography in *Hamatocaulis vernicosus* (Calliergonaceae, Bryophyta). *Plant Systematics and Evolution* **268**, 131–145.

Hedenäs, L., Huttunen, S., Shevock, J.R., Norris, D.H. (2009). *Homalothecium californicum* (Brachytheciaceae), a new endemic species to the California Floristic Province, Pacific Coast of North America. *The Bryologist* **112**, 593–604.

Heinrichs, J., Groth, H., Lindner, M., Feldberg, K., Rycroft, D.S. (2004a). Molecular, morphological, and phytochemical evidence for a broad species concept of *Plagiochila bifaria* (Hepaticae). *The Bryologist* **107**, 28–40.

Heinrichs, J., Groth, H., Lindner, M. et al. (2004b). Intercontinental distribution of *Plagiochila corrugata* (Plagiochilaceae, Hepaticae) inferred from nrDNA ITS sequences and morphology. *Botanical Journal of the Linnean Society* **146**, 469–481.

Heinrichs, J., Lindner, M., Groth, H. et al. (2006). Goodbye or welcome Gondwana? – insights into the phylogenetic biogeography of the leafy liverwort *Plagiochila* with a description of *Proskauera*, gen. nov. (Plagiochilaceae, Jungermanniales). *Plant Systematics and Evolution* **258**, 227–250.

Heinrichs, J., Hentschel, J., Feldberg, K., Bombosch, A., Schneider, H. (2009a). Phylogenetic biogeography and taxonomy of disjunctly distributed bryophytes. *Journal of Systematics and Evolution* **47**, 497–508.

Heinrichs, J., Klugmann, F., Hentschel, J., Schneider, H. (2009b). DNA taxonomy, cryptic speciation and diversification of the Neotropical-African liverwort, *Marchesinia brachiata* (Lejeuneaceae, Porellales). *Molecular Phylogenetics and Evolution* **53**, 113–121.

Hentschel, J., Zhu, R.-L., Long, D.G. et al. (2007). A phylogeny of *Porella* (Porellaceae, Jungermanniopsida) based on nuclear and chloroplast DNA sequences. *Molecular Phylogenetics and Evolution* **45**, 693–705.

Hill, M.O., Bell, N., Bruggeman-Nannenga, M.A. et al. (2006). An annotated checklist of the mosses of Europe and Macaronesia. *Journal of Bryology* **28**, 198–267.

Hillebrand, H., Watermann, F., Karez, R., Berninger, U.G. (2001). Differences in species richness patterns between unicellular and multicellular organisms. *Oecologia* **126**, 114–124.

Hovestadt, T., Poethke, H.J. (2005). Dispersal and establishment: spatial patterns and species-area relationships. *Diversity and Distributions* **11**, 333–340.

Huttunen, S., Hedenäs, L., Ignatov, M.S., Devos, N., Vanderpoorten, A. (2008). Origin and evolution of the northern hemisphere disjunction in the moss genus *Homalothecium* (Brachytheciaceae). *American Journal of Botany* **95**, 720–730.

Ingerpuu, N., Vellak, K., Kukk, T., Pärtel, M. (2001). Bryophyte and vascular plant species richness in boreo-nemoral moist forests and mires. *Biodiversity and Conservation* **10**, 2153–2166.

Kimmerer, R.W., Driscoll, M.J.L. (2000). Bryophyte species richness on insular boulder habitats: the effect of area, isolation, and microsite diversity. *The Bryologist* **103**, 748–756.

Kuusinen, M., Penttinen, A. (1999). Spatial pattern of the threatened epiphytic bryophyte *Neckera pennata* at two scales in a fragmented boreal forest. *Ecography* **22**, 729–735.

Lara, F., Garilleti, R., Mazimpaka, V. (2003). Noticias sobre el estado de *Orthotrichum handiense* en Fuerteventura (Islas Canarias). *Boletín de la Sociedad Española de Briología* **22–23**, 11–16.

Limpens, J., Berendse, F., Blodau, C. et al. (2008). Peatlands and the carbon cycle: from local processes to global implications – a synthesis. *Biogeosciences* **5**, 1475–1491.

Longton, R.E., Schuster, R.M. (1983). Reproductive biology. In Schuster, R.M. (ed.), *New Manual of Bryology*, pp. 386–462. Nichinan: Hattori Botanical Laboratory.

Löbel, S., Snäll, T., Rydin, H. (2006). Metapopulation processes in epiphytes inferred patterns of regional distribution and local abundance fragmented forest landscapes. *Ecology* **94**, 856–868.

Machado, A. (2002). La biodiversidad de las islas Canarias. In Pineda, F.D., Miguel, J.M., Casado, M.A., Montalvo, J. (eds.), *La diversidad biológica de España*, pp. 89–100. Madrid: Pearson Educación S.A.

Marino, P., Raguso, R., Goffinet, B. (2009). The ecology and evolution of fly-dispersed dung mosses (Splachnaceae): manipulating insect behavior through odour and visual cues. *Symbiosis* **47**, 61–76.

McDaniel, S.F., Shaw, A.J. (2003). Phylogeographic structure and cryptic speciation in the Trans-Antarctic moss

Pyrrhobryum mnioides. *Evolution* **57**, 205–215.

McDowall, R.W. (2004). What biogeography is: a place for process. *Journal of Biogeography* **31**, 345–351.

Miles, C.J., Longton, R.E. (1990). The role of spores in reproduction in mosses. *Biological Journal of the Linnean Society* **104**, 149–173.

Mishler, B. (2009). Species are not uniquely real biological entities. In Ayala, F., Arp, R . (eds.), *Contemporary Debates in Philosophy of Biology*, pp. 110–122. Chichester: Wiley-Blackwell.

Morat, P. (1993). Our knowledge of the flora of New Caledonia: endemism and diversity in relation to vegetation types and substrates. *Biodiversity Letters* **1**, 72–81.

Mouquet, N., Loreau, M. (2002). Coexistence in metacommunities: the regional similarity hypothesis. *American Naturalist* **159**, 420–426.

Muñoz, J., Felicísimo, A.M., Cabezas, F., Burgaz, A.R., Martínez, I. (2004). Wind as a long-distance dispersal vehicle in the southern hemisphere. *Science* **304**, 1144–1147.

Mutke, J., Barthlott, W. (2005). Patterns of vascular plant diversity at continental to global scales. *Biologiske Skrifter* **55**, 521–537.

Nakanishi, K. (2001). Floristic diversity of bryophyte vegetation in relation to island area. *Journal of Hattori Botanical Laboratory* **91**, 301–316.

Nekola, J.C., White, P.S. (1999). Special Paper: The distance decay of similarity in biogeography and ecology. *Journal of Biogeography* **26**, 867–878.

O'Shea, B.J. (2006). Checklist of the mosses of sub-Saharan Africa (version 5, 12/06). *Tropical Bryology Research Reports* **6**, 1–252.

Ochyra, R., Buck, W.R. (2003). *Arctoa fulvella*, new to Tierra del Fuego, with notes on trans-American bipolar bryogeography. *The Bryologist* **106**, 532–538.

Ochyra, R., Smith, R.L., Bednarek-Ochyra, H. (2008). *Illustrated Moss Flora of Antarctica*. Cambridge: Cambridge University Press.

Oguri, E., Yamaguchi, T., Shimamura, M., Tsubota, H., Deguchi, H. (2008). Phylogenetic and morphological reevaluation of *Leucobryum boninense* (Leucobryaceae), endemic to the Bonin Islands. *The Bryologist* **111**, 260–270.

Oliver, M.J., Velten, J., Mishler, B.D. (2005). Desiccation tolerance in bryophytes: a reflection of the primitive strategy for plant survival in dehydrating habitats? *Integrative and Comparative Biology* **45**, 788–799.

Peintinger, M., Bergamini, A., Schmid, B. (2003). Species-area relationships and nestedness of four taxonomic groups in fragmented wetlands. *Basic and Applied Ecology* **4**, 385–394.

Pócs, T. (1998). Bryophyte diversity along the Eastern Arc. *Journal of East African Natural History* **87**, 75–84.

Porley, R., Hodgetts, N. (2005). *Mosses and Liverworts*. London: HarperCollins.

Proctor, M.C.F. (2009). Physiological ecology. In Shaw, A., Goffinet, B. (eds.), *Bryophyte Biology* (2nd edition), pp. 237–268. Cambridge: Cambridge University Press.

Pugnaire, F.I. , Valladares, F. (eds.). (2007). *Functional Plant Ecology* (2nd edition). New York, NY: Taylor & Francis Group.

Renzaglia, K.S., Schuette, S., Duff, R.J. et al. (2007). Bryophyte phylogeny: advancing the molecular and morphological frontiers. *The Bryologist* **110**, 179–213.

Rice, S.K., Collins, D., Anderson, A.M. (2001). Functional significance of variation in bryophyte canopy structure. *American Journal of Botany* **88**, 1568–1576.

Rycroft, D.S., Groth, H., Heinrichs, J. (2004). Reinstatement of *Plagiochila maderensis* (Jungermanniopsida: Plagiochilaccae) based on chemical evidence and nrDNA ITS sequences. *Journal of Bryology* **26**, 37–45.

Sáinz Ollero, H., Moreno Saiz, J.C. (2002). Flora vascular endémica española. In Pineda, F., Miguel, J., Casado, M., Montalvo, J. (eds.), *Diversidad biológica de España,* pp. 175–196. Madrid: Pearson Educación S.A.

Schofield, W.B. (1988). Bryophyte disjunctions in the northern hemisphere: Europe and North America. *Biological Journal of the Linnean Society* **98**, 211–224.

Schofield, W.B. (1992). Bryophyte distribution patterns. In Bates, J., Farmer, A. (eds.), *Bryophytes and Lichens in a Changing Environment,* pp. 103–130. Oxford: Oxford University Press.

Schofield, W.B., Crum, H.A. (1972). Disjunctions in bryophytes. *Annals of the Missouri Botanical Garden* **59**, 174–202.

Schuster, R.M. (1983). Phytogeography of the bryophyta. In Schuster, R.M. (ed.), *New Manual of Bryology,* Vol. 1, pp. 463–623. Nichinan: The Hattori Botanical Laboratory, Nichinan.

Shaw, A.J. (2001). Biogeographic patterns and cryptic speciation in bryophytes. *Journal of Biogeography* **28**, 253–261.

Shaw, A.J., Goffinet, B. (2000). *Bryophyte Biology.* Cambridge: Cambridge University Press.

Shaw, A.J., McDaniel, S.F., Werner, O., Ros, R.M. (2002). New frontiers in bryology and lichenology (invited essay). Phylogeography and phylodemography. *The Bryologist* **105**, 373–383.

Shaw, A.J., Werner, O., Ros, R.M. (2003). Intercontinental mediterranean disjunct mosses: morphological and molecular patterns. *American Journal of Botany* **90**, 540–550.

Shaw, A.J., Cox, C.J., Goffinet, B. (2005). Global patterns of moss diversity: taxonomic and molecular inferences. *Taxon* **54**, 337–352.

Shaw, A.J., Holz, I., Cox, C.J., Goffinet, B. (2008). Phylogeny, character evolution, and biogeography of the Gondwanic moss family Hypopterygiaceae (Bryophyta). *Systematic Botany* **33**, 21–30.

Smith, A.J.E. (2004). *The Moss Flora of Britain and Ireland.* Cambridge: Cambridge University Press.

Sotiaux, A., Enroth, J., Olsson, S., Quandt, D., Vanderpoorten, A. (2009). When morphology and molecules tell us different stories: a case-in-point with *Leptodon corsicus*, a new and unique endemic moss species from Corsica. *Journal of Bryology* **31**, 186–196.

Spratt, B.G., Staley, J.T., Fisher, M.C. (2006). Introduction: species and speciation in micro-organisms. *Philosophical Transactions of the Royal Society B* **365**, 1897–1898.

Staples, G.W., Imada, C.T. (2006). Checklist of Hawaiian anthocerotes and hepatics. *Tropical Bryology* **28**, 15–47.

Staples, G.W., Imada, C.T., Hoe, W.J., Smith, C.W. (2004). A revised checklist of Hawaiian mosses. *Tropical Bryology* **25**, 35–69.

Stech, M., Dohrmann, J. (2004). Molecular relationships and biogeography of two Gondwanan *Campylopus* species, *C. pilifer* and *C. introflexus* (Dicranaceae). *Monographs in Systematic Botany* **98**, 415–432.

Stech, M., Sim-Sim, M., Esquível, M.G. et al. (2008). Explaining the 'anomalous' distribution of *Echinodium* (Bryopsida: Echinodiaceae): independent evolution in Macaronesia and Australasia. *Organisms, Diversity and Evolution* **8**, 282–292.

Sundberg, S., Rydin, H. (2002). Habitat requirements for establishment of *Sphagnum* from spores. *Journal of Ecology* **90**, 268–278.

Szweykowski, J., Buczkowska, K., Odrzykoski, I.J. (2005). *Conocephalum salebrosum* (Marchantiopsida, Conocephalaceae) – a new Holarctic liverwort species. *Plant Systematics and Evolution* **253**, 133–158.

Söderström, L., Jonsson, B.G. (1989). Spatial pattern and dispersal in the leafy hepatic *Ptilidium pulcherrimum*. *Journal of Bryology* **15**, 793–802.

Tan, B., Pócs, T. (2000). Biogeography and conservation of bryophytes. In Shaw, A. J., Goffinet, B. (eds.), *Biology of Bryophytes*, pp. 403–448. Cambridge: Cambridge University Press.

Tangney, R.S., Wilson, J.B., Mark, A.F. (1990). Bryophyte island biogeography: a study in Lake Manapouri, New Zealand. *Oikos* **59**, 21–26.

Taylor, T.N., Taylor, E.L., Krings, M. (2009). *Paleobotany: The Biology and Evolution of Fossil Plants*. New York, NY: Academic Press.

van Zanten, B.O. (1978). Experimental studies on transoceanic long-range dispersal of moss spores in the southern hemisphere. *Journal of Hattori Botanical Laboratory* **44**, 455–482.

van Zanten, B.O., Pócs, T. 1981. Distribution and dispersal of bryophytes. *Advances in Bryology* **1**, 479–562.

van Zanten, B.O., Gradstein, S.R. (1988). Experimental dispersal geography of neotropical liverworts. *Beihefte zur Nova Hedwigia* **90**, 41–94.

Vanderpoorten, A., Goffinet, B. (2009). *Introduction to Bryophytes*. Cambridge: Cambridge University Press.

Vanderpoorten, A., Long, D.G. (2006). Budding speciation and neotropical origin of the Azorean endemic liverwort, *Leptoscyphus azoricus*. *Molecular Phylogenetics and Evolution* **40**, 73–83.

Vanderpoorten, A., Devos, N., Goffinet, B., Hardy, O.J., Shaw, A.J. (2008). The barriers to oceanic island radiation in bryophytes: insights from the phylogeography of the moss *Grimmia montana*. *Journal of Biogeography* **35**, 654–663.

Virtanen, R., Oksanen, J. (2007). The effects of habitat connectivity on cryptogam richness in boulder metacommunity. *Biological Conservation* **135**, 415–422.

von Konrat, M., Hagborg, A., Söderström, L. et al. (2008). Early land plants today: Global patterns of liverwort diversity, distribution, and floristic knowledge. In Mohamed, H., Baki, B., Nasrulaq-Boyce, A., Lee, P.K.Y. (eds.), *Bryology in the New Millennium*, pp. 425–438. Kuala Lumpur: University of Malaya.

Werner, O., Ros, R.M., Guerra, J., Shaw, A.J. (2003). Molecular data confirm the presence of *Anacolia menziesii* (Bartramiaceae, Musci) in southern Europe and its separation from

Anacolia webbii. *Systematic Botany* **28**, 483–489.

Werner, O., Patiño, J., González-Mancebo, J.M., Gabriel, R.M.A., Ros, R.M. (2009). The taxonomic status and the geographical relationships of the Macaronesian endemic moss *Fissidens luisieri* (Fissidentaceae) based on DNA sequence data. *The Bryologist* **112**, 315–324.

Zander, R.H. (2007). When biodiversity study and systematics diverge. *Biodiversity* **8**, 43–48.

12

Dispersal limitation or habitat quality – what shapes the distribution ranges of ferns?

Hanno Schaefer

Ecology and Evolutionary Biology, Imperial College London, Silwood Park Campus, Ascot, UK; and Organismic and Evolutionary Biology, Harvard University, Cambridge, MA, USA

12.1 Introduction

Ferns are the second largest vascular plant group on earth with more than 9000 living species currently placed in four classes: (1) whisk ferns – Psilotopsida, *c.* 92 species, (2) horsetails – Equisetopsida, *c.* 15 species, (3) marattioid ferns – Marattiopsida, *c.* 150 species, and (4) leptosporangiate ferns – Polypodiopsida, *c.* 9000 species (Smith et al., 2006). Their origin dates back to the Late Devonian or early Carboniferous more than 350 million years ago (Pryer et al., 2004). They reproduce by haploid spores, which grow into a free-living gametophyte, usually a photosynthetic prothallus with microscopic male and female organs. The male sexual organs, the anteridia, release mobile sperms that swim to the female sexual organs, the archegonia (often on the same prothallus), and fertilise an egg that remains attached to the prothallus. The resulting zygotes divide by mitoses and grow into the diploid sporophytes, usually with characteristic rhizome and fronds

Biogeography of Microscopic Organisms: Is Everything Small Everywhere?, ed. Diego Fontaneto. Published by Cambridge University Press.

(Lloyd, 1974). Most of the ferns are perennial hemicryptophytes (less commonly tree ferns, rarely annuals) that produce up to millions of tiny, long-lived, mostly wind-dispersed spores every year (Smith et al., 2006). The notable exceptions are some genera of Polypodiaceae (e.g. *Grammitis, Jungermannia*), which produce relatively few, chlorophyllous spores per frond that live only days or weeks (Schaefer, 2001a) and some water- or bird-dispersed heterosporous ferns (Marsileaceae, Salviniaceae). Chlorophyllous spores are thought to be less well adapted to long-distance dispersal (Lloyd and Klekowski, 1970) but this might be compensated in some groups of tropical epiphytic ferns by the production of gametophytic gemmae (Dassler and Farrar, 2001). In contrast to flowering plants, ferns do not depend on pollinators and are rarely attacked by herbivores (Barrington, 1993).

Among pteridologists, the hypothesis 'Everything is everywhere, but the environment selects' (Baas Becking 1934), which was originally developed for microbial organisms with latent life stages, was entertained especially by Tryon (e.g. Tryon, 1970, 1972, 1985). He thought that most ferns' enormous dispersal ability should allow them to colonise almost any suitable habitat. And indeed, most fern families have a worldwide distribution (Smith et al., 2006) and managed to colonise even isolated archipelagos like Hawai'i and the Azores multiple times (Geiger et al., 2007; Schaefer, 2001b). Apparently, Tryon never explicitly referred to Baas Becking but he stated that 'studies of fern geography can largely eliminate dispersal capacity as a variable, or as a limiting factor, and attention can be focused on other aspects of geographic and evolutionary processes' (Tryon, 1972), which corresponds very well to the original hypothesis. However, there is still a debate among pteridologists as to whether dispersal or habitat are more important in shaping the ranges of ferns (e.g. Barrington, 1993; Kato, 1993; Wolf et al., 2001; Haufler, 2007).

Spore sizes in ferns are in general so small (<60 μm) that unlimited wind dispersal seems plausible (Tryon, 1970). However, recent studies have shown that the idea of relatively unlimited dispersal for ferns might be too simple: Gradstein and van Zanten (1999) tested spores of ferns that had been attached to passenger aeroplanes to find out if high-altitude wind currents are a safe way of spore dispersal. They found evidence for UV damage in most species, so strong high-altitude air currents might be effective to transport spores over long distances but the spores lose the ability to germinate. Even more important, research of the past few decades has produced strong evidence that similar to spermatophytes, in many (most?) fern species more than one germinating fern spore is required to avoid inbreeding problems (Schneller, 1988; Schneller et al., 1990; Haufler, 2007). Intragametophytic selfing (fusion of sperm and egg from a single gametophyte), intergametophytic selfing (fusion of sperm and egg from different gametophytes, which are products of the same sporophyte – analogous to inbreeding in angiosperms), and intergametophytic crossing (fusion of sperm and egg from different gametophytes, each produced by a different sporophyte) (Lloyd, 1974) were shown to produce significantly different results in laboratory studies (e.g. Schneller, 1988).

Intragametophytic selfing is the only mechanism that would allow establishment of a new population after a single-spore colonisation event. This has been postulated for some species (e.g. Vogel et al., 1999) but is thought to be an exception, especially among diploid taxa (Trewick et al., 2002). These breeding system constraints led to the conclusion that selfing ability might be one of the key limitations in the establishment of new fern populations following dispersal events (Flinn, 2006; Wolf et al., 2001).

A comprehensive analysis of fern phylogenies, distribution ranges and available habitats is therefore needed to find out what shaped the modern ranges of our ferns. Here, we (1) summarise published studies that went in this direction and (2) combine these results with data of the fern flora of the isolated Azores archipelago, which is an ideal study system to test dispersal limitations.

12.2 Evidence from modern fern phylogenies and chorological analyses

Extant fern lineages differ considerably in age: while most of the lineages had diversified already in the Late Cretaceous, the diversification of one of the biggest families, the Polypodiaceae, took place in the Eocene, some 40 million years ago (Schneider et al., 2004; Schuettpelz and Pryer, 2009). Yet, a comparison of the distribution ranges of all 37 currently accepted (and mostly monophyletic) fern families shows that regardless of age, most of them are pantropical or even subcosmopolitan (Smith et al., 2006). Rare exceptions are the Thyrsopteridaceae (endemic to the Juan-Fernandez archipelago) and the exclusively Asian Dipteridaceae (Smith et al., 2006) but the former are monotypic and the latter were more widespread in Mesozoic times (Skog, 2001). Even many genera, and some species (or species groups) like bracken, *Pteridium aquilinum*, seem to have extremely broad distribution ranges compared to angiosperms (Smith, 1972).

However, even though they share most families and many genera of ferns, the regional species pools of the Asian, African and American continent are clearly distinct: in one of the relatively few phylogeny-based analyses available so far, Janssen et al. (2007) found that the Asian and Neotropical Polypodiaceae have few common elements but the African Polypodiaceae (and maybe also some other African fern groups) are a mix of Neotropical and Asian lineages that colonised the African continent in different time windows. Little and Barrington (2003) found in their analysis of the pantropical genus *Polystichum* (with about 200 accepted species) clearly distinct clades on the American, African and Asian continents. A species group that based on morphology had been thought to occur in the West Indies and East Asia turned out to represent in fact two independent lineages with some convergent adaptations.

The hypothesis that all fern species can be found everywhere is also challenged by detailed analyses of regional fern floras: it turns out that many species are widespread, but rare and confined to few populations with only small numbers of individuals (e.g. many European *Asplenium* species), others are confined to 'endemic centres' that are explained by geographic isolation (Barrington, 1993). Furthermore, disjunctions are common at the regional scale. Whether they are a result of habitat structure with specialist ferns restricted to rare soil or rock types or a result of dispersal limitations is often hard to decide. In a recent study, Wild and Gagnon (2005) analysed the distribution of three rare ferns in Canada and found no evidence for habitat limitations: a big proportion of suitable habitat was not colonised by the study species. The most likely explanation seems to be dispersal limitation. In contrast, an analysis of the fern flora of Japan (Guo et al., 2003) found that probably habitat availability and not dispersal is the key factor for the shape of modern fern distribution ranges within the Japanese archipelago.

12.3 Evidence from the fern flora of the Azores

The Azores archipelago is a group of nine volcanic islands in the Northern Atlantic, more than 1000 km off the European mainland (Portuguese coast) and more than 800 km from the closest island (Madeira). Its angiosperm flora is species-poor (149 natives plus about 650 naturalised species) and mainly of European origin (Schaefer, 2003; Carine and Schaefer, 2010). Its fern flora is similarly well known (Schaefer, 2001b, 2003) and consists of 42 native and 24 naturalised species in 19 families (Fig 12.1).

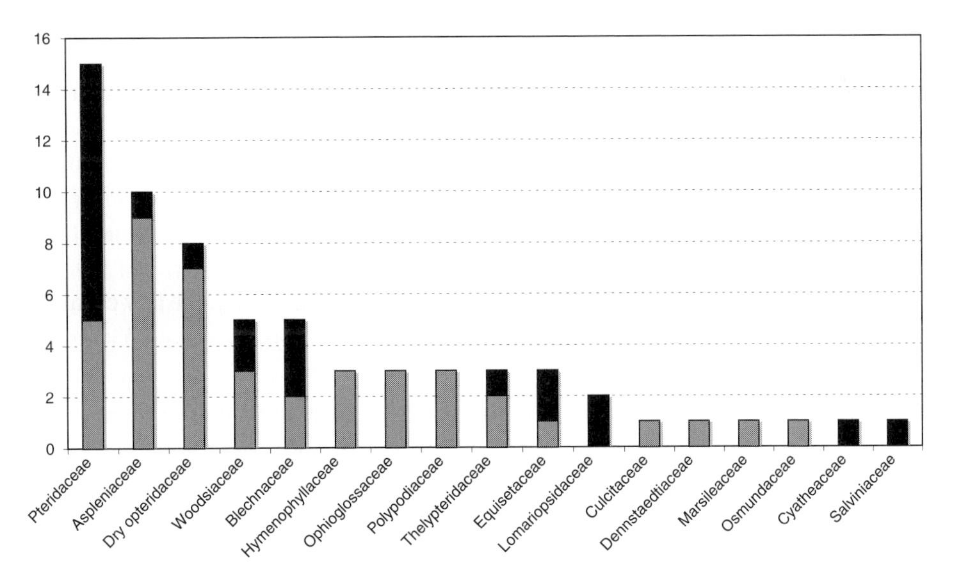

Fig 12.1 The fern flora of the Azores: number of species per family (grey – native species; black – naturalised species); based on Schaefer (2003).

Fig 12.2 Endemic ferns of the Azores: (A) *Grammitis azorica*; (B) *Polypodium azoricum*; (C) *Marsilea azorica* (all photos by H. Schaefer).

The native fern flora is the result of *c.* 40 independent colonisation events (Schaefer, 2001b). Even though the predominantly westerly winds of the region should help American lineages to reach the islands, the native ferns are mainly of European origin and the species composition is very similar to the fern flora of Madeira (Schaefer, 2003; Vanderpoorten et al., 2007 – but the latter analysis does not distinguish between native and naturalised species, rendering its results questionable). Whether the native ferns reached the Azores through stepping-stone colonisation via Madeira or directly from the continents through parallel dispersal events has never been tested. Only six fern species are thought to be Azorean endemics: *Asplenium azoricum, Dryopteris azorica, Dryopteris crispifolia, Grammitis azorica* (Fig 12.2A), *Polypodium azoricum* (Fig 12.2B) and *Marsilea azorica* (Fig 12.2C). Five of these endemic ferns occur on two or more islands of the archipelago indicating that dispersal is not limited at shorter distances (< 600 km) but (so far) they seem to be unable to spread to Madeira (850 km away), which would have suitable habitats for all of them (Schaefer, 2003; Rumsey et al., 2004). The only single island endemic among the Azorean ferns is *Marsilea azorica*, an aquatic, heterosporous clover fern that is known from only a single small pond on Terceira Island. Clover ferns are probably dispersed by waterfowl, but their dispersal ability was found to be very low (Vitalis et al., 2002) and most of these birds are only rare vagrants in the Azores. The small number of endemics and the absence of endemic radiations among the Azorean ferns seem to support the Baas Becking hypothesis. However, a similar pattern was found for the Azorean angiosperms and is perhaps related

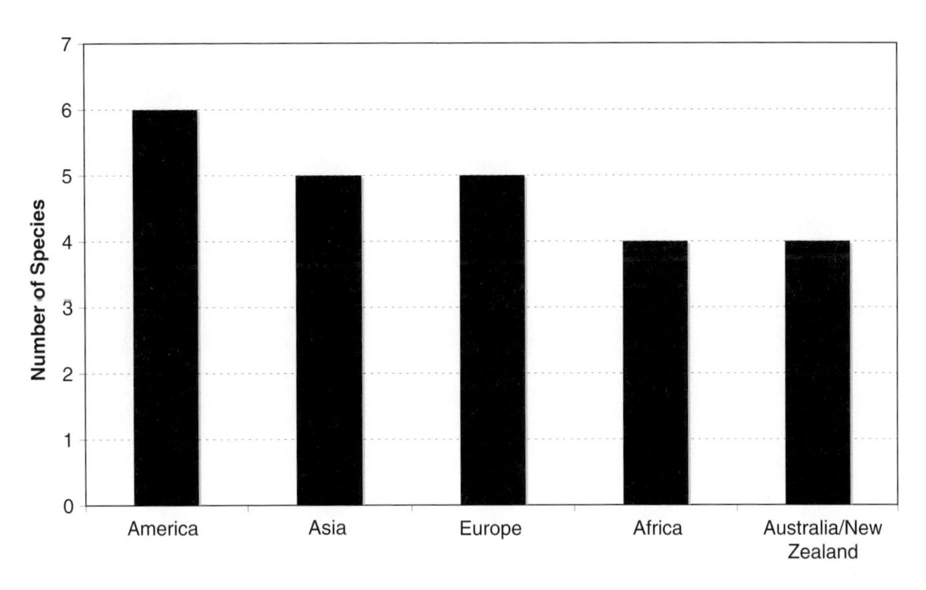

Fig 12.3 Origin of the naturalised fern flora of the Azores; based on Schaefer (2003).

to the palaeoclimatic history of the region rather than to dispersal or habitat factors (Carine and Schaefer, 2010).

If ferns were not dispersal limited, one would expect to find very few invasive fern species. Long-distance wind dispersal should have brought the spores of most species to all suitable habitats so they should have been able to establish populations long before nurseries started to send ferns around the globe. However, there are quite a few invasive fern species in the Azores as well as in other continental and island ecosystems worldwide (e.g. some 30 species on Hawai'i; Wilson, 1996). The 24 naturalised ferns of the Azores are mostly invasive (here defined as 'spreading into natural habitats') and common. Probably all of them were introduced as ornamental plants by nineteenth century gardeners and soon escaped. Almost identical proportions were introduced from the American continent, Africa, Asia, continental Europe and Australia/New Zealand (Fig 12.3). Even though the proportion of naturalised species in ferns is much lower than in angiosperms (*c.* 30% vs. *c.* 70%), these species seem to be highly competitive in disturbed and natural habitats of the Azorean islands and most of them managed to spread throughout the archipelago within less than a century (Schaefer, 2001b). One of them, *Deparia petersenii* (Fig 12.4A), was first reported from the Azores in 1907 but was soon so well established that it was considered as a native and even an Azorean endemic species for decades and described as *Diplazium allorgei* before it was discovered that the species is actually native to Asia and only escaped from cultivation in the Azores. The tree fern *Sphaeropteris cooperi* (Fig 12.4B) was first collected outside cultivated areas on Sao Miguel Island by B. Carreiro in 1895

Fig 12.4 Invasive ferns of the Azores: (A) *Deparia petersenii*; (B) *Sphaeropteris cooperi*; (C) *Cyrtomium falcatum* (all photos by H. Schaefer).

(specimen in the herbarium of Coimbra). Since then it spread to all other islands of the group and is invasive in all types of lowland forest and former pastures (Schaefer, 2003). This shows clearly that the habitat is suitable for all these species but they did not manage to reach the Azores before gardeners transported them across the Atlantic.

12.4 Discussion and conclusions

In general, ferns are extremely widespread and at family level mostly pantropical to subcosmopolitan (Smith, 1972; Smith et al., 2006). This seems to support the theory of unlimited dispersal in this plant group but other factors like high lineage age or high number of species attributed to each family could also explain this pattern. At smaller scales, however, and when island fern floras are analysed, it turns out that (a) the regional species composition (excluding introduced species) differs considerably, often with high proportions of rare and localised (but not necessarily endemic) species and (b) when fern species are introduced to areas outside their current range or isolated archipelagos (e.g. the Azores, New Zealand or Hawai'i), they can very often become invasive within less than a century (Schaefer, 2003). This is strong evidence for the existence of dispersal barriers. Once they have been crossed (e.g. by human-mediated transport), the habitat quality does not seem to be a limiting factor as long as the climatic conditions allow the species to become established.

All in all, the currently available data point to considerable dispersal limitations in ferns, perhaps comparable to those observed in most angiosperm groups.

Whether these are caused by the wind dispersal itself, by limitations related to the breeding system of the different fern species or by a combination of factors is unclear. More work on historical biogeography of the different fern lineages is needed but already it seems clear that for ferns the Baas Becking hypothesis can be rejected.

References

Baas Becking, L.G.M. (1934). *Geobiologie of inleiding tot de milieukunde.* The Hague: Van Stockum and Zoon.

Barrington, D.S. (1993). Ecological and historical factors in fern biogeography. *Journal of Biogeography* **20**, 275–280.

Birks, H.J.B. (1976). The distribution of European Pteridophytes: a numerical analysis. *New Phytologist* **77**, 257–287.

Carine, M.A., Schaefer, H. (2010). The Azores diversity enigma: why are there so few Azorean endemic flowering plants and why are they so widespread? *Journal of Biogeography* **37**, 77–89.

Dassler, C.L., Farrar, D.R. (2001). Significance of gametophyte form in long-distance colonization by tropical, epiphytic ferns. *Brittonia* **53**, 352–369.

Flinn, K.M. (2006). Reproductive biology of three fern species may contribute to differential colonization success in post-agricultural forests. *American Journal of Botany* **93**, 1289–1294.

Geiger, J.M.O., Ranker, T.A., Ramp Neale, J.M., Klimas, S.T. (2007). Molecular biogeography and origins of the Hawaiian fern flora. *Brittonia* **59**, 142–158.

Gradstein, R., van Zanten, B. (1999). High altitude dispersal of spores: an experimental approach. XVI International Botanical Congress, St. Louis. Abstract 15.4.3.

Guo, Q., Kato, M., Ricklefs, R.E. (2003). Life history, diversity and distribution: a study of Japanese pteridophytes. *Ecography* **26**, 129–138.

Haufler, C.H. (2007). Genetics, phylogenetics and biogeography: considering how shifting paradigms and continents influence fern diversity. *Brittonia* **59**, 108–114.

Janssen, T., Kreier, H.-P., Schneider, H. (2007). Origin and diversification of African ferns with special emphasis on Polypodiaceae. *Brittonia* **59**, 159–181.

Kato, M. (1993). Biogeography of ferns: dispersal and vicariance. *Journal of Biogeography* **20**, 265–274.

Little, D., Barrington, D.S. (2003). Major evolutionary events in the origin and diversification of the fern genus *Polystichum* (Dryopteridaceae). *American Journal of Botany* **90**, 508–514.

Lloyd, R.M. (1974). Reproductive biology and evolution in the pteridophyta. *Annals of the Missouri Botanical Garden* **61**, 318–331.

Lloyd, R.M., Klekowski, E.J. (1970). Spore germination and viability in Pteridophyta: evolutionary significance of chlorophyllous spores. *Biotropica* **2**, 129–137.

Pryer, K.M., Schuettpelz, E., Wolf, P.G. et al. (2004). Phylogeny and evolution of ferns (monilophytes) with a focus on the early leptosporangiate divergences. *American Journal of Botany* **91**, 1582–1598.

Rumsey, F., Russel, S., Schaefer, H., Rasbach, H. (2004). Distribution, ecology and cytology of *Asplenium azoricum* Lovis, Rasbach and Reichstein (Aspleniaceae, Pteridophyta) and its hybrids. *American Fern Journal* **94**, 113–125.

Schaefer, H. (2001a). The Grammitidaceae, Pteridophyta, of Macaronesia. *Feddes Repertorium* **112**, 509–523.

Schaefer, H. (2001b). Distribution and status of the pteridophytes of Faial island, Azores (Portugal). *Fern Gazette* **16**, 213–237.

Schaefer, H. (2003). Chorology and diversity of the Azorean Flora. *Dissertationes Botanicae* **374**. J. Cramer, Stuttgart, 130 pp. + CD rom (580 pp.).

Schneider, H., Schuettpelz, E., Pryer, K.M. et al. (2004). Ferns diversified in the shadow of angiosperms. *Nature* **428**, 553–557.

Schneller, J.J. (1988). Remarks on reproductive biology of homosporous ferns. *Plant Systematics and Evolution* **161**, 91–94.

Schneller, J.J., Haufler, C.H., Ranker, T.A. (1990). Antheridiogen and natural gametophyte populations. *American Fern Journal* **80**, 143–152.

Schuettpelz, E., Pryer, K.M. (2009). Evidence for a Cenozoic radiation of ferns in an angiosperm-dominated canopy. *Proceedings of the National Academy of Sciences USA* **106**, 27.

Skog, J.E. (2001). Biogeography of Mesozoic leptosporangiate ferns related to extant ferns. *Brittonia* **53**, 236–269.

Smith, A.R. (1972). Comparison of fern and flowering plant distributions with some evolutionary interpretations for ferns. *Biotropica* **4**, 4–9.

Smith, A.R., Pryer, K.M., Schuettpelz, E. et al. (2006). A classification for extant ferns. *Taxon* **55**, 705–731.

Trewick, S.A., Morgan-Richards, M., Russell, S.J. et al. (2002). Polyploidy, phylogeography and Pleistocene refugia of the rockfern *Asplenium ceterach*: evidence from chloroplast DNA. *Molecular Ecology* **11**, 2003–2012.

Tryon, R. (1970). Development and evolution of fern floras of Oceanic islands. *Biotropica* **2**, 76–84.

Tryon, R. (1972). Endemic areas and geographic speciation in tropical American ferns. *Biotropica* **4**, 121–131.

Tryon, R. (1985). Fern speciation and biogeography. *Proceedings of the Royal Society of Edinburgh* **86B**, 353–360.

Vanderpoorten, A., Rumsey, F., Carine, M.A. (2007). Does Macaronesia exist? Conflicting signal in the bryophyte and pteridophyte floras. *American Journal of Botany* **94**, 625–639.

Vitalis, R., Riba, M., Colas, B., Grillas, P., Olivieri, I. (2002). Multilocus genetic structure at contrasted spatial scales of the endangered water fern *Marsilea strigosa* Willd. (Marsileaceae, Pteridophyta). *American Journal of Botany* **89**, 1142–1155.

Vogel, J.C., Rumsey, F.J., Russell, S.J. et al. (1999). Genetic structure, reproductive biology and ecology in isolated populations of *Asplenium ciskii* (Aspeniaceae, Pteridophyta). *Heredity* **83**, 604–612.

Wild, M., Gagnon, D. (2005). Does lack of available suitable habitat explain the patchy distributions of rare calcicole fern species? *Ecography* **28**, 191–196.

Wilson, K.A. (1996). Alien ferns in Hawai'i. *Pacific Science* **50**, 127–141.

Wolf, P.G., Schneider, H., Ranker, T.A. (2001). Geographic distributions of homosporous ferns: does dispersal obscure evidence of vicariance? *Journal of Biogeography* **28**, 263–270.

13

Ubiquity of microscopic animals? Evidence from the morphological approach in species identification

TOM ARTOIS[1], DIEGO FONTANETO[2], WILLIAM D. HUMMON[3], SANDRA J. MCINNES[4], M. ANTONIO TODARO[5], MARTIN V. SØRENSEN[6] AND ALDO ZULLINI[7]

[1] *Hasselt University, Centre for Environmental Sciences, Diepenbeek, Belgium*
[2] *Department of Invertebrate Zoology, Swedish Museum of Natural History, Stockholm, Sweden; Division of Biology, Imperial College London, Ascot, UK*
[3] *Department of Biological Sciences, Ohio University, Athens, OH, USA*
[4] *British Antarctic Survey, Cambridge, UK*
[5] *Dipartimento di Biologia, Università di Modena & Reggio Emilia, Modena, Italy*
[6] *Natural History Museum of Denmark, Zoological Museum, Copenhagen, Denmark*
[7] *Dipartimento di Biotecnologie & Bioscienze, Università di Milano-Bicocca, Milan, Italy*

Biogeography of Microscopic Organisms: Is Everything Small Everywhere?, ed. Diego Fontaneto.
Published by Cambridge University Press.

13.1 Introduction

Zoologists always hope to find unusual and interesting new animals in exotic places. Over the last few centuries, scientific expeditions in remote places outside Europe and North America have indeed discovered new species and even higher taxa of vertebrates, insects and other macroscopic animals, completely different from the ones previously known in the home country. In contrast, scientists working on microscopic animals, looking at samples from remote areas, have often found organisms that could be ascribed to familiar species. Microscopic animals have thus been considered not interesting in biogeography, as their distribution may not be limited by geography.

Are microscopic animals really widely distributed? Is their cosmopolitanism an actual biological property or only a common misconception based on false assumptions and unreliable evidence? Is the scenario more complex than the claimed clear-cut difference between micro- and macroscopic animals? This chapter will review all the faunistic knowledge gathered so far on the global distribution of free-living microscopic animals smaller than 2 mm (gastrotrichs, rotifers, tardigrades, micrognathozoans, cycliophorans, loriciferans, kinorhynchs and gnathostomulids). Moreover, we will deal with microscopic free-living species in other groups of animals such as nematodes and flatworms, which have both micro- and macroscopic species. The focus will be on species identification from traditional taxonomy based on morphology, whereas Chapter 14 will deal with more recent evidence gathered from analyses on molecular phylogeny and phylogeography from the same groups.

13.2 Gastrotrichs

(M. Antonio Todaro and William D. Hummon)

Gastrotrichs are microscopic free-living, acoelomate, aquatic worms of a meiobenthic lifestyle. In marine habitats they are mainly interstitial, whereas in fresh waters they are ubiquitous in the periphyton and epibenthos, and to a limited extent also in the plankton. Their total length ranges between 70 μm for the freshwater *Heterolepidoderma lamellatum* and 3.5 mm for the marine *Megadasys pacificus*. The simultaneous presence of ventral ciliation, adhesive tubes and terminal mouth make the Gastrotricha easily distinguishable from other microscopic biota (Fig 13.1). The body is enwrapped by a two-layered cuticle, which in many species forms protective ornamentations such as plates, scales and spines, whose ample variety of shape and size is extensively used for taxonomic purposes (e.g. taxon/species identification).

The phylum counts about 765 species grouped into two orders: Macrodasyida, with some 310 strap-shaped species (in 32 genera of eight families), all but two of

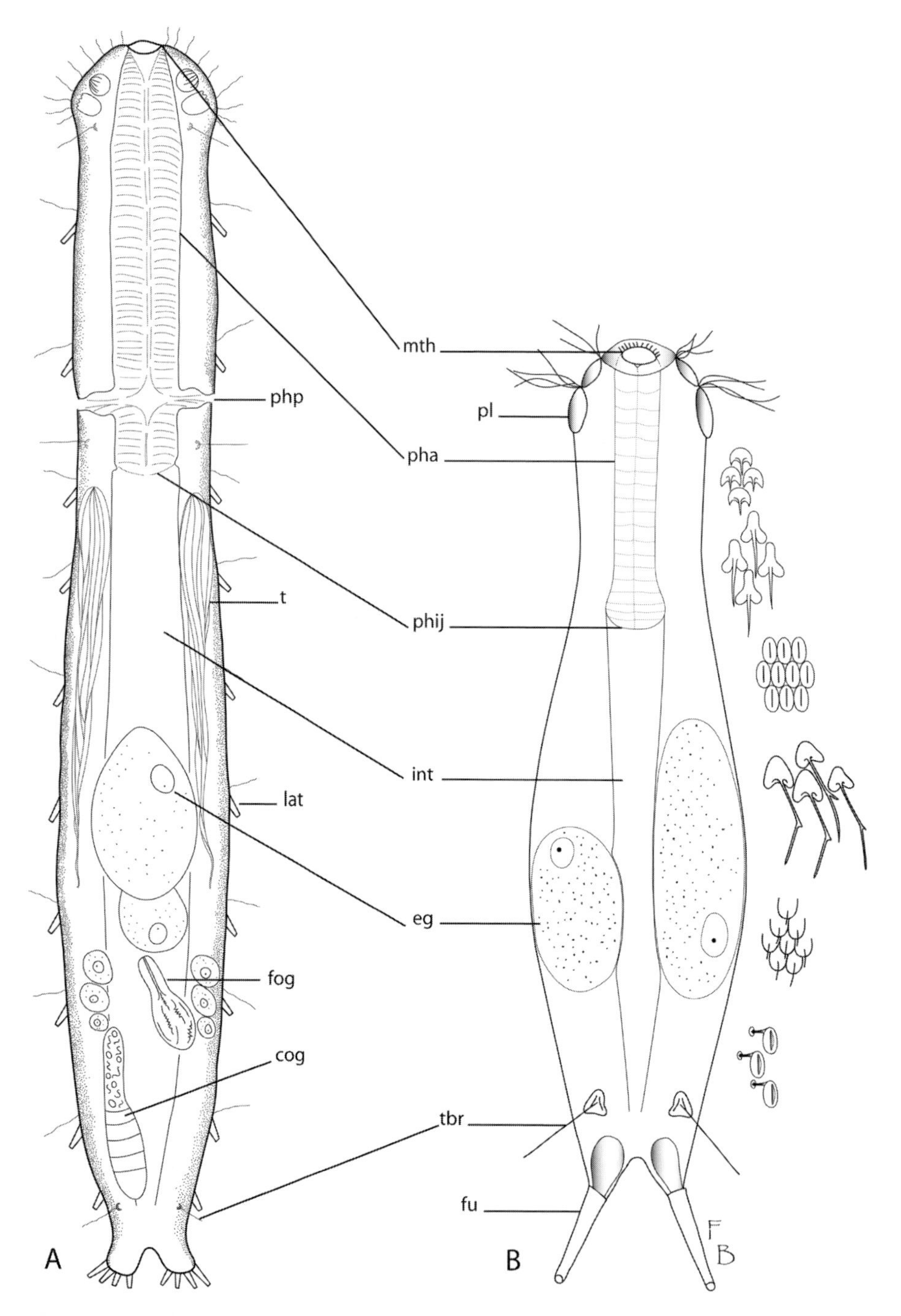

Fig 13.1 Drawing of generalised gastrotrichs, seen from dorsal side: (A) hermaphroditic Macrodasyida, (B) parthenogenetic Chaetonotida (showing on the side different types of cuticular covering), not in scale. cog: caudal organ; eg: egg; fog: frontal organ; fu: furca; int: intestine; lat: lateral adhesive tubes; mth: mouth; pha: pharynx; phij: pharyngeo-intestinal junction; php: pharyngeal pores; pl: pleurae; t: testis; tbr: tactile bristle.

which are marine or estuarine, and Chaetonotida with some 455 tenpin-shaped species (in 31 genera of eight families), 70% of which are freshwater (Balsamo et al., 2009; Hummon and Todaro, 2010). The continuing discovery and description of new species and the many areas of the world still unexplored with regard to the gastrotrich fauna, suggest that these statistics should be considered as very conservative (e.g. Todaro et al., 2005; Hummon, 2008; Kanneby et al., 2009).

Macrodasyidan gastrotrichs are hermaphroditic practising internal, putatively cross-fertilisation; the same appears to be true for members of the marine chaetonotidan families Muselliferidae, Neodasyidae and Xenotrichulidae (six genera and *c.* 30 spp.; Todaro and Hummon, 2008). By contrast, most chaetonotidans, including all the freshwater families and also the marine representatives of the family Chaetonotidae (*c.* 420 spp.), reproduce largely by obligate, apomictic parthenogenesis (Hummon, 1984). In fact, spermatozoa produced after the parthenogenetic phase (Weiss, 2001) have never been proved to be functional and their ultrastructure suggests otherwise (e.g. Balsamo, 1992). Resting eggs are known for several freshwater species, but so far they are unreported for marine taxa. High gastrotrich diversity, found in temporary inland water bodies, suggests that the occurrence of resting eggs is more widespread across the spectrum of freshwater taxa than is currently believed. Among a handful of chaetonotidans cultured in the laboratory, mature gastrotrichs lay 1–10 eggs singly, by rupture of the body wall, eggs being attached to sediment grains, where cleavage usually begins (Balsamo and Todaro, 1988). Development is direct and adulthood is reached within a few days from hatching, with the lifespan extending from 10–20 days to a couple of months. Almost nothing is known about macrodasyidans, but slower growth and longer lifespan (6–12 months) should not come as a surprise for large marine species, such as species of *Dolichodasys*, where the presence of mature specimens only in certain seasons is suggested (Todaro and Hummon, unpublished data).

The phylum Gastrotricha is cosmopolitan in distribution; this is also true of the two orders, almost all families and taxonomically reliable genera. Given the short life cycle, the small number of offspring, the absence of a pelagic larval stage and the comparatively limited swimming ability of the adults, gastrotrich species would be expected to have restricted geographic ranges. Dispersal via long-shore currents over a long geological time combined with oceanic dispersal and continental drift may be invoked to explain biogeographic patterns of Gastrotricha distribution in many marine systems; however, indirect dispersal by phoretic, rafting and ballast may play a relevant role too. Of the 45 species found by Todaro et al. (1995) along the northern Gulf of Mexico the proportion of amphi-Atlantic/cosmopolitan, regional and endemic species found amounted to 60, 22 and 18% respectively, most endemics being macrodasyidans. Of the 56 putative species found by Todaro and Rocha (2004, 2005) from the Brazilian beaches of the State of São Paulo, 31 (55%) could be endemics, again mostly macrodasyidans. A key role of the asexual reproduction in the widespread distribution of chaetonotidian species in marine habitas is best testified to by the high occurrence among cosmopolitan taxa of the initially parthenogenetic Chaetonotidae and by the widespread geography of three marine species belonging to otherwise hermaphroditic higher taxa, i.e. *Draculiciteria tesselata, Heteroxenotrichula pygmaea* and *Urodasys viviparus*. Meanwhile, one of the species found along the Gulf of Mexico has been reclassified as a new species (*Xenotrichula lineata*, redescribed as *X. paralineata*, see Hummon and Todaro, 2007), so that the contrasts between North and South America indicate that a latitudinal gradient may have influenced the biogeographic patterns of marine gastrotrichs. Hermaphroditic species currently thought as cosmopolitan gastrotrichs may be reconsidered as having narrower geographic boundaries as more precise studies are completed, e.g. some species of *Neodasys* and *Musellifer* (see Hochberg, 2005; Leasi and Todaro, 2010). In this framework, a paradigmatic example is provided by the chaetonotidan *Xenotrichula intermedia*. Todaro et al. (1996), using morphometric characters and mitochondrial DNA (CO-I), demonstrated that trans-Atlantic morphologically indistinguishable populations are in fact genetically distinct, bearing different haplotypes with genetic divergence among populations up to 11%. Also, individuals of the Mediterranean Sea and the Arabian Gulf, that appear almost identical when surveyed with conventional microscopy (i.e. DIC and/or SEM), show clearly different arrangements of the muscular system when studied under confocal microscopy (Leasi and Todaro, 2009).

Small size of the adults and resting eggs makes freshwater gastrotrichs good candidates for passive dispersal, aerial and especially phoretic. For instance, single parthenogenetic females hatching from resting eggs and transported over wide areas and even around the world by avian fauna, could initiate a new population. By contrast, many species have been reported as having a wide distribution with a large number of certain cases of cosmopolitanism. Balsamo et al. (2008), in summarising data for freshwater taxa, reported that approximately 33% of European species appear to be cosmopolitan. Detailed faunistic comparisons at several spatial scales were made by Kisielewski (1991, 1999) through extensive studies carried out in Poland, Brazil and Israel. About 34% of the species found in the South American countries were also known in Europe, while in Israel the percentage of European species was higher, 55%, probably due to the closer zoogeographic relationships and shorter distance between the two areas. The high diversity of endemic freshwater genera in the Brazilian fauna, evidenced by Kisielewski (1991), may be due to the scarcity of comparable works carried out in other tropical-equatorial regions of the world. The great faunistic homogeneity of the European freshwater gastrotrich fauna, on the other hand, seems confirmed by preliminary results of the first extensive survey in Sweden, which found several Italian *Ichtydium* species previously unknown outside the Mediterranean peninsula (Kanneby et al., 2009, and own unpublished data).

To summarise, in spite of the unsatisfactory knowledge of their taxonomy, Gastrotricha do exhibit specific patterns of diversity and distribution.

Notwithstanding possible problems and biases in the interpretations of gastrotrich distribution, some blueprints emerge: the percentage of widespread taxa is high, especially among the highly speciose chaetonotids; some species have a more restricted geographic range, but even these species are relatively widespread within one biogeographic region. Future molecular and detailed microscopy investigations will reveal whether morphological identification based on current microscopical techniques alone may be misleading, as the case of *Xenotrichula intermedia* seems to suggest.

13.3 Rotifers

(Diego Fontaneto)

Rotifers (see Wallace et al., 2006 for a recent review of the group) are common microscopic aquatic animals, smaller than 2 mm and usually between 50 μm and 800 μm in length. They can be easily distinguished from other microscopic organisms by the corona of cilia on the head (Fig 13.2) and by the typical hard jaws called trophi; these trophi are so variable that they are widely used as a species-specific taxonomic feature. There are three major groups of rotifers: monogononts, bdelloids and seisonids. Monogonont rotifers may occur in any kind of water and are the richest group, with 1500 species (Wallace et al., 2006; Segers, 2007); they reproduce by cyclical parthenogenesis and the outcome of sexual recombination is usually a dormant embryo called a resting egg, which is considered the dispersal propagule. Bdelloid rotifers only reproduce via obligate parthenogenesis, indeed only females are present (Ricci and Fontaneto, 2009); they occur in any aquatic habitat, from proper water bodies to soil, lichens, mosses and even to deserts; *c.* 450 species have been described. They are able to enter a dormant stage in any period of their life; in their dry dormant stage they can be passively dispersed. Resting eggs in monogononts and dormant stages in bdelloids can remain viable for a long time, surviving desiccation, high temperatures and frost, and thus may act as very effective indirect dispersal propagules. The third group of rotifers, seisonids, is represented by four marine species living only as epibionts on crustaceans of the genus *Nebalia* (Ricci et al., 1993). They do not have resting stages and very few data are available on their distribution; thus, they will not be mentioned here.

The presence of small and drought-resistant dormant stages, perfectly tailored for passive dispersal, aerial or phoretic, make both monogonont and bdelloid rotifers potential candidates for a widespread or even cosmopolitan distribution. Moreover, parthenogenetic females hatch from resting eggs in monogononts and recover from dormancy in bdelloids, so that a single individual can potentially found a new population. These biological properties combined with presumed

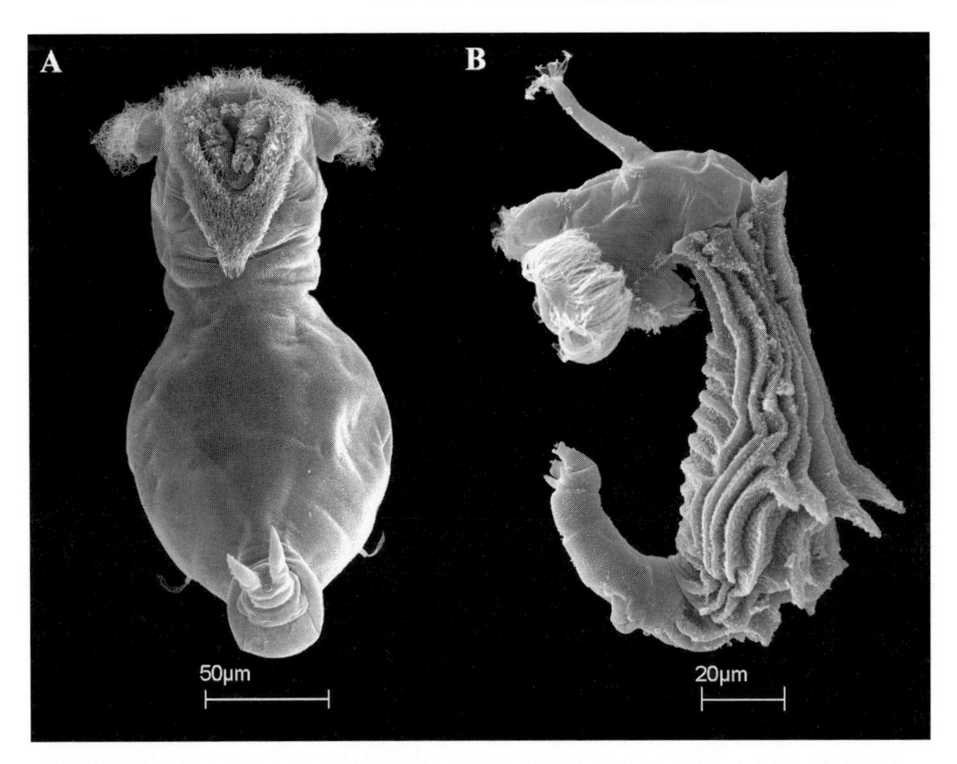

Fig 13.2 Scanning electron microsopy micrographs of (A) a monogonont rotifer, *Notommata collaris*, in ventral view and (B) a bdelloid rotifer, the potential 'flagship' species, *Pleuretra hystrix* in lateral view.

ubiquity, traditionally were believed to render rotifers useless for biogeographic studies (Rousselet, 1909; Hutchinson, 1967; Ball, 1976). Indeed, many species have been reported from different continents, thus supporting their ubiquity (Fontaneto et al., 2007; Segers, 2007).

Other than a large number of cosmopolitan species, another theoretical expectation of rotifer ubiquity is that a locally large representation of the global diversity should be found (Fenchel and Finlay, 2004). This expectation appears to be fulfilled by rotifers: in that any temperate or tropical water body is expected to host between 150 and 250 species respectively, that is 7–12.5% of global diversity (Dumont and Segers, 1996). In a study focused on bdelloids, 20% of all known species were found in a few samples collected in one afternoon in a small valley in Italy (Fontaneto et al., 2006).

According to this evidence, rotifers seem to be potentially and actually ubiquitous. Nevertheless, the fact that some species may be cosmopolitan does not immediately imply that all rotifers are cosmopolitan. Early studies from the 1970s provided increasing evidence of spatial patterns in rotifer biogeography and restricted distributions, comparable to those commonly found in larger animals. For example, Green (1972) showed a latitudinal zonation in planktonic rotifers; Pejler (1977) confirmed restricted distributions in monogononts of the family Brachionidae, highlighting a striking number of endemic species in Lake Baikal for the genus *Notholca*; and De Ridder (1981) demonstrated limited distribution for almost 50% of periphytic and benthic rotifers. More recently, the suspicion that the apparent cosmopolitanism was due more to inadequate knowledge than to reality arose (Dumont, 1983; Segers, 1996). More detailed studies on specific taxa of monogononts found compelling evidence of endemic species for several genera in all the major biogeographic regions: with 6–22% of all *Lecane* spp. in each region being endemic (Segers, 1996); hotspots of biodiversity and endemism exist for *Trichocerca* in the northern hemisphere, especially north-east Nearctic (Segers, 2003); in the case of *Keratella*, hotspots of relict endemics have been identified in the east Palaearctic and in temperate and cold regions of the southern hemisphere, together with a recent radiation in the Nearctic (Segers and De Smet, 2008).

Recently, Segers (2008) analysed the non-marine fauna (96.5% of all rotifers) and found that in monogononts 44% of the species are endemic to one biogeographic region and only 23% of the species may be considered as truly cosmopolitan (defined as present in five or more of the eight biogeographic regions); the same scenario was also shown for bdelloids, with 51% endemic and only 13% truly cosmopolitan species.

It is of course difficult to ascertain the absence of rotifer species from any specific area, especially with the dearth of available information. Moreover, many species are known only from the locality where they have been described; even in Europe, the most well-known area, *c.* 25% of the *c.* 300 European bdelloids have never been collected since their original description (Fontaneto and Melone, 2003). Thus, the high number of species supposedly endemic for only one biogeographic region may reflect a biased picture, resulting from a lack of knowledge rather than representing the actual distribution pattern. Faunistic data for rotifers are scarce, and it has been suggested that rotifer distribution follows the distribution of rotiferologists studying them and not the actual distribution of rotifers (Wallace et al., 2006). Notwithstanding this problem, some easily recognised 'flagship' species (*sensu* Foissner, 2006), are limited in distribution. For instance, the bdelloid *Pleuretra hystrix*, with a characteristic spiny lorica (Fontaneto and Melone, 2003), is known only from arctic or alpine samples; it has never been found in the Antarctic or sub-Antarctic area, despite the more intensive southern polar faunistic studies for bdelloids (Kaya et al., 2010).

It is interesting to note, however, that even some of the biological assumptions for widespread distribution in rotifers are not verified but originate from potentially unreliable generalisations. For instance, resting eggs are known only for very few species, and it is only assumed, but not proven, that all monogononts are able to produce them: thus, it is possible that only few species have such potential for passive dispersal. Moreover, resting eggs of some species do not survive desiccation, and hatching from dormant resting eggs is triggered by rather specific environmental cues, so that even if dispersal may be possible, colonisation of distant habitats is prevented.

There are also other caveats when considering rotifer biogeography. For instance, rotifer taxonomy is all but adequate and it is almost exemplary of the taxonomic impediment; many new species are still to be discovered and very few taxonomists are actively working on rotifers (Dumont, 1980; Segers and De Smet, 2008). Moreover, species identification from morphology alone may be misleading, as suggested by the increasing evidence of complexes of cryptic species in all rotifers analysed so far (Gómez et al., 2002; Schroeder and Walsh, 2007; Fontaneto et al., 2009).

These difficulties may have a significantly negative impact on our ability to understand the diversity of rotifers and consequently their distribution. Nevertheless, in spite of the unsatisfactory knowledge of their taxonomy, rotifers do exhibit patterns of diversity and distribution. Notwithstanding possible problems and biases in the interpretations of rotifer distribution, one pattern is clear: the percentage of widespread taxa is high; some species are more restricted geographically, but even the species with more restricted distributions are relatively widespread within one biogeographic region. Moreover, almost all genera and higher taxa are really cosmopolitan, one feature that is completely different from macroscopic organisms, even if the definition of taxa higher than species in different groups may not be directly comparable.

13.4 Tardigrades

(Sandra J. McInnes)

Tardigrades, common name for the phylum Tardigrada, are also known as water bears, moss piglets and bear animalcules (Fig 14.3). They were first recorded in the 1770s (Goeze, 1773; Spallanzani, 1776) and have been subject to several monographs and systematic revisions (Thulin, 1928; Marcus, 1929, 1936; Ramazzotti, 1962, 1972; Ramazzotti and Maucci, 1983). They are 'aquatic' in that they require a coating of water to permit locomotion and respiration. As a common component of the limno-terrestrial meiofauna they are found in a variety of habitats including bryophytes, cushion-forming plants, lichens, algae and soils, to truly aquatic habitats, and from polar to tropical environments. Marine tardigrades range from the tide line to the abyssal depths.

Tardigrades are typically 100–500 μm but range from 50 to 1700 μm. They are cylindrical with four pairs of lobopodal limbs that terminate in 'claws'; they have a nervous system that may include eyespots and sensory structures, a digestive tract with mouth and anus, and a reproductive system (Kinchin, 1994; Nelson, 2002). Two classes are recognised; the Heterotardigrada, incorporating the Arthrotardigrada (marine) and Echiniscoidea (marine and terrestrial); and the Eutardigrada, incorporating the Parachela (mainly terrestrial) and Apochela (terrestrial). A third class (Mesotardigrada) with a single species is considered dubious (e.g. Nelson, 2002; Nelson and McInnes, 2002) as neither the type material nor the type locality have survived. The original description of this species (*Thermozodium esakii*) was limited and modern classification now suggests a potential relationship with Carphaniidae, which is placed in the order Echiniscoidea (Binda and Kristensen, 1986). The palaeontological record for tardigrades is limited to subfossil records from Cretaceous amber (Cooper, 1964; Bertolani and Grimaldi, 2000), Eocene lake sediments (Cromer et al., 2006, 2008), and Pliocene peat cores (Jankovska, 1991; Miller and Heatwole, 2003). A current comprehensive taxonomic summary of all known extant tardigrade taxa can be found in Guidetti and Bertolani (2005) with updates in Degma and Guidetti (2007) and at: http://www.tardigrada.modena.unimo.it/miscellanea/Actual%20checklist%20of%20Tardigrada.pdf

Reproductive strategies within the Tardigrada include both sexual and parthenogenesis (Nelson, 1982a; Bertolani, 1982, 1987; Rebecchi and Bertolani, 1988, 1994; Bertolani and Rebecchi, 1999), with parthenogenesis conveying an advantage for the invasion of new habitats. Reproduction produces eggs (40–60 μm in diameter), which may be smooth and laid singly or en masse inside an exuvium or are ornamented and laid free. Both individually free eggs and those encased in exuvia have the potential for transportation. As with many of the micrometazoans, changes in environmental conditions can induce the tardigrade to enter a latent state (i.e. cryptobiosis). Cryptobiosis provides resistance to environmental extremes (i.e. cold, heat, drought, chemicals and ionising radiation), which has a significant impact on the ecological role of the organism (see Wright et al., 1992; Kinchin, 1994; Wright, 2001) and offers the potential for relatively long-range dispersal (Kristensen, 1987; Pugh and McInnes, 1998).

The combination of small size, parthenogenesis and potential for cryptobiosis had led to the assumption that limno-terrestrial tardigrades should be cosmopolitan. However, the analysis of the most likely form of transport – wind – has barely been explored. A simple experiment run by Sudzuki (1972) showed that tardigrades and other microinvertebrates were rarely dispersed by wind speeds less than 2 m/s over 2 months. In the Antarctic, Janiec (1996) and Nkem et al. (2006) found that most microinvertebrates are transported with sediment or habitat (moss, lichen) near ground level and over relatively short distances. Kristensen (1987) mentioned that *Echiniscus* sp. were 'common in raindrops or 'air plankton' after Föhn storms in Greenland'.

Tardigrada have a limited suite of morphological characters and are morphologically conserved, factors which make traditional taxonomic classification challenging. For example, in some of the early reports bi-polar tardigrade species were recorded (e.g. Richters, 1905; Murray, 1906), which in part may be a consequence of the then relatively immature state of tardigrade taxonomy, microscopy limitations and observer expectations. Subsequent researchers have used the predominantly northern hemisphere taxonomic literature to create several cosmopolitan species or group-complexes (e.g. *Echiniscus arctomys, Pseudechiniscus suillus, Macrobiotus harmsworthi, Mac. hufelandi, Minibiotus intermedius, Hypsibius convergens, H. dujardini, Diphascon* (*Diphascon*) *pingue, Milnesium tardigradum*). More recent traditional taxonomic literature has indicated that some of these group-complexes are 'hiding' a number of species that are more restricted in their distribution (e.g. Bertolani and Rebecchi, 1993; Claxton, 1998).

The confusion over group-complexes and potential errors in published taxonomic records would imply that the phylum Tardigrada was not a good subject for biogeographic studies. However, there seems to be a general understanding that while there are cosmopolitan genera, most of these exhibiting parthenogenesis and capable of cryptobiosis (Pilato, 1979), not all genera are cosmopolitan and many of these have a lower capacity for cryptobiosis and/or reproduce sexually (Kristensen, 1987). Of the 64 limno-terrestrial genera only 11 are considered endemic (Europe: *Macroversum* (Murryidae), *Necopinatum* (Necopinatidae), *Carphania* (Carphaniidae), *Pseudohexapodibius* (Macrobiotidae); Africa: *Paradiphascon* (Hypsibiidae); Asia: *Famelobiotus* (Macrobiotidea); Australasia: *Milnesioides, Limmenius* (Milnesiidae); North America: *Haplohexapodibius* (Calohypsibiidae), *Proechiniscus* (Echiniscidae); South America: *Minilentus* (Macrobiotidae)). Endemism at the species level is relatively high (between 25–58%) for the major continents (McInnes and Pugh, 2007). Despite the potential identification problems, Tardigrada can be used to explore biogeography (Pilato and Binda, 2001) and the current biogeographic distribution patterns of the non-marine tardigrades show evidence of palaeogeographic events such as the break-up of Pangaea, and the division of east and west Gondwana (McInnes and Pugh, 2007).

13.5 Micrognathozoans, cycliophorans, loriciferans, kinorhynchs and gnathostomulids

(Martin V. Sørensen)

Other microscopic animals exist which may provide further empirical evidence supporting or denying cosmopolitanism for microorganisms.

Micrognathozoa is a recently described animal group with affinities to Rotifera and Gnathostomulida (Sørensen, 2003). Only one species, *Limnognathia maerski*, is currently known, and our extremely scarce knowledge about its biogeography suggests that it has a peculiar bipolar and patchy distribution. The species was described from a spring on Disko Island, Greenland (Kristensen and Funch, 2000), where it occurs in relatively high numbers in the short Arctic summer. Among the thousands of springs on Disko Island, it has only been recorded in one other spring on the island, and only at a single occasion (R. M. Kristensen, personal communication). Outside Disko Island, the species has been recorded once from a spring area in Wales, UK (J. M. Schmid-Araya, personal communication), and from the Subantarctic Crozet Islands (De Smet, 2002). On the latter, the species is quite abundant. The Subantarctic recording of the species, in particular, is puzzling. The great distance between this locality and the Greenlandic type locality suggests that the species possess a great migratory potential, but this is contradicted by the otherwise very few recordings of the species. The most likely explanation would probably be that *L. maerski* is a relatively widespread species throughout the cold and temperate regions of both hemispheres, but that it often occurs in very low numbers, and therefore are rarely encountered. In general, *L. maerski* does not appear to be a species adapted for distribution over great ranges. The adults are short lived, without cryptobiotic capabilities, and apparently intolerant to great abiotic changes in their environment. However, since they can survive the Arctic winter, one would expect that at least the eggs are freeze tolerant, and therefore could be tolerant to dehydration and other stresses that would be lethal for the hatched specimen. One kind of micrognathozoan egg actually resembles sculptured rotifer resting eggs, but hatching from previously frozen eggs has not yet been observed, hence this adaptation is still speculative. However, if the sculptured eggs turn out to be comparable with rotifer resting eggs, they would be more suitable for dispersal than the adults.

Cycliophora is another recently discovered microscopic animal group with only few known species (Funch and Kristensen, 1995). The animals live as commensals on the mouthparts of lobsters, and are characterised by an extremely complex life cycle that involves asexual feeding stages, parthenogenetic Pandora larvae, short-lived females, Prometheus larvae that produce dwarf males, and the sexually produced chordoid larvae (Funch and Kristensen, 1997; Obst and Funch, 2003). Currently two species are described: *Symbion pandora* from the Norwegian lobster, *Nephrops norvegicus* (Funch and Kristensen, 1995), and *Symbion americanus* from the American lobster *Homarus americanus* (Obst et al., 2006). A yet to be described species has been recorded from the European lobster, *Homarus gammarus* in the Mediterranean Sea (Nedved, 2004; Baker and Giribet, 2007; R. M. Kristensen, personal communication). *Symbion pandora* has been recorded on the Norwegian lobster all along the Atlantic European West Coast from Norway to Portugal and around Faroe Island (Funch and Kristensen, 1997; Neves et al., 2010). The American species, *S. americanus*, is distributed along the North American East Coast, and has been found in lobster populations at several localities between Nova Scotia and Maryland (Obst et al., 2006; Baker et al., 2007). *Symbion americanus* appears to be a complex of cryptic species that may co-occur on the same lobster specimen (Baker et al., 2007). Knowledge on the distribution of these organisms is still scanty, but it seems that each morphologically recognisable species is restricted to specific regions (Northern Europe, east coast of North America and Mediterranean Sea, respectively), and that the cryptic species on *H. americanus* do not show any geographic patterning and their distributions overlap (Baker et al., 2007).

The restricted distribution is clearly tied to the host specificity, cycliophoran species are constrained to the distribution of the host. Furthermore, no stage in the cycliophoran life cycle is adapted for dispersal over greater ranges. The feeding stage is sessile, while the female, the Prometheus larva and the Pandora larva are only capable of moving over very short distances – probably not longer than from one mouthpart bristle to another. Usually the chordoid larva is considered the 'dispersal stage' in the life cycle, as it has the most developed locomotory ciliation (Funch and Kristensen, 1997). However, 'dispersal' in this context means dispersal to another mouth appendage, or eventually, another host in the same population. Nothing indicates that the chordoid larva can disperse over greater distances. Until now species of cycliophorans have been recorded from Europe and northeast America only, and other preliminary reports of potential cycliophorans from other parts of the world have always turned out to be sessile protists after closer examination.

Loricifera (Kristensen, 1983) is another recently described phylum of microscopic marine animal. The loriciferans are among the smallest known metazoans, but they are morphologically complex, and may have an extremely complicated life cycle with various larval and pre-adult stages. They are found in the interstices of sand and shell gravel, but may also be present in more muddy sediments. The first loriciferans were described from coastal areas in West Europe and along the North American East Coast (Kristensen, 1983; Higgins and Kristensen, 1986), but more recent studies have demonstrated a relatively high loriciferan diversity on banks and seamounts (Heiner, 2004, 2008; Gad, 2005a). In addition, the number of recordings and discovery of new species from the deep sea is currently increasing, and there are indications that the deep sea holds a diverse loriciferan fauna (Kristensen and Shirayama, 1988; Gad, 2005b, 2005c).

Loricifera have now been recorded in most parts of the world, and so the group can be considered cosmopolitan (e.g. Todaro and Kristensen, 1998; Gad and Martinez Arbizu, 2005; Heiner et al., 2009). The species, however, all have relatively limited distributions, and usually a species is restricted to a specific region, such as the West European Coast or North American East Coast. However, the distribution of the animals is extremely patchy, which limits the chances of finding a species outside its type locality, or sometimes even the chances of refinding it on a locality where it previously has been recorded. The latter is probably best exemplified by one of the first loriciferan specimens ever recorded. The specimen was collected in 1975 in Øresund, Denmark, but was lost during preparation. Knowing that the individual could represent a new animal phylum, the locality was sampled intensively for several years to obtain more specimens; but even today, 35 years later, no further loriciferans have been recorded from Øresund.

It is unclear why loriciferans occur in such a patchy distribution and why they do not, at least in coastal areas, form more continuous populations. One explanation could be very specific requirements to the sediment and abiotic conditions in their habitat, but this is only speculation. What is certain, however, is that no stage in the loriciferan life cycle is specialised for dispersal over great distances. Neither the eggs, the primary larva (Higgins larva, see Fig 13.3A), nor the adult appear to leave the sediment; even still, in the sediment they move very slowly. Species that inhabit seamounts or submarine banks are likewise trapped on these localities. For example, a species that is adapted to inhabit the calcareous shell sediment on Faroe Bank (see Heiner, 2004, 2008) would have difficulties leaving the bank and would therefore need to migrate through the surrounding muddy sediments to reach another bank in the area. Our knowledge on deep sea loriciferans is

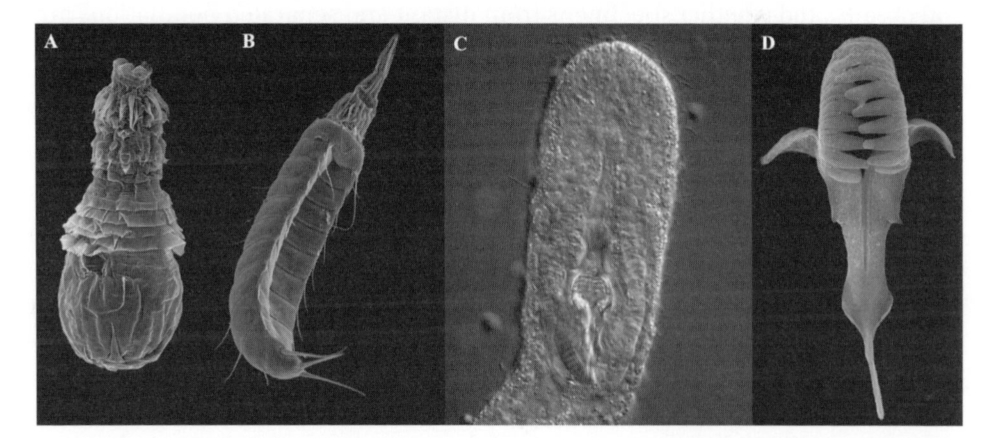

Fig 13.3 Scanning electron microscopy micrographs of (A) Higgins larva of an as yet undescribed species of Loricifera from the Western Pacific and (B) the kinorhynch species *Campyloderes* cf. *macquariae* from the Faroe Island, North Atlantic; this species could be the only known example of a cosmopolitan kinorhynch. Light (C) and scanning electron (D) micrographs of the gnathostomulid *Rastrognathia macrostoma*, showing its head with the prominent pharyngeal hard parts inside (C), and the isolated hard parts (D).

still extremely limited, but it is not unlikely that loriciferans living at great depths would show broader distributional ranges.

The Kinorhyncha represents another phylum of marine, microscopic animals. They are composed of a head with numerous appendages (so-called scalids), a short neck region and a trunk with 11 segments. Kinorhynchs show the highest diversity in muddy sediments, but are also present in sandy sediments (Sørensen and Pardos, 2008). They are known from marine localities throughout the world and have the highest levels of diversity in coastal areas. The latter, however, may be a sampling artefact due to coastal areas being sampled more frequently than less accessible oceanic and deep sea localities, and thus does not reflect the true distribution of the species.

The known kinorhynch distribution in some ways resembles the distributional patterns for loriciferans; their occurrence is patchy (although more consistent than the loriciferans), and the distribution of a species appears to be regional and in restricted areas of a few hundred kilometres squared. Our knowledge on kinorhynch distribution is still scarce. Only a few regions, such as the North American East Coast, the European West Coast and the Mediterranean, can be considered relatively well-investigated in terms of systematic kinorhynch studies, but even within these regions, the discovery of new species is not unusual (e.g. GaOrdóñez et al., 2008). When sampling in previously unstudied regions, discovering new species is much more likely than finding known ones (e.g. Sørensen, 2008; Sørensen and Rho, 2009). There are only a few examples of either cosmopolitan kinorhynch species distribution, or a distribution that spans across several oceans, and whether specimens from distant and separated populations are conspecific is questionable. One of the few kinorhynchs that, according to previous recordings, could be considered cosmopolitan is *Echinoderes dujardinii*. The species is widely distributed through the Mediterranean and along the European West Coast, but it has also been reported from Japan and the North American West Coast. In light of this rather wide distribution, Higgins (1977) revised the species and concluded that the identity of the Japanese specimens should be considered doubtful, and reports of the American specimens, in actuality, were based on a new species that he described as *E. kozloffi*. Consequently, *E. dujardinii* turned out to be a strictly European species, with a restricted and continuous distribution along the European West Coast.

There are only one or two kinorhynch species that are currently candidates as true cosmopolitans, but the taxonomy of these species, or species complexes, is puzzling and currently under investigation. The species *Campyloderes vanhoeffeni* and *C. macquariae* (Fig 13.3B) are both described from localities close to Antarctica (Zelinka, 1913; Johnston, 1938). In the descriptions the two species appear literally identical and they should most certainly be considered as a single species. Interestingly, *C. macquariae/vanhoeffeni* has over the years been reported from various localities around the world, including New Caledonia, Korea, Galapagos, the Pacific Coast of Central America and the North Atlantic (see complete list in Sørensen and Pardos, 2008). Specimens from additional localities are currently being examined (Neuhaus and Sørensen, work in progress), and the numerous recordings suggest that this could be the only example of a cosmopolitan kinorhynch species. However, specimens from the various populations tend to differ in minor details, and the consistency of this variation still requires further studies to determine if they indicate the presence of several distinct and regionally restricted species, or whether *C. macquariae/vanhoeffeni* is a true cosmopolitan.

In general kinorhynchs do not possess any dispersal mechanisms that would suggest cosmopolitanism or very wide distributional ranges. They have no locomotory cilia, and move very slowly through the sediment, and thus are unable to enter the epibenthic or pelagic zones. In fact, their highly hydrophobic cuticular surface makes even a short stay in the open water rather dangerous, as contact with a small air bubble would mean immediate adhesion and subsequent transport to the surface where they would be trapped. Additionally their eggs are not efficient dispersal stages because only one or a few eggs are laid at a time, and they are immediately coated with detritus and sediment (Kozloff, 2007). This makes it even more unlikely that the eggs could enter the water column and be dispersed by the water currents.

Gnathostomulida is a phylum of microscopic marine worms that inhabit the interstices in sandy sediments. The animals appear in many ways very simple, but are equipped with a rather complex pharyngeal apparatus (Sterrer, 1972; Sørensen and Sterrer, 2002). About 100 valid species are currently known, of which some show very narrow distributional ranges, whereas others must be considered cosmopolites (e.g. Sterrer, 1998). As is the case with many other meiofaunal organisms, Gnathostomulida is an under-sampled group and thus their biodiversity and the distributional ranges of known species are probably greater than we currently know. Nevertheless, we have indications that distributional ranges differ greatly between the species. For example, *Rastrognathia macrostoma* represents a species of the monotypic genus *Rastrognathia* (Kristensen and Nørrevang, 1977); specimens are always found in relatively high numbers at its type locality north of Zealand, Denmark, but this species or any other undescribed *Rastrognathia* species have never been recorded anywhere else. Its pharyngeal hard parts are very prominent (Fig 13.3C), and if specimens had been collected elsewhere, they would not have been confused with anything else; this indicates that *R. macrostoma* most probably has a very limited distribution. Several other gnathostomulid species are only known from single localities, and even some genera, for example *Problognathia*, *Valvognathia* and *Ratugnathia* are known from very restricted areas (Sterrer and Farris, 1975; Kristensen and Nørrevang, 1978; Sterrer, 1991).

Examples of more widely distributed species are found within the genus *Gnathostomula*. Along the European West Coast, *Gnathostomula paradoxa* is one of the most common and widely distributed gnathostomulid species. At the same latitudes, but along the North American East Coast, *G. armata* tends to be widespread and common, and further south, along the Carolinas, Florida and in the Caribbean, *G. peregrina* tends to be the dominant species.

Finally, some species appear to be true cosmopolitans; this is especially true among the filospermoid genera *Haplognathia*, *Pterognathia* and *Cosmognathia*, where the likelihood of cosmopolitanism is high. For example, the species *Haplognathia rosea* and *H. ruberrima* have been reported from the Southwest Pacific, the Central Pacific, the Caribbean, the Northwest and Northeast Atlantic, and Scandinavian waters, whereas *P. ctenifera* also is known from various localities in the Pacific, West Atlantic and Caribbean (Sterrer, 1968, 1991, 1997, 1998, 2001). These distributions indicate that all three species could be considered cosmopolitan. It is likely that molecular studies would indicate significant genetic distances between the populations and that all three would represent complexes of several cryptic species; this remains to be tested.

Currently it remains puzzling why some gnathostomulid species show great regionality whereas others have worldwide distributions. Nothing indicates that they should be specialised migrators, and since all species are strictly interstitial they would be unable to travel through muddy sediments. They are furthermore direct developers; as such there are no planktonic larvae that could serve as dispersal stages. Oviposition and egg development have only been described from a single species, *Gnathostomula jenneri*; here the egg is laid in the sediment and immediately attached to the substratum (Riedl, 1969), which disables the egg from spreading.

13.6 Nematodes

(Aldo Zullini)

Nematodes (Fig 13.4) are probably the most numerous metazoan group living on our planet. In non-desert soil there are, on average, about 2 million individuals per square metre, and at the sea bottom their abundance, following a conservative estimate, is about 100 thousand individuals per square metre (Lambshead, 2004); there are at least 10^{20} nematodes in the world. Assuming nematodes are on average 1 mm in body length, a queue of all these individuals would span a distance equivalent to 10 light years. The number of known species, in contrast, is not as impressive: there are about 27 000 nominal species, half of them free-living, and half plant and zooparasites. However, many authors think that the existing nematode species may be 10^5 or even 10^6 (Hugot et al., 2001). The biogeography of the

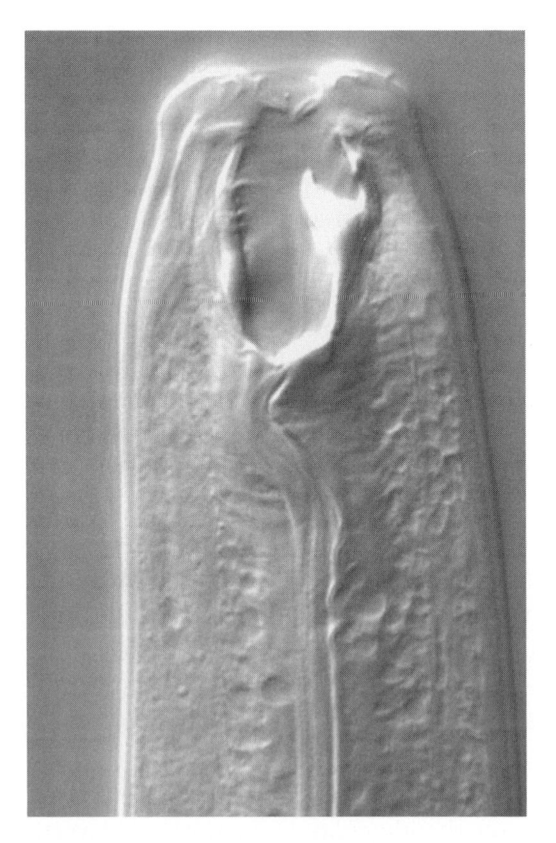

Fig 13.4 Anterior end of a soil nematode (*Mylonchulus* sp.), a predaceous species about 1.5 mm long. (Photo L. Poiras).

parasitic species depends on the hosts; therefore they obey macrofaunal dispersal logic. The biogeography of marine nematodes is still poorly known (Lambshead, 2004; Tchesunov, 2006) and goes beyond the scope of the present work. We consider the free-living soil and freshwater species only, remembering that a clearcut taxonomic distinction between these two nematode groups does not exist.

At a microscale level it is well known that soil nematodes exhibit an aggregate distribution tied to their limited dispersal and to the soil resource patchiness (Ettema and Wardle, 2002). But at the large geographic level, the existence of recognisable species patterns is still unclear. Much work was conducted on nematode dispersal; given that many soil species and some freshwater species are able to withstand dryness and low temperatures in cryptobiosis (anhydrobiosis) for many months or years, it has been postulated that nematodes in this state can be dispersed at great distances (Womersley et al., 1998; Wharton, 2004). Applying mathematical models, Carroll and Viglierchio (1981) found that dust-devil events can redeposit a significant number of nematodes within 0.5–1.5 km from a given vortex, whereas deposition at longer distances (13–40 km downwind) are much rarer. They calculated that the nematode eggs are less transportable. The dispersion of nematodes by wind is possible for dry soils and this has been studied by catching transported soil in traps. In Texas, Orr and Newton (1971) placed traps 2 m above ground to catch the wind-transported sand; in total 28 genera of free-living and plant parasitic nematodes were collected. In the most ancient European desert (44°50′N, 21°10′E, near Belgrade) the sand caught by traps 1 m above ground contained from 2 to 9 nematodes per 100 ml, mainly Dorylaimida, Tylenchomorpha and Rhabditomorpha (Krnjaic' and Krnjaic', 1973). In the peanut-cropping area of Senegal, traps (pots opened at ground level) to collect wind-transported sand, captured nematodes (mainly mycophagous and bacteriophagous species, Dorylaimida and Tylenchomorpha being less numerous) especially during the dry season (Baujard and Martiny, 1994). On the coastal dunes in the Netherlands 30 nematodes per 100 g dry soil were found in the sand traps (de Rooij-van der Goes et al., 1997).

The McMurdo Dry Valleys, Antarctica, is the driest and coldest desert on Earth: air temperature averages −20 °C, and the area practically lacks snow and ice cover since precipitation is < 5 cm water equivalent annually and the wind speed can exceed 300 km h^{-1}. Wind-transported microfauna was collected (at 77°S, 162°E) by traps and from debris deposited on ice: 10 nematodes (of three species: *Eudorylaimus antarcticus*, *Plectus antarcticus* and *Scottnema lindsayae*, all Antarctic endemites) along with 82 tardigrades and 105 rotifers were found, on average, in 100 g of wind-transported soil. No microarthropods were present (Nkem et al., 2006). It is interesting to note that the inland nunataks of Ellsworth Land (at 75°–77°S, 70°–73°W) is the only known large area without nematodes: the sole existing metazoans are rotifers and tardigrades (Freckman and Virginia, 1997; Convey and McInnes, 2005). Similar nunataks with rotifer and tardigrade diversity, but devoid of nematodes, were found in an area at 73°–75°S, 11°–14°W (Sohlenius and Boström, 2005).

Water is another important dispersal mean for nematodes. Agriculturally polluted irrigation canals in south central Washington contain 150–16 000 nematodes per cubic metre of water; these were mainly free-living but there were also 5–12% of plant parasites: it was calculated that $2–16 \times 10^9$ nematodes per day were carried past a given point (Faulkner and Bolander, 1966, 1970). Densities of 50–222 nematodes per cubic metre were found in some irrigation canals in southern Italy (Roccuzzo and Ciancio, 1991). In northern Italy near Milan, about 300 million nematodes per day are drifted by the River Adda: 38 of the 40 identified species are typical of fresh water, the commonest nematode being *Paroigolaimella bernensis* (Zullini, unpublished data). Nematodes have been found to be dispersed via tap water: data from nine different papers record 1–41 nematode species present in drinking water, and about 0–156 individuals per litre (Dózsa-Farkas, 1965; Mott and Harrison, 1983; Lupi et al., 1994).

Phoresy seems to be much less important for free-living nematode dispersal. Their association with other animals can be internal (e.g. reproductive tract or trachea) or external. Nematodes involved are mainly Rhabditomorpha (*Caenorhabditis* species included), Diplogasteromorpha and Tylenchomorpha (Gaugler and Bilgrami, 2004). Internal phoresy (in the genital chamber) on certain Diptera and Coleoptera is physiologically obligatory for some, e.g. *Paroigolaimella coprophila* (Kiontke, 1996). An example of external phoresy is the transport of *Rhabditophanes schneideri* on the legs and pedipalps of pseudoscorpions (Curčic et al., 2004). More important is the transport via large mammals. It was observed that the Great Plains (large region extending from Canada to the Mexican border) hosts an almost identical nematode faunal diversity from north to south along its 3200 km. This region was, until 1870, inhabited by about 50–100 million American bisons; these roamed the Plains for thousands of years. They would often wallow in mud to protect themselves from swarms of flies: in this way they could carry around 5–20 kg of soil containing, of course, a lot of nematodes. These bisons would transport millions of tonnes of soil; such movement of mud has made for a uniform nematode diversity across the vast region (Thorne, 1968). At present, moreover, the anthropogenic transport of plants and soil is more frequent than ever.

All the factors mentioned above involving nematode dispersal can explain why many nematode species are cosmopolitan or nearly so. Nematode cosmopolitanism was in fact assumed as a general rule, but the bulk of new data suggests an alternate view, even if the existence of small-area endemism (like those observed for insects and terrestrial gastropods) has not yet been proved.

Sohlenius (1980), examining data from 81 sites encompassing the principal biomes, found that the abundance of soil nematodes was higher in temperate than in tropical or in cold regions (Fig 13.5A). His data were only indicative, given the few sites examined in tropical soils and the different methods used to extract nematodes. Examining this work and all the existing biogeographic literature, Procter (1984, 1990) published two papers asserting the following points:

(1) Nematode species richness, in contrast with the general rule valid for most animals and plants, reaches its maximum at higher latitudes. For example, a comparison of 18 soil samples gives an average of 81 species (range 33–162) in the tundra habitats, 56 species (31–95) in temperate grassland and forest habitats and 18 species (12–24) in tropical habitats.

(2) Genus *Plectus* is dominant, or nearly so, in high-latitude faunas, whereas it is insignificant in temperate and tropical areas. Other dominant genera in high latitudes include *Tylenchus*, *Dorylaimus*, *Eudorylaimus* and *Teratocephalus*. The same genera dominate at high altitudes.

(3) Nematode densities and biomass are both high at high altitudes, whereas they are very low in the tropical forests.

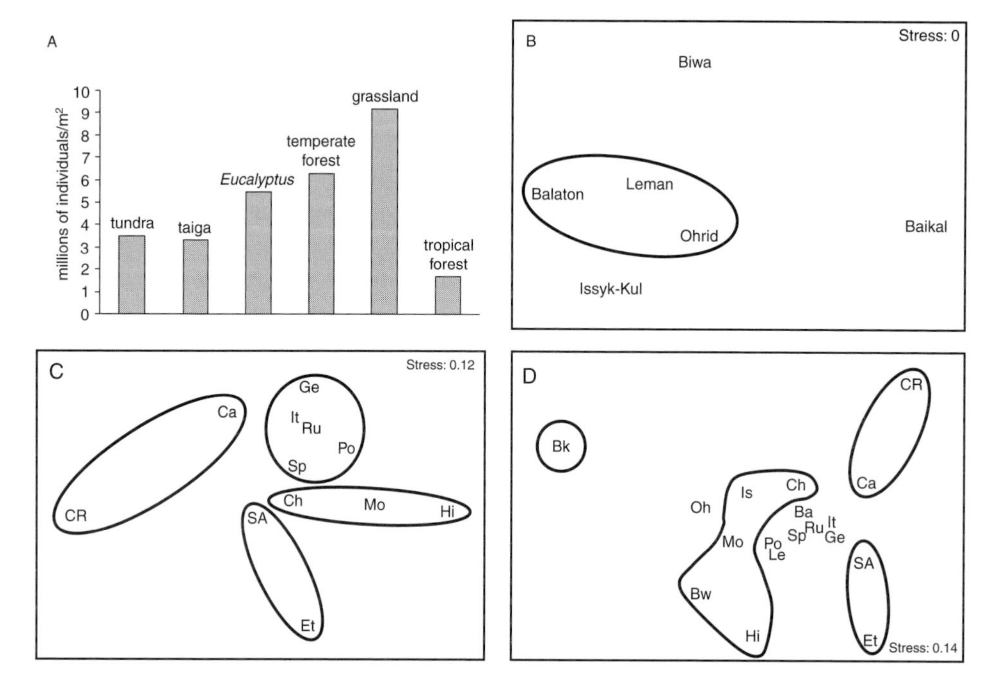

Fig 13.5 (A) Average density of soil nematodes (million individuals per square metre) in different biomes. Drawn from Sohlenius (1980) data. (B) MDS (multidimensional scaling) scatterplot of six lakes well known from the nematological point of view. (C) MDS scatterplot of freshwater regional nematofauna. America (Ca: Canada + USA; CR: Costa Rica), Europe (Ge: Germany + Austria; It: Italy; Ru: European Russia; Po: Poland; Sp: Spain), Asia (Ch: China; Mo: Mongolian waters < 5 per thousand salinity; Hi: Himalayas), Africa (SA: South Africa; Et: Ethiopia). (D) MDS scatterplot of freshwater regional nematofauna (abbreviations as before) encircling groups of five Asian faunas, two American and two African faunas, plus the nematofauna of six lakes (Bk: Baikal; Ba: Balaton; Bw: Biwa; Is: Issyk-Kul; Le: Léman; Oh: Ohrid). European fauna not encircled.

(4) At high latitudes many nematodes species (e.g. of *Plectus*, *Eudorylaimus* and *Dorylaimus*) are unusually large.

(5) Antarctic regions have lower nematode richness than do Arctic regions.

Since nematodes tolerate unfavourable conditions such as freezing and desiccation by means of an intermittent activity, Procter (1984) considers nematodes an invertebrate analogue of lichens (but see Chapter 10 for a detailed discussion on lichen biogeography). Moreover many nematodes, being often parthenogenetic (e.g. *Plectus*), can reproduce in habitats being very poor in nutrients, permitting only low population densities. The compelling short distance between individuals, usually necessary for sexual encounters, is not a limiting factor for parthenogenetic nematodes. Procter's final interpretation of the above-mentioned facts was that most microfauna (mites, springtails, etc.), usually very active in tropical soils, are hindered in the cold climates owing to the arthropods' higher ecological requirements. Consequently nematodes experience little competition and therefore can maintain relatively high diversities and densities.

The rule that nematodes are larger in cold climates is often, but not always, confirmed: for example, the largest known species of *Tobrilus* (*sensu lato*) was found on Mount Kenya (*T. elephas*: almost 7 mm long), and many relatively large nematodes were found in the Himalayas from 5200 to 6100 m above sea level (Andrássy, 1964; Zullini, 1973). Boag and Yeates (1998), examining the existing literature on 134 soil samples at different sites across the world, found that soil nematode biodiversity is lower near the poles than in temperate and tropical regions. In particular, they found minimum values of species at 20°–30° latitude (north plus south) and at 60°–80° latitude, thus contradicting the suggestion of Procter (1984) about the scarce tropical nematode biodiversity.

The assessment of geographic patterns in nematodes is difficult for three reasons. First of all, the free-living species of many regions are understudied; only in Europe, and in some parts of India, is nematode diversity sufficiently known. Second, there are generally poor taxonomic descriptions of many species erected before 1950. The last reason is the yet unclear taxonomic relevance of nematode morphological variability: for example, *Dorylaimus stagnalis* and *Monhystera stagnalis* are species reported from lakes and rivers all over the world, but it is probable that these names include, in fact, a large complex of similar, but different species. Molecular analyses will perhaps resolve this and many other similar problems.

Despite these problems, it seems that some geographic patterns are evident. In 1964 Andrássy named Africa 'das Land der Actinolaimen'. In fact this family (whose species inhabit fresh water, moss and humid soil) containing 132 valid species, includes only two real cosmopolitan species. Africa hosts the largest number of Actinolaimidae (30%), followed by India (25%) and by Neotropic regions (23%). In total, 84% of the species of this family are known for the Gondwanian regions only. In 1970 Dao compared the two, at that time, best-known temperate and tropical soil faunas: the nematodes of the Netherlands (362 species) and Venezuela (113 species). Excluding the doubtful species and the Venezuelan non-tropical records (nematodes from above 1200 m), only 5% of the considered species were really eurytopic, i.e. common to both temperate and tropical areas. Loof (1971), in the same years, examined almost 25 000 nematodes from Spitzbergen (78°–79°N) and concluded that some genera (*Tylenchus*, *Teratocephalus*, *Cervidellus*, *Plectus*, *Ereptonema*, *Prismatolaimus*) are much more dominant on Spitzbergen than in the temperate regions. A surprising fact was the complete absence of Rhabditidae,

Diplogastridae, Mononchidae and Aporcelaimidae (less surprising was the absence of some families of plant parasitic nematodes). The Spitzbergen nematofauna was found to be very similar to that of northern Canada, Greenland and Novaya Zemlya.

The biogeography of freshwater species has received much less attention and thus entails many unsolved problems. One puzzling case is given by the nematodes living in freshwater habitats in the Galápagos archipelago: two of the 18 collected species were new, six were widely distributed in the southern hemisphere, and the remaining 10 were cosmopolitan. Since Galápagos are remote volcanic (therefore oceanic) islands 960 km away from the South American coast, the most likely vector of these nematodes, obviously in resistant stages, is the passive and very occasional transport by birds (Eyualem-Abebe and Coomans, 1995). The ability of freshwater nematodes to withstand harsh conditions, including long-distance transport, is illustrated by Hodda et al. (2006). The most dominant species in Alpine lakes are also cosmopolitan, but a survey on these lakes found that only three species (*Eumonhystera filiformis, E. longicaudatula* and *Tripyla glomerans*) were found in all the studied lakes (Traunspurger et al., 2006). Eyualem-Abebe et al. (2008) published an interesting paper on the global diversity of freshwater nematodes. They pointed out that this group of nematodes is only 7% of all the nominal nematode species and discussed the geographic distribution of the nematode orders including free-living species.

To find possible geographic patterns in freshwater nematodes, 102 papers and species lists from 1913 to 2003 were examined. After some taxonomic corrections, 717 nominal species were found; only strictly freshwater nematodes were considered, excluding the Tylenchomorpha (most of them being plant-parasitic). Comparing nematodes of six biogeographically important and well-studied lakes (Baikal: the most ancient lake and the richest in endemics; Balaton: a remnant of Paratethys; Biwa: the largest Japanese freshwater lake; Leman: the largest Alpine lake; Issyk-Kul + Sonkul: the largest endorheic lake; Ohrid: the most ancient European lake) by means of multidimensional scaling, a good propinquity (stress = 0) connects the three European lakes, whereas Baikal is placed well apart from all other lakes (Fig 13.5B). A well-studied lake for nematodes is the Königssee (Traunspurger, 1991), but in the following diagrams it is omitted since its placements practically coincide with the Léman: this means that these two Alpine lakes, whose distance apart is exactly 500 km, are extremely similar in their nematode fauna. Comparing the freshwater nematodes at the regional level (excluding the data of the above-mentioned lakes, and including rivers and other lakes), four distinct groups, at a continental level, emerge (Fig 13.5C). By including all data together (regional data plus the data of the six lakes) the following pattern emerges (Fig 13.5D): European faunas are well grouped together, excluding Ohrid (pliocenic lake rich in unusual species) and Baikal, of which the latter is placed in an isolated position (Zullini, unpubl. data).

All these data prove that both the soil and the freshwater species (or freshwater groups of species) are not distributed randomly in the world.

It is one thing to discover natural patterns and it is another to tentatively intepret them. In our case, interpretation is based upon the distinction between environmental (ecological) and historical (evolutionary) factors: for the small eukaryotes, in particular, declaring that some 'species' are cosmopolitan might be approximately equivalent to saying that a genus or family of birds is cosmopolitan (Martiny et al., 2006). The concept of 'small' animal, and of the related microfauna concept, usually refers to the body length (generally defined as < 2 mm), but many soil and freshwater nematodes are outside the 2 mm length, and yet still remain within the frame of the microfauna. Therefore the maximum threshold (2 mm) should refer to the maximum body width and not to the body length. The geographic distribution of nematode species attacking roots of cultivated plants has been investigated (e.g. Navas et al., 1993; Coomans, 1996). Biogeographic analyses on soil free-living species were made by Ferris et al. (1976, 1981) adopting a plate tectonics perspective and using cladistic analysis. They analysed the family Leptonchidae because these hyphal and omnivorous feeders do not have cryptobiotic ability and their geographic distribution is rather clear. The genus *Tyleptus*, in particular, was regarded as a Gondwanian genus, except for one species present in Venezuela and in North America, which probably migrated northwards via the Central America land bridge. All other genera of this family radiated primarily in Gondwana areas, except *Funaria* which is of Laurasian origin. The fact that one of the cladograms (about *Tyleptus*) published by these authors presents species/continental area correspondences, could raise strong objections by many biogeographers, as a species level is generally considered a too low systematic level (= recent origin) to be correlated with the ancient tectonic splits. However, we must remember that species evolution in nematodes has a different pace: for example, the free-living soil nematodes *Caenorhabditis elegans* and *C. briggsae* (whose genome sequences were completed in 2002 and 2003, respectively) are morphologically almost indistinguishable, despite the fact that their most recent common ancestor existed about a 100 million years ago (Hillier et al., 2007).

In conclusion, it is still not clear what kind of geographic pattern is really true and important for the continental free-living nematodes, nor if there is a geographic parallelism between the soil and the freshwater species, nor if vicariance or dispersion are the main biogeographic factors. We can only foresee that new field collections and DNA data, that many laboratories are collecting with a constantly increasing speed, will permit us to define a picture of nematode distribution and history; one which is vaster and more correct than any we can conceive at present.

13.7 Flatworms

(Tom Artois)

Flatworms (Platyhelminthes) are a species-rich group of acoelomate, soft-bodied, protostome animals (over 20 000 species described). Traditionally they were subdivided into the parasitic Neodermata (*c*. 15 000 species) and the free-living or symbiotic 'Turbellaria' (*c*. 6500 species), this group included all the flatworms that do not replace the ciliated epidermis with a non-ciliated one (the so-called neodermis) during ontogenetic development. Phylogenetic analyses, based on molecular as well as morphological data, have made clear that 'Turbellaria' is not monophyletic (see Willems et al., 2006 and references therein). However, the term turbellarian is still commonly used as a vernacular name to indicate all non-parasitic free-living flatworms.

Until recently, four large taxa were recognised within Platyhelminthes: Acoela, Nemertodermatida, Catenulida and Rhabditophora (including Neodermata). However, it is clear now that Acoela and Nemertodermatida do not belong to the monophyletic Platyhelminthes, but are basal bilaterians (Hejnol et al., 2010). Here we will restrict the term microturbellaria to all species of the Catenulida and Rhabditophora that are free-living or symbiotic, and are less than a few millimetres in length (Fig 13.6). Hence Neodermata, triclads (planarians) and polyclads are not treated. Defined as such, microturbellaria includes about 3300 species. Microturbellaria all are simultaneous hermaphrodites, which

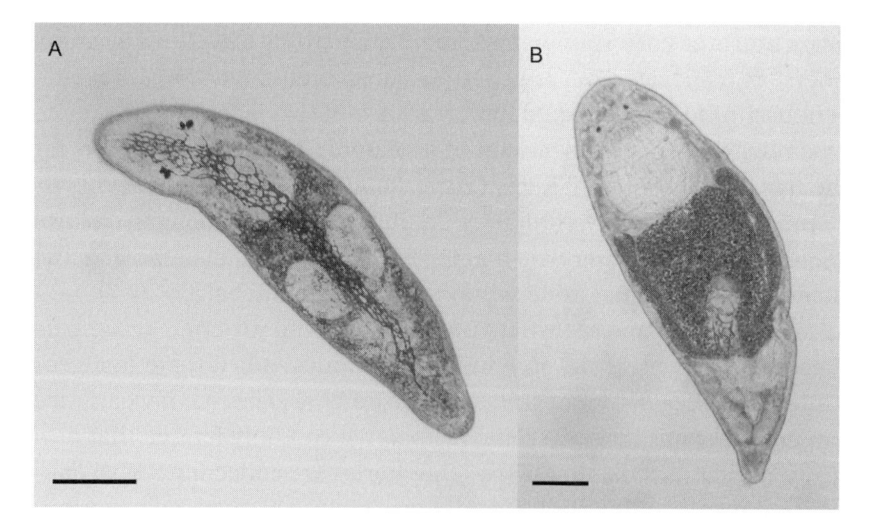

Fig 13.6 Two microturbellarians from India, December 2008. (A) *Trigonostomum franki* (marine) and (B) a new species of Dalyelliidae (freshwater). Scale bars 0.1 mm.

(probably) mostly cross-fertilise, although self-fertilisation can occur (Sekera, 1906). They have a direct development, without any free-living larvae. An overview of the world's distribution of freshwater turbellaria was recently provided by Schockaert et al. (2008). As yet, a worldwide review of marine turbellaria is not available.

Microturbellarians can be found in all types of wet environments all over the world: streams, ponds, salt marshes, sandy and rocky beaches and different kinds of sublittoral environments. They either live interstitially between the sand grains, epiphytically on algae and/or plants, or epizoically, e.g. on cnidarians and bryozoans. Whereas most turbellarians need wet habitats, about 50 species can occur in moist terrestrial environments. Very few species are described from dry environments (desiccated mosses), but recent collections have shown that at least rhabdocoel flatworms are ubiquitous in these temporarily dry environments (Van Steenkiste et al., 2010). These terrestrial rhabdocoels survive periods of extreme drought by encystment, often after the deposition of one or two eggs. A few days after excystment, the worms start laying eggs (for details, see Van Steenkiste et al., 2010). Encystment by adult worms, to survive adverse conditions, has also been reported in freshwater lecithoepitheliates and the freshwater species *Bothrioplana semperi* (see Luther, 1960). A number of limnic species produce resting eggs, which can survive periods of drought or cold. For example, mature specimens of five species of rhabdocoels were collected a few days after inundation of sediment containing resting eggs from a dried-out pool in Botswana (Artois et al., 2004). Such resting eggs have been reported from many taxa: Catenulida (Hyman, 1951), Macrostomida (Graff, 1913; Ingole, 1987), and Rhabdocoela (Graff, 1913; Luther, 1955; Young, 1974; Heitkamp, 1988).

Very little is known about the dispersal capacities of microturbellaria. Cases of worms in rafting material (such as coconuts, seaweeds, drifting algae and plants, driftwood) have been reported (Gerlach, 1977); this could be a major source of dispersal of the adult (and juvenile) worms, and of eggs. In polar areas, drifting sea ice could be an important carrier of flatworms. For acoels it has been known for a long time that they can survive in sea ice (see Gradinger et al., 1999; Janssen and Gradinger, 1999 and references therein; Friedrich and Hendelberg, 2001), and recently rhabditophorans have been discovered in the sea ice endofauna of the Weddell Sea (Melnikov, 1997; M. Kramer, personal communication). Worms as well as eggs could be displaced by animals that regularly visit moist habitats, such as waterfowl (Steinböck, 1931). Resting stages of limnic species, such as cysts and resting eggs, could be important stages for airborne or phoretic dispersal (Reisinger and Steinböck, 1927; Young and Young, 1976; Vanschoenwinkel et al., 2008a, 2008b, 2009). In a recent study, Vanschoenwinkel et al. (2008c) found a relatively high number of viable turbellarian propagules in the faeces of wild boars, and in mud from rubbing trees used by these animals, which indicates that large mammal activity could be an important means of dispersal. Moreover, flooding has been shown to be important for the dispersal of microturbellarian resting stages (Vanschoenwinkel et al., 2008a). For interstitial marine microturbellarians, sand displacement by wave action and sea currents can be a means of egg and individual worm dispersal, which, together with the sand, are swept from their original location and deposited in a new location (Boaden, 1964, 1968). Moreover, and somewhat surprisingly, microturbellarians (and other meiofauna) are readily displaced while suspended in the water column (Palmer, 1988). They can enter the water column by accident, as a consequence of heavy wave actions, but they can also enter it actively. In an experiment carried out by Hagerman and Rieger (1981), specimens of several flatworm taxa, along with some other meiofauna taxa (predominantly nematodes), were retrieved from a meiofauna trap that was suspended in the water column. As such they can be dispersed by water currents, and colonise new areas. When mature, flatworms often carry one or more eggs, or have viable sperm stored from previous copulations, which can be used to fertilise eggs. Moreover, self-fertilisation can occur, although seldom does (Sekera, 1906). By these means, one individual can be responsible for the colonisation of a new locality. In more recent times, anthropogenic introduction of microturbellarians in new localities has occurred, for instance by the use of large bodies of sediment as counter ballast in sailing ships, or by the (international) trade of fauna and flora (Young and Young, 1976; Gerlach, 1977; Faubel and Gollasch, 1996).

Although the above-mentioned means of dispersal undoubtedly occur, it is far from clear how much these processes have influenced present-day distribution of microturbellaria. According to some authors, there are almost no boundaries to the spread of small interstitial animals, whereas others consider continental drift to be the major historical cause of the present-day distribution of these animals; the processes discussed above are only responsible for intracontinental rather than intercontinental dispersal (reviewed by Sterrer, 1973). At least the latter option seems conceivable for freshwater turbellarians; for these animals, large marine waters constitute an unbridgeable biogeographic barrier.

Based on the data sets compiled in the framework of the FADA-project (Balian et al., 2008; data set not publicly available yet), we calculated that about 50–80% (depending on the taxon considered) of the freshwater species are known from their type locality and the nearby vicinity only. Based on the data available in the Turbellarian taxonomic database (Tyler et al., 2006–2009) and on our own field experiences, the situation is comparable for the marine taxa. These high numbers of species with restricted distributions are reflective of our lack of knowledge as to the real distributions, rather than indicative of high levels of endemicity. In Europe, for instance, most species generally show a much wider distribution than species known from other continents; occurring in comparable habitats on the entire continent (including Russia). The obvious reason is sampling bias, since most flatworm taxonomists in the past have been European, sampling predominantly in their own 'backyard'. Only Lake Baikal seems to be a real hotspot of microturbellarian endemicity (Schockaert et al., 2008).

On the other hand, there are many examples of species with a very disparate distribution. For instance, *Plagiostomum acoluthum* (Prolecithophora) is known from Hawai'i and Brazil (Karling et al., 1972), *Annalisella bermudensis* (Rhabdocoela) is recorded from Bermuda, Curaçao and Zanzibar (Artois and Tessens, 2008); these are but a few of numerous examples. More specifically, many species are known from localities in the European Northern Atlantic (Scandinavian and German Coast), the Mediterranean and Black Sea, and often the North American Atlantic Coast. Not coincidentally, these are the most densely sampled areas. Therefore, one could easily state that distribution maps of species of microturbellaria actually reflect the distributions of taxonomists, and the expeditions they have done. Without doubt, denser sampling will show that many species actually have a much wider distribution than is thought now.

In almost all taxa there are species that have an extremely wide or even cosmopolitan distribution. The most renowned example is *Gyratrix hermaphroditus* (Rhabdocoela). This 'species' occurs in freshwater, marine and brackish water habitats from the north pole to the south pole. However, populations (sometimes sympatric) can differ in the number of chromosomes, the morphology of the chromosomes and the detailed morphology and dimensions of the hard parts of the copulatory organ. This variation can even be found within populations. To what extent these differences indicate reproductive isolation and/or evolutionary differentiation is unknown, but it is clear that *Gyratrix hermaphroditus* represent a complex of species (see Curini-Galletti and Puccinelli, 1998; Timoshkin et al., 2004 and references therein). Most other 'cosmopolitan species' often belong to taxa in which species identification is extremely difficult (e.g. some species of *Mesostoma*, several taxa within Typhloplanidae, some species of *Macrostomum*, many species of Catenulida, etc.). A detailed molecular and morphological study of these taxa is certainly necessary to indicate whether these species are indeed cosmopolitan, or whether they form complexes of many cryptic taxa. The molecular study of cryptic biodiversity, however, has only very recently been started (e.g. Casu and Curini-Galletti, 2006).

It is clear that, at present, too few data are available to make any definitive inferences about patterns of distributions in microturbellaria. On one hand, it is clear that many species probably have a much wider distribution than is now known. On the other hand, many cosmopolitan species could represent complexes of sibling species, each with a much narrower distribution. Only intensive sampling, a combination of molecular and morphology-based taxonomy and experimental laboratory and field studies on the dispersal capacities of microturbellarians, will give a realistic image of their distribution patterns.

References

Andrássy, I. (1964). Süsswasser-Nematoden aus den grossen Gebirgsgegenden Ostafrikas. *Acta Zoologica, Budapest* **10**, 1–59.

Artois, T., Tessens, B. (2008). Polycystididae (Rhabditophora: Rhabdocoela: Kalyptorhynchia) from the Indian Ocean, with the description of twelve new species. *Zootaxa* **1849**, 1–27.

Artois, T., Willems, W., De Roeck, E., Jocqué, M., Brendonck, L. (2004). Freshwater Rhabdocoela (Platyhelminthes) from ephemeral rock pools from Botswana, with the description of four new species and one new genus. *Zoological Science* **21**, 1063–1072.

Baker, J.M., Giribet, G. (2007). A molecular phylogenetic approach to the phylum Cycliophora provides further evidence for cryptic speciation in *Symbion americanus*. *Zoologica Scripta* **36**, 353–359.

Baker, J.M., Funch, P., Giribet, G. (2007). Cryptic speciation in the recently discovered American cycliophoran *Symbion americanus*; genetic structure and population expansion. *Marine Biology* **151**, 2183–2193.

Balian, E.V., Segers, H., Lévêque, C., Martens, K. (2008). Freshwater animal diversity assessment: an introduction to the Freshwater Animal Diversity Assessment (FADA) project. *Hydrobiologia* **595**, 3–8.

Ball, I.R. (1976). Nature and formulation of biogeographical hypothesis. *Systematic Zoology* **24**, 407–430.

Balsamo, M. (1992) Hermaphroditism and parthenogenesis in lower Bilateria: Gnathostomulida and Gastrotricha. In Dallai, R. (ed.), *Sex Origin and Evolution*, pp. 309–327. Modena: Mucchi editore.

Balsamo, M., Todaro, M.A. (1988). Life history traits of two chaetonotids (Gastrotricha) under different experimental conditions. *Invertebrate Reproduction and Development* **14**, 161–176.

Balsamo, M., d'Hondt, J.-L., Kisielewski, J., Pierboni, L. (2008). Global diversity of gastrotrichs (Gastrotricha) in fresh waters. *Hydrobiologia* **595**, 85–91.

Balsamo, M., d'Hondt, J.-L., Pierboni, L., Grilli, P. (2009). Taxonomic and nomenclatural notes on freshwater Gastrotricha. *Zootaxa* **2158**, 1–19.

Baujard, P., Martiny, B. (1994). Transport of nematodes by wind in the peanut cropping area of Senegal, West Africa. *Fundamental and Applied Nematology* **17**, 543–550.

Bertolani, R. (1982). Cytology and reproductive mechanisms in tardigrades. In Nelson, D.R. (ed.), *Proceedings of the Third International Symposium on the Tardigrada*, pp. 93–114. August 3–6, 1980, Johnson City, Tennessee. Johnson City, TN: East Tennessee State University Press.

Bertolani, R. (1987) Sexuality, reproduction, and propagation in tardigrades. In Bertolani, R. (ed.), *Biology of Tardigrades. Selected Symposia and Monographs UZI* **1**, 93–101.

Bertolani, R., Grimaldi, D. (2000). A New Eutardigrade (Tardigrada: Milnesiidae) in amber from the Upper Cretaceous (Turonian) of New Jersey. In Grimaldi, D. (ed.), *Studies on Fossils in Amber, with Particular Reference to the*

Cretaceous of New Jersey, pp. 103–110. Leiden: Backhuys Publishers.

Bertolani, R., Rebecchi, L. (1993). A revision of the *Macrobiotus hufelandi* group (Tardigrada, Macrobiotidae), with some observations on the taxonomic characters of eutardigrades. *Zoologica Scripta* **22**, 127–152.

Bertolani, R., Rebecchi, L. (1999). Tardigrada. In Knobil, E., Neill, J.D. (eds.), *Encyclopedia of Reproduction*, Vol. 4, pp. 703–718. San Diego, CA: Academic Press.

Binda, M.G., Kristensen, R.M. (1986). Notes on the genus *Oreella* (Oreellidae) and the systematic position of *Carphania fluviatilis* Binda, 1978 (Carphanidae fam. nov., Heterotardigrada). *Animalia* **13**, 9–20.

Boaden, P.J.S. (1964). Grazing in the interstitial habitat: a review. In Crisp, D.J. (ed.), Grazing in terrestrial and marine environments. *British Ecological Society Symposium* **4**, 299–303.

Boaden, P.J.S. (1968). Water movement – a dominant factor in interstitial ecology. *Sarsia* **34**, 125–136.

Boag, B., Yeates, G.W. (1998). Soil nematode biodiversity in terrestrial ecosystems. *Biodiveristy and Conservation* **7**, 617–630.

Carroll, J.J., Viglierchio, D.R. (1981). On the transport of nematodes by the wind. *Journal of Nematology* **13**, 476–482.

Casu, M., Curini-Galletti, M. (2006). Genetic evidence for the existence of cryptic species in the mesopsammic flatworm *Pseudomonocelis ophiocephala* (Rhabditophora: Proseriata). *Biological Journal of the Linnean Society* **87**, 553–576.

Claxton, S.K. (1998). A revision of the genus *Minibiotus* (Tardigrada: Macrobiotidae) with descriptions of eleven new species from Australia. *Records of the Australian Museum* **50**, 125–160.

Convey, P., McInnes, S. (2005). Exceptional tardigrade-dominated ecosystems in Ellsworth Land, Antarctica. *Ecology* **86**, 519–527.

Coomans, A. (1996). Phylogeny of the Longidoridae. *Russian Journal of Nematology* **4**, 51–60.

Cooper, K.W. (1964). The first fossil tardigrade: *Beorn leggi* Cooper, from Cretaceous amber. *Psyche* **71**, 41–48.

Cromer, L., Gibson, J.A.E., Swadling, K.M., Hodgson, D.A. (2006). Evidence for a lacustrine faunal refuge in the Larsemann Hills, East Antarctica, during the Last Glacial Maximum. *Journal of Biogeography* **33**, 1314–1323.

Cromer, L., Gibson, J.A.E., McInnes, S.J., Agius, J.T. (2008). Tardigrade remains from lake sediments. *Journal of Paleolimnology* **39**, 143–150.

Curčic, B.P.M., Sudhaus, W., Dimitrijevic, R.N. (2004). Phoresy of *Rhabditophanes* schneideri (Bütschli) (Rhabditida: Alloionematidae) on pseudoscorpiones (Arachnida: Pseudoscorpiones). *Nematology* **6**, 313–317.

Curini-Galletti, M.C., Puccinelli, I. (1998). The *Gyratix hermaphroditus* species complex (Kalyptorynchia: Polycystididae) in marine habitats of eastern Australia. *Hydrobiologia* **383**, 287–298.

Dao, F. (1970). Climatic influence on the distribution pattern of plant parasitic and soil inhabiting nematodes. *Mededelingen Landbouwhogenschool Wageningen* **70**, 1–181.

De Ridder, M. (1981). Some considerations on the geographic distribution of rotifers. *Hydrobiologia* **85**, 209–225.

de RooiJ-van der Goes, P.C.E.M., van Dijk, C., van der Putten, W.H, Jungerius, P.D. (1997). Effects of sand movement by wind on nematodes and soil-borne fungi in coastal foredunes. *Journal of Coastal Conservation* **3**, 133–142.

De Smet, W.H. (2002). A new record of *Limnognathia maerski* Kristensen & Funch, 2000 (Micrognathozoa) from the subantarctic Crozet Islands, with redescription of the trophi. *Journal of Zoology* **258**, 381–393.

Degma, P., Guidetti, R. (2007). Notes to the current checklist of Tardigrada. *Zootaxa* **1579**, 41–53.

Dózsa-Farkas, K. (1965). Untersuhungen über die Fauna des Budapester Leitungswassers, mit besonderer Berücksichtigung der Nematoden. *Opuscola Zoologica, Budapest* **5**, 173–181.

Dumont, H.J. (1980). Workshop on taxonomy and biogeography. *Hydrobiologia* **73**, 205–206.

Dumont, H.J. (1983). Biogeography of rotifers. *Hydrobiologia* **104**, 19–30.

Dumont, H., Segers, H. (1996). Estimating lacustrine zooplankton species richness and complementarity. *Hydrobiologia* **341**, 125–132.

Ettema, C.H., Wardle, D. (2002). Spatial soil ecology. *Trends in Ecology and Evolution* **17**, 177–183.

Euyalem-Abebe, Coomans, A. (1995). Freshwater nematodes of the Galápagos. *Hydrobiologia* **299**, 1–51.

Eyualem-Abebe, Decraemer, W., De Ley, P. (2008). Global diversity of nematodes (Nematoda) in freshwater. *Hydrobiologia* **595**, 67–78.

Faubel, A., Gollasch, S. (1996). *Cryptostylochus hullensis* sp. nov. (Polycladida, Acotylea, Platyhelminthes): a possible case of transoceanic dispersal on a ship's hull. *Helgoländer Meeresuntersuchungen* **50**, 533–537.

Faulkner, L.R., Bolander, W.J. (1966). Occurrence of large nematode populations in irrigation canals in South Central Washington. *Nematologica* **12**, 591–600.

Faulkner, L.R., Bolander, W.J. (1970). Agriculturally-polluted irrigation water as a source of plant-parasitic nematode infestation. *Journal of Nematology* **2**, 368–374.

Fenchel, T., Finlay, B.J. (2004). The ubiquity of small species: patterns of local and global diversity. *Bioscience* **54**, 777–784.

Ferris, V.R., Goseco, C.G., Ferris, J.M. (1976). Biogeography of free-living soil nematodes from the perspective of plate tectonics. *Science* **193**, 508–510.

Ferris, V.R., Ferris, J.M., Goseco, C.G. (1981). Phylogenetic and biogeographic hypotheses in Leptonchidae (Nematoda: Dorylaimida) and a new classification. *Proceedings of the Helminthological Society, Washington* **48**, 163–171.

Foissner, W. (2006). Biogeography and dispersal of micro-organisms: a review emphasizing protists. *Acta Protozoologica* **45**, 111–136.

Fontaneto, D., Melone, G. (2003). Redescription of *Pleuretra hystrix*, an endemic alpine bdelloid rotifer. *Hydrobiologia* **497**, 153–160.

Fontaneto, D., Ficetola, G.F., Ambrosini, R., Ricci, C. (2006). Patterns of diversity in microscopic animals: are they comparable to those in protists or in larger animals? *Global Ecology and Biogeography* **15**, 153–162.

Fontaneto, D., Herniou, E.A., Barraclough, T.G., Ricci C. (2007). On the global distribution of microscopic animals: new worldwide data on bdelloid rotifers. *Zoological Studies* **46**, 336–346.

Fontaneto, D., Kaya, M., Herniou, E.A., Barraclough, T.G. (2009). Extreme levels of hidden diversity in microscopic animals (Rotifera) revealed by DNA taxonomy. *Molecular Phylogenetics and Evolution* **53**, 182–189.

Freckman, D.W., Virginia, R.A. (1997). Low-diversity Antarctic soil nematode communities: distribution and response to disturbance. *Ecology* **78**, 363–369.

Friedrich, C., Hendelberg, J. (2001). On the ecology of Acoela living in the Arctic Sea ice. *Belgian Journal of Zoology* **131 (Supplement 1)**, 213–216.

Funch, P., Kristensen, R.M. (1995). Cycliophora is a new phylum with affinities to Entoprocta and Ectoprocta. *Nature* **378**, 711–714.

Funch, P., Kristensen, R.M. (1997). Cycliophora. In Harrison, F.W., Woollacott, R.M. (eds.), *Microscopic Anatomy of Invertebrates*, Vol. 13. *Lophophorates, Entoprocta, and Cycliophora*, pp. 409–474. New York, NY: Wiley-Liss.

Gad, G. (2005a). Successive reduction of the last instar larva of Loricifera, as evidenced by two new species of *Pliciloricus* from the Great Meteor Seamount (Atlantic Ocean). *Zoologischer Anzeiger* **243**, 239–271.

Gad, G. (2005b). Giant Higgins-larvae with paedogenetic reproduction from the deep sea of the Angola Basin – evidence for a new life cycle and for abyssal gigantism in Loricifera? *Organisms, Diversity and Evolution* **5**, 59–75.

Gad, G. (2005c). A parthenogenetic, simplified adult in the life cycle of *Pliciloricus pedicularis* sp. n. (Loricifera) from the deep sea of the Angola Basin (Atlantic). *Organisms, Diversity and Evolution* **5**, 77–103.

Gad, G., Martinez Arbizu, P. (2005). First description of an Arctic Loricifera – a new *Rugiloricus*-species from the Laptev Sea. *Marine Biology Research* **1**, 313–325.

GaOrdóñez, D., Pardos, F., Benito, J. (2008). Three new *Echinoderes* (Kinorhyncha, Cyclorhagida) from North Spain, with new evolutionary aspects in the genus. *Zoologischer Anzeiger* **247**, 95–111.

Gaugler, R., Bilgrami, A.L. (2004). Nematode Behaviour. Wallingford: CABI Publishing.

Gerlach, S.A. (1977). Means of meiofauna dispersal. *Microfauna Meeresboden* **61**, 89–103.

Goeze, J.A.E. (1773). Über den kleinen Wasserbär. In Bonnet, K. (ed.), *Abhandlungen aus der Insektologie*, pp. 367–375. Halle: JJ Gebauers Wittwe und Joh Jac Gebauer.

Gómez, A., Serra, M., Carvalho, G.R., Lunt, D.H. (2002). Speciation in ancient cryptic species complexes: evidence from the molecular phylogeny of *Brachionus plicatilis* (Rotifera). *Evolution* **56**, 1431–1445.

Gradinger, R., Friedrich, C., Spindler, M. (1999). Abundance, biomass and composition of the sea ice biota of the Greenland sea pack ice. *Deep Sea Research II* **46**, 1457–1472.

Graff, L. von (1913). Platyhelminthes. Turbellaria II. Rhabdocoelida. *Tierreich* **35 II–XX**, 1–484.

Green, J. (1972). Latitudinal variations in associations of planktonic rotifers. *Journal of Zoology* **167**, 31–39.

Guidetti, R., Bertolani, R. (2005). Tardigrade taxonomy: an updated check list of the taxa and a list of characters for their identification. *Zootaxa* **845**, 1–46.

Hagerman, G.M., Rieger, R.M. (1981). Dispersal of benthic meiofauna by wave and current action in Bogue Sound, N.C., USA. *PSZN Marine Ecology* **2**, 245–270.

Heiner, I. (2004). *Armorloricus kristenseni* (Nanaloricidae, Loricifera), a new species from the Faroe Bank (north Atlantic). *Helgoland Marine Research* **58**, 192–205.

Heiner, I. (2008). *Rugiloricus bacatus* sp. nov. (Loricifera – Pliciloricidae) and a ghost-larva with paedogenetic reproduction. *Systematics and Biodiversity* **6**, 225–247.

Heiner, I., Boesgaard, T.M., Kristensen, R.M. (2009). First time discovery of Loricifera from Australian waters and marine caves. *Marine Biology Research* **5**, 529–546.

Heitkamp, U. (1988). Life-cycles of microturbellarians of pools and their strategies of adaptation to their habitats. *Progress in Zoology* **36**, 449–456.

Hejnol, A., Obst, M., Stamatakis, A. et al. (2010). Assessing the root of bilaterian animals with scalable phylogenomic methods. *Proceedings of the Royal Society B* **276**, 4261–4270.

Higgins, R.P. (1977). Redescription of *Echinoderes dujardinii* (Kinorhyncha) with descriptions of closely related species. *Contributions to Zoology* **248**, 1–26.

Higgins, R.P., Kristensen, R.M. (1986). New Loricifera from Southeastern United States Coastal Waters. *Smithsonian Contributions to Zoology* **438**, 1–70.

Hillier, L.W., Miller, R.D., Baird, S.E. et al. (2007). Comparison of *C. elegans* and *C. briggsae* genome sequence reveals extensive conservation of chromosome organization and synteny. *PLoS Biology* **5**, 1603–1616.

Hochberg, R. (2005). Musculature of the primitive gastrotrich *Neodasys* (Chaetonotida): Functional adaptations to the interstitial environment and phylogenetic significance. *Marine Biology* **146**, 315–323.

Hodda, M., Ocaña, A., Traunspurger, W. (2006). Nematodes from extreme freshwater habitats. In Abebe, E., Traunspurgerm, W., Andrássy, I. (eds.), *Freshwater Nematodes: Ecology and Taxonomy*, pp. 179–210. Wallingford: CABI Publishing.

Hugot, J.P., Baujard, P., Morand, S. (2001). Biodiversity in helminths and nematodes as a field of study: an overview. *Nematology* **3**, 199–208.

Hummon, M.R. (1984). Reproduction and sexual development in a freshwater gastrotrich. 1. Oogenesis of parthenogenic eggs (Gastrotricha). *Zoomorphologie* **104**, 33–41.

Hummon, W.D. (2008). Gastrotricha of the North Atlantic Ocean: 1. Twenty four new and two redescribed species of Macrodasyida. *Meiofauna Marina* **16**, 117–174.

Hummon, W.D., Todaro, M.A. (2007). A new species of Xenotrichulidae (Gastrotricha) from southern and southeastern USA. *Cahiers de Biologie Marine* **48**, 297–302.

Hummon, W.D., Todaro, M.A. (2010). Analytic taxonomy and notes on marine, brackish-water and estuarine Gastrotricha. *Zootaxa* **2392**, 1–32.

Hutchinson, G.E. (1967). *A Treatise on Limnology*. Vol. II. *Introduction to Lake Biology and the Limnoplankton*. New York, NY: John Wiley & Sons.

Hyman, L.H. (1951). *The Invertebrates. Platyhelminthes and Rhynchocoela. The Acoelomate Bilateria*, Vol II. New York, NY: McGraw-Hill.

Ingole, B.S. (1987). Occurrence of resting eggs in *Macrostomum orthostylum* (M. Braun, 1885) (Turbellaria: Macrostomida). *Zoologischer Anzeiger* **219**, 19–24.

Janiec, K. (1996). Short distance wind transport of microfauna in maritime Antarctic (King George Island, South Shetland Islands). *Polish Polar Research* **17**, 203–211.

Jankovska, V. (1991). Unbekannte Objekte in Pollenpräparaten – Tardigrada. In Kovar-Eder, J. (ed.), *Palaeovegetational Development in Europe and Regions Relevant to its Palaeofloristic Evolution*, pp. 19–23. Pan-European Palaeobotanical Conference, Vienna, Austria, September.

Janssen, H.H., Gradinger, R. (1999). Turbellaria (Archoophora: Acoela) from Antarctic sea ice endofauna: examination of their micromorphology. *Polar Biology* **21**, 410–416.

Johnston, T.H. (1938). Echinoderida. *Scientific Reports, Ser. C.- Zoology and Botany*. Sydney: David Harold Paisley, Government Printer.

Kanneby, T., Todaro, M.A., Jondelius, U. (2009). One new species and records of Ichthydium Ehrenberg, 1830 (Gastrotricha: Chaetonotida) from Sweden with a key to the genus. *Zootaxa* **2278**, 26–46.

Karling, T.G., Mack-Fira, V., Dörjes, J. (1972). First report on marine microturbellarians from Hawaii. *Zoologica Scripta* **1**, 251–269.

Kaya, M., De Smet, W.H., Fontaneto, D. (2010). Survey of moss-dwelling bdelloid rotifers from middle Arctic Spitsbergen (Svalbard). *Polar Biology* **33**, 833–842.

Kinchin, I.M. (1994). *The Biology of Tardigrades*. Chapel Hill, NC: Portland Press.

Kiontke, K. (1996). The phoretic association of *Diplogaster coprophila* Sudhaus & Rehfeld, 1990 (Diplogastridae) from cow dung with its carriers, in particular flies of the family Sepsidae. *Nematologica* **42**, 354–366.

Kisielewski, J. (1991). Inland-water Gastrotricha from Brazil. *Annales Zoologici (Warsaw)* **43** (Suppl. 2), 1–168.

Kisielewski, J. (1999). A preliminary study of the inland-water Gastrotricha of Israel. *Israel Journal of Zoology* **45**, 135–157.

Kozloff, E.N. (2007). Stages of development, from first cleavage to hatching, of an Echinoderes (Phylum Kinorhyncha: Class Cyclorhagida). *Cahiers de Biologie Marine* **48**, 199–206.

Kristensen, R.M. (1983). Loricifera, a new phylum with Aschelminthes characters from the meiobenthos. *Zeitschrift für Zoologische Systematik und Evolutionsforschung* **21**, 163–180.

Kristensen, R.M. (1987). Generic revision of the Echiniscidae (Heterotardigrada), with a discussion of the origin of the family. In Bertolani, R. (ed.), *Biology of Tardigrades. Selected Symposia and Monographs UZI*, **1**, 261–335.

Kristensen, R.M., Funch, P. (2000). Micrognathozoa: A new class with complicated jaws like those of Rotifera and Gnathostomulida. *Journal of Morphology* **246**, 1–49.

Kristensen, R.M., Nørrevang, A. (1977). On the fine structure of *Rastrognathia macrostoma* gen. et sp. n. placed in Rastrognathiidae fam. n. (Gnathostomulida). *Zoologica Scripta* **6**, 27–41.

Kristensen, R.M., Nørrevang, A. (1978). On the fine structure of *Valvognathia pogonostoma* gen. et sp.n. (Gnathostomulida, Onychognathiidae) with special reference to the jaw apparatus. *Zoologica Scripta* **7**, 179–186.

Kristensen, R.M., Shirayama, Y. (1988). *Pliciloricus hadalis* (Pliciloricidae), a new loriciferan species collected from the Izu-Ogasawara Trench, Western Pacific. *Zoological Science* **5**, 875–881.

Krnjaic', D.J., Krnjaic', S. (1973). Dispersion of nematodes by wind. *Bollettino del Laboratorio di Entomologia Agraria F. Silvestri, Portici* **30**, 66–70.

Lambshead, J.D. (2004). Marine nematode biodiversity. In Chen, Z.X., Chen, S.Y., Dickson, D.W. (eds.), *Nematology, Advances and Perspectives*, Vol. I, pp. 438–468. Cambridge, MA: Tsinghua University Press and CABI Publishing.

Leasi, F., Todaro, M. A. (2009). Meiofaunal cryptic species revealed by confocal microscopy: the case of *Xenotrichula intermedia* (Gastrotricha). *Marine Biology* **156**, 1335–1346.

Leasi, F., Todaro, M.A. (2010). The gastrotrich community of a north Adriatic Sea site, with a redescription of *Musellifer profundus* (Chaetonotida: Muselliferidae). *Journal of the Marine Biological Association UK* **90**, 645–653.

Loof, P.A.A. (1971). Freeliving and plant parasitic nematodes from Spitzbergen, collected by Mr. H. Van Rossen. *Mededelingen Landbouwhogenschool Wageningen* **71–7**, 1–86.

Lupi, E., Ricci, V., Burrini, D. (1994). Occurrence of nematodes in surface water used in a drinking water plant. *Journal Water SRT-Aqua* **43**, 107–112.

Luther, A. (1955). Die Dalyelliiden (Turbellaria, Neorhabdocoela): eine Monographie. *Acta Zoologica Fennica* **87**, 1–337.

Luther, A. (1960). Die Turbellarien Ostfennoskandiens. I. Acoela, Catenulida, Macrostomida, Lecithoepitheliata, Prolecithophora und Proseriata. *Fauna Fennica* **7**, 1–155.

Marcus, E. (1929). Tardigrada. In Bronn, H.G. (ed.), *Klassen und Ordnungen des Tierreichs*, Vol. 5, pp. 1–608. Leipzig: Akademische Verlagsgesellschaft.

Marcus, E. (1936). Tardigrada. In Schultze, F. (ed.), *Das Tierreich*, Vol. 66, pp. 1–340. Berlin: Walter de Gruyter.

Martiny, J.B.H., Bohannan, B.J.M., Brown, J.H. et al. (2006). Microbial biogeography: putting microorganisms on the map. *Nature Reviews* **4**, 102–112.

McInnes, S.J., Pugh, P.J.A. (2007). An attempt to revisit the global biogeography of limno-terrestrial Tardigrada. *Journal of Limnology* **66**, 90–96.

Melnikov, I.A. (1997). The Arctic Sea Ice Ecosystem. Amsterdam: Gordon and Breach Science Publishers.

Miller, W.R., Heatwole, H.F. (2003). Tardigrades of the sub-Antarctic: 5000 year old eggs from Marion Island. Abstract: 9th International Symposium on Tardigrada, Florida, USA.

Mott, J.B., Harrison, A.D. (1983). Nematodes from river drift and surface drinking water supplies in southern Ontario. *Hydrobiologia* **102**, 27–38.

Murray, J. (1906). Scottish National Antarctic Expedition: Tardigrada of the South Orkneys. *Transactions of the Royal Society of Edinburgh* **45**, 323–338.

Navas, A., Baldwin, J.G., Barrios, L., Nombela, G. (1993). Phylogeny and biogeography of Longidorus

(Nematoda: *Longidoridae*) in Euromediterranea. *Nematologia Mediterranea* **21**, 71–88.

Nedved, O. (2004). Occurrence of the phylum Cycliophora in the Mediterranean. *Marine Ecology – Progress Series* **277**, 297–299.

Nelson, D.R. (1982a) Developmental biology of the Tardigrada In Harrison, F., Cowden, R. (eds.), *Developmental Biology of Freshwater Invertebrates*, pp. 363–368. New York, NY: Alan R. Liss.

Nelson, D.R. (2002). Current status of the Tardigrada: evolution and ecology. *Integrative and Comparative Biology* **42**, 652–659.

Nelson, D.R., McInnes, S.J. (2002). Tardigrades. In Rundle, S.D, Robertson, A.L., Schmid-Araya, J.M. (eds.), *Freshwater Meiofauna: Biology and Ecology*, pp. 177–215. Leiden: Buckhuys.

Neves, R.C., Cunha, M.R., Funch, P., Kristensen, R.M., Wanninger, A. (2010). Comparative myoanatomy of cycliophoran life cycle stages. *Journal of Morphology* **271**, 596–611.

Nkem, J.N., Wall, D.H., Virginia, R.A. et al. (2006). Wind dispersal of soil invertebrates in the McMurdo Dry Valleys, Antarctica. *Polar Biology* **29**, 346–352.

Obst, M., Funch, P. (2003). Dwarf male of *Symbion pandora* (Cycliophora). *Journal of Morphology* **255**, 261–278.

Obst, M., Funch, P., Kristensen, R.M. (2006). A new species of Cycliophora from the mouthparts of the American lobster, *Homarus americanus* (Nephropidae, Decapoda). *Organisms, Diversity and Evolution* **6**, 83–97.

Orr, C.C., Newton, O.H. (1971). Distribution of nematodes by wind. *Plant Disease* **55**, 61–63.

Palmer, M.A. (1988). Dispersal of marine meiofauna: a review and conceptual model explaining passive transport and active emergence with implications for recruitment. *Marine Ecology – Progress Series* **48**, 81–91.

Pejler, B. (1977). On the global distribution of the family Brachionidae (Rotatoria). *Archiv fuer Hydrobiologie (Suppl.)* **53**, 255–306.

Pilato, G. (1979). Correlations between cryptobiosis and other biological characteristics in some soil animals. *Bollettino di Zoologica*, **46**, 319–332.

Pilato, G., Binda, M.G. (2001). Biogeography and limnoterrestrial tardigrades: are they truly incompatible binomials? *Zoologischer Anzeiger* **240**, 511–516.

Procter, D.L.C. (1984). Towards a biogeography of free-living soil nematodes. I. Changing species richness, diversity and densities with changing latitude. *Journal of Biogeography* **11**, 103–117.

Procter, D.L.C. (1990). Global overview of the functional roles of soil-living nematodes in terrestrial communities and ecosystems. *Journal of Nematology* **22**, 1–7.

Pugh, P.J.A., McInnes, S.J. (1998). The origin of Arctic terrestrial and freshwater tardigrades. *Polar Biology* **19**, 177–182.

Ramazzotti, G. (1962). Il Phylum Tardigrada. *Memorie dell'Istituto Italiano di Idrobiologia* **16**, 1–595.

Ramazzotti, G. (1972). Il Phylum Tardigrada. II edizione. *Memorie dell'Istituto Italiano di Idrobiologia* **19**, 101–212.

Ramazzotti, G., Maucci, W. (1983). Il Phylum Tardigrada. III edizione riveduta e aggiornata. *Memorie dell'Istituto Italiano di Idrobiologia* **41**, 1–1012.

Rebecchi, L., Bertolani, R. (1988). New cases of parthenogenisis and polyploidy in the genus *Ramazzottius* (Tardigrada, Hypsibiidae) and a hypothesis concerning their origin. *Invertebrate Reproduction and Development* **14**, 187–196.

Rebecchi, L., Bertolani, R. (1994). Maturative pattern of ovary and testis in eutardigrades of freshwater and terrestrial habitats. *Invertebrate Reproduction and Development* **26**, 107–117.

Reisinger, E., Steinböck, O. (1927). Foreløbig meddelelse om vor zoologiske Rejse i Grønland 1926. *Meddelelser om Grønland, København* **74**, 33–42.

Ricci, C., Fontaneto, D. (2009). The importance of being a bdelloid: ecological and evolutionary consequences of dormancy. *Italian Journal of Zoology* **76**, 240–249.

Ricci, C., Melone, G., Sotgia, C. (1993). Old and new data on Seisonidea (Rotifera). *Hydrobiologia* **255**/256, 495–511.

Richters, F. (1905). Moss dwellers. *Scientific American Supplement* **60 (1556)**, 24937.

Riedl, R.J. (1969). Gnathostomulida from America – This is the first record of the new phylum from North America. *Science* **163**, 445–452.

Roccuzzo, G., Ciancio, A. (1991). Notes on nematodes found in irrigation water in southern Italy. *Nematologia Mediterranea* **19**, 105–108.

Rousselet, C.F. (1909). On the geographical distribution of the Rotifera. *Journal of the Quekett Microscopical Club Ser. 2.* **10**, 465–470.

Schockaert, E.R., Hooge, M., Sluys, S. et al. (2008). Global diversity of free-living flatworms (Platyhelminthes, "Turbellaria") in freshwater. *Hydrobiologia* **595**, 41–48.

Schroeder, T., Walsh, E.J. (2007). Cryptic speciation in the cosmopolitan *Epiphanes senta* complex (Monogononta, Rotifera) with the description of new species. *Hydrobiologia* **593**, 129–140.

Segers, H. (1996). The biogeography of littoral *Lecane* Rotifera. *Hydrobiologia* **323**, 169–197.

Segers, H. (2003). A biogeographical analysis of rotifers of the genus Trichocerca Lamarck. 1801 (Trichocercidae, Monogononta, Rotifera), with notes on taxonomy. *Hydrobiologia* **500**, 103–114.

Segers, H. (2007). A global checklist of the rotifers (Phylum Rotifera). *Zootaxa* **1564**, 1–104.

Segers, H. (2008). Global diversity of rotifers (Rotifera) in freshwater. *Hydrobiologia* **595**, 49–59.

Segers, H., De Smet, W.H. (2008). Diversity and endemism in Rotifera: a review, and *Keratella* Bory de St Vincent. *Biodiversity Conservation* **17**, 303–316.

Sekera, E. (1906). Über die Verbreitung der Selbstbefruchtung bei den Rhabdocoeliden. *Zoologischer Anzeiger* **30**, 142–153.

Sohlenius, B. (1980). Abundance, biomass and contribution to energy flow by soil nematodes in terrestrial ecosystems. *Oikos* **34**, 186–194.

Sohlenius, B., Boström, S. (2005). The geographic distribution of metazoan microfauna on East Antarctic nunataks. *Polar Biology* **28**, 439–448.

Sørensen, M.V. (2003). Further structures in the jaw apparatus of *Limnognathia maerski* (Micrognathozoa), with notes on the phylogeny of the Gnathifera. *Journal of Morphology* **255**, 131–145.

Sørensen, M.V. (2008). A new kinorhynch genus from the Antarctic deep sea and a new species of Cephalorhyncha from Hawaii (Kinorhyncha: Cyclorhagida:Echinoderidae). *Organisms, Diversity and Evolution* **8**, 230–232.

Sørensen, M.V., Pardos, F. (2008). Kinorhynch systematics and biology – an introduction to the study of kinorhynchs, inclusive identification keys to the genera. *Meiofauna Marina* **16**, 21–73.

Sørensen, M.V., Rho, H.S. (2009). *Triodontoderes anulap* gen. et sp. nov. – A new cyclorhagid kinorhynch genus and species from Micronesia. *Journal of the Marine Biology Association UK* **89**, 1269–1279.

Sørensen, M.V., Sterrer, W. (2002). New characters in the gnathostomulid mouth parts revealed by scanning electron microscopy. *Journal of Morphology* **253**, 310–334.

Spallanzani, L. (1776). II tardigrado Volume II, Opuscolo IV, Sezione II, 222–253. Opuscoli di Fisica Animale e Vegetabile. Modena.

Steinböck, O. (1931). Marine Turbellaria. *Zoology of the Faroes* **8**, 1–26.

Sterrer, W. (1968). Beiträge zur Kenntnis der Gnathostomulida I. Anatomie und Morphologie des Genus *Pterognathia* Sterrer. *Arkiv för Zoologi, Ser. 2* **22**, 1–125.

Sterrer, W. (1972). Systematics and evolution within the Gnathostomulida. *Systematic Zoology* **21**, 151–173.

Sterrer, W. (1973). Plate tectonics as a mechanism for dispersal and speciation in interstitial sand fauna. *Netherlands Journal of Sea Research* **7**, 200–222.

Sterrer, W. (1991). Gnathostomulida from Fiji, Tonga and New Zealand. *Zoologica Scripta* **20**, 107–128.

Sterrer, W. (1997). Gnathostomulida from the Canary Islands. *Proceedings of the Biological Society of Washington* **110**, 186–197.

Sterrer, W. (1998). Gnathostomulida from the (sub)tropical northwestern Atlantic. *Studies on the Natural History of the Caribbean Region* **74**, 1–178.

Sterrer, W. (2001). Gnathostomulida from Australia and Papua New Guinea. *Cahiers de Biologie Marine* **42**, 363–395.

Sterrer, W., Farris, R. (1975). *Problognathia minima* n. g., n. sp., representative of a new family of Gnathostomulida, Problognathidae n. fam. from Bermuda. *Transactions of the American Microscopical Society* **94**, 357–367.

Sudzuki, M. (1972). An analysis of colonization in freshwater micro-organisms. II. Two simple experiments on the dispersal by wind. *Japanese Journal of Ecology* **22**, 222–225.

Tchesunov, A.V. (2006). *Biology of Marine Nematodes*. Moscow: KMK Scientific Press Ltd. (in Russian).

Thorne, G. (1968). *Nematodes of the Northern Great Plains*. I. *Tylenchida* (*Nemata, Secernentea*), pp. 1–111. Brookings, SD: Agricultural Experiment Station, South Dakota State University.

Thulin, G. (1928). Über die Phylogenie und das System der Tardigraden. *Hereditas* **11**, 207–266.

Timoshkin, O.A., Kawakatsu, M., Korgina, E.M., Vvedenskaya, T.L. (2004). Preliminary analysis of the stylets of the *Gyratrix hermaphroditus* Ehrenberg, 1831 species complex (Platyhelminthes, Neorhabdocoela, Kalyptorhynchia) from lakes of central Russia, Pribaikalye and Kamachatka, lakes Baikal and Biwa. In Timoshkin, O.A, Sitnikova, T.Ya.,

Rusinek, O.T. et al. (eds.), *Index of Animal Species Inhabiting Lake Baikal and its Catchment Area*. Vol. 1. *Lake Baikal*, Book 2, pp. 1321–1343. Novosibirsk: Nauka.

Todaro, M.A., Hummon, W.D. (2008). An overview and a dichotomous key to genera of the phylum Gastrotricha. *Meiofauna Marina* **16**, 3–20.

Todaro, M.A., Kristensen, R.M. (1998). A new species and first report of the genus *Nanaloricus* (Loricifera, Nanaloricida, Nanaloricidae) from the Mediterranean Sea. *Italian Journal of Zoology* **65**, 219–226.

Todaro, M.A., Rocha, C.E.F. (2004). Diversity and distribution of marine Gastrotricha along the northern beaches of the state of Sao Paulo (Brazil), with description of a new species of *Macrodasys* (Macrodasyida, Macrodasyidae). *Journal of Natural History* **38**, 1605–1634.

Todaro, M.A., Rocha, C.E.F. (2005). Further data on marine gastrotrichs from the State of São Paulo and the first records from the State of Rio de Janeiro (Brazil). *Meiofauna Marina* **14**, 27–31.

Todaro, M.A., Fleeger, J.W., Hummon, W.D. (1995). Marine gastrotrichs from the sand beaches of the northern Gulf of Mexico: Species list and distribution. *Hydrobiologia* **310**, 107–117.

Todaro, M.A., Fleeger, J.W., Hu, Y.P., Hrincevich, A.W., Foltz, D.W. (1996). Are meiofauna species cosmopolitan? Morphological and molecular analysis of *Xenotrichula intermedia* (Gastrotricha: Chaetonotida). *Marine Biology* **125**, 735–742.

Todaro, M.A., Balsamo, M., Kristensen, R.M. (2005). A new genus of marine chaetonotids (Gastrotricha), with a description of two new species from Greenland and Denmark. *Journal of the Marine Biological Association UK* **85**, 1391–1400.

Traunspurger, W. (1991). *Das Meiobenthos des Königssees: systematische und ökologische Untersuhungen unter besonderer Berücksichtigung der Nematoda. Fischbiologie des Königssees. Nahrunsangebot und Nahrungswahl.* Band I. Nationalpark Berchtesgaden: Forschungsbericht 22.

Traunspurger, W., Michiels, I.C., Eyualem-Abebe (2006). Composition and distribution of free-living freshwater nematodes: global and local perspectives. In Eyualem-Abebe, Traunspurger, W., Andrássy, I. (eds.), *Freshwater Nematodes: Ecology and Taxonomy*, pp. 46–76. Wallingford: CABI Publishing.

Tyler, S., Schilling, S., Hooge, M., Bush, L.F. (comp.) (2006–2009). *Turbellarian Taxonomic Database*. Version 1.5 http://turbellaria.umaine.edu.

Vanschoenwinkel, B., Gielen, S., Vandewaerde, H., Seaman, M., Brendonck, L. (2008a). Relative importance of different dispersal vectors for small aquatic invertebrates in a rock pool metacommunity. *Ecography* **31**, 567–577.

Vanschoenwinkel, B., Gielen, S., Seaman, M., Brendonck, L. (2008b). Any way the wind blows – frequent wind dispersal drives species sorting in ephemeral aquatic communities. *Oikos* **117**, 125–134.

Vanschoenwinkel, B., Waterkeyn, A., Vandecaetsbeek, T. et al. (2008c). Dispersal of freshwater invertebrates by large terrestrial mammals: a case study with wild boar (*Sus scrofa*) in Mediterranean wetlands. *Freshwater Biology* **53**, 2264–2273.

Vanschoenwinkel, B., Gielen, S., Seaman, M., Brendonck, L. (2009).

Wind mediated dispersal of freshwater invertebrates in rock pool metacommunity: differences in dispersal capacities and modes. *Hydrobiologia* **635**, 363–372.

Van Steenkiste, N., Davison, P., Artois, T. (2010). *Bryoplana xerophila* n.g. n.sp., a new limnoterrestrial microturbellarian (Platyhelminthes, Typhloplanidae, Protoplanellinae) from epilithic mosses, with notes on its ecology. *Zoological Science* **27**, 285–291.

Wallace, R.L., Snell, T.W., Ricci, C., Nogrady, T. (2006). Rotifera vol. 1: biology, ecology and systematics (2nd edition). In Segers, H., Dumont, H.J. (eds.), *Guides to the Identification of the Microinvertebrates of the Continental Waters of the World*, 23. Gent: Kenobi Productions and The Hague: Backhuys Academic Publishing BV.

Weiss, M.J. (2001). Widespread hermaphroditism in freshwater gastrotrichs. *Invertebrate Biology* **120**, 308–341.

Wharton, D.A. (2004). Survival strategies. In Gaugler, R., Bilgrami, A.L. (eds.), Nematode Behaviour, pp. 371–399. Wallingford: CABI Publishing.

Willems, W.R., Wallberg, A., Jondelius, U. et al. (2006). Filling a gap in the phylogeny of flatworms: relationships within the Rhabdocoela (Platyhelminthes), inferred from 18S ribosomal DNA sequences. *Zoologica Scripta* **35**, 1–17.

Womersley, C.Z., Wharton, D., Higa, L.M. (1998). Survival biology. In Perry, R.N., Wright, D.J. (eds.), *The Physiology and Biochemistry of Free-living and Plant-parasitic Nematodes*, pp. 271–302. Wallingford: CABI Publishing.

Wright, J.C. (2001). Cryptobiosis 300 years on from van Leuwenhoek: what have we learned about tardigrades? *Zoologischer Anzeiger* **240**, 563–582.

Wright, J.C., Westh, P., Ramløv, H. (1992). Cryptobiosis in Tardigrada. *Biological Reviews of the Cambridge Philosophical Society* **67**, 1–29.

Young, J.O. (1974). The occurence of diapause in the egg stage of the life-cycle of *Phaenocora typhlops* (Vejdovsky) (Turbellaria: Neorhabdocoela). *Journal of Animal Ecology* **43**, 719–731.

Young, J.O., Young, B.M. (1976). First records of eight species and new records of four species of freshwater microturbellaria from East Africa, with comments on modes of dispersal of the group. *Zoologischer Anzeiger* **96**, 93–108.

Zelinka, C. (1913). *Der Echinoderen der Deutschen Südpolar-Expedition, 1901–1903*. Band 14. Berlin: Reimer.

Zullini, A. (1973). Su alcuni nematodi di alta quota del Nepal. *Khumbu Himal* **4**, 401–412.

14

Molecular approach to micrometazoans. Are they here, there and everywhere?

NOEMI GUIL

Department of Biodiversity and Evolutionary Biology, National Museum of Natural History (CSIC), Madrid, Spain

14.1 Introduction

The 'Everything is everywhere, but the environment selects' hypothesis (EiE hereafter) was originally proposed to explain the apparent ubiquity of microorganisms based on evidence from bacteria (Beijerinck, 1913). Recently, this has been proposed also for protists (e.g. Fenchel and Finlay, 2004) and further extended to micrometazoans (animals smaller than 2 mm) (Foissner, 2006). This hypothesis assumes that microorganisms disperse worldwide due to their microscopic sizes and dormancy capabilities, and that their distributions are restricted only by environmental limitations. High local:global diversity ratios for species assemblages and high gene flow between populations are thus expected. Micrometazoans share a common evolutionary history and the multicellular condition with macroscopic animals, while they are similar in terms of resources used, microscopic size and dormancy capability to microscopic unicellular organisms, which are supposed to be without biogeographies. Even though micrometazoans may provide interesting evidence for the EiE hypothesis, their diversity and phylogeography has not received much attention. However, results

Biogeography of Microscopic Organisms: Is Everything Small Everywhere?, ed. Diego Fontaneto. Published by Cambridge University Press.

so far (which we will deal with along the chapter) give us some indications of ecological and historical–geographic influence on micrometazoan distributions.

Little is known about distributional patterns and phylogeography in micrometazoans – as yet they neither support nor reject the EiE hypothesis. However, the few studies using a molecular approach are providing useful results on micrometazoan patterns, cryptic species and phylogeography. The micrometazoans that have been studied in the framework of the EiE hypothesis are those that may potentially achieve global distributions due to their dormancy capabilities and production of resting stages (Rebecchi et al., 2007; and Chapter 13 for a detailed review of biogeography in other micrometazoans). Accordingly, this review will focus on rotifers, nematodes (free-living forms, generally microscopic; Eyualem Abebe et al., 2008) and tardigrades (Fig 14.1) (Wright et al., 1992). A quick analysis of the information included in the ISI Web of Knowledge (accessed March 2010) on spatial patterns, biodiversity and phylogeography of micrometazoans reveals that within these groups, our knowledge is biased towards rotifers (Fig 14.2A). Moreover, there are more marine-based studies available for tardigrades and free-living nematodes, and more freshwater studies for rotifers (Fig 14.2B). Focusing only on molecular studies for these three phyla, the lack of information becomes dramatic (rotifers, 25 papers; tardigrades, 11 papers; nematodes, 10 papers) (Fig 14.2C).

In this chapter, I will review the available molecular information on micrometazoans within the EiE framework and discuss assumptions, conclusions and future perspectives which could shed some light about where micrometazoans are.

14.2 Dispersal and colonisation

The EiE hypothesis assumes that organisms with microscopic size have high passive dispersal rates and consequently are widely distributed (Fenchel and Finlay, 2004). To achieve highly effective dispersal, dormant stages have to survive during dispersal, recover afterwards and be able to establish in the new place. Here, I review the evidence for each of these steps.

14.2.1 Direct evidence of dispersal

Active dispersal in micrometazoans may be effective at very short distances, but passive dispersal is possible over very long distances. Main vectors discussed for transporting zooplankton are wind, rainfall and animals. Waterfowl have been found as vectors of zooplankton (Figuerola et al., 2005; Frisch et al., 2007), including nematodes and rotifers, transporting them both externally (on feet and feathers) and internally (in the stomach) (Derycke et al., 2005; Eyualem Abebe et al., 2008; Segers and De Smet, 2008). Freshwater rotifers disperse efficiently in water bodies as they have been found in artificial mesocosms left open in the field (Cáceres

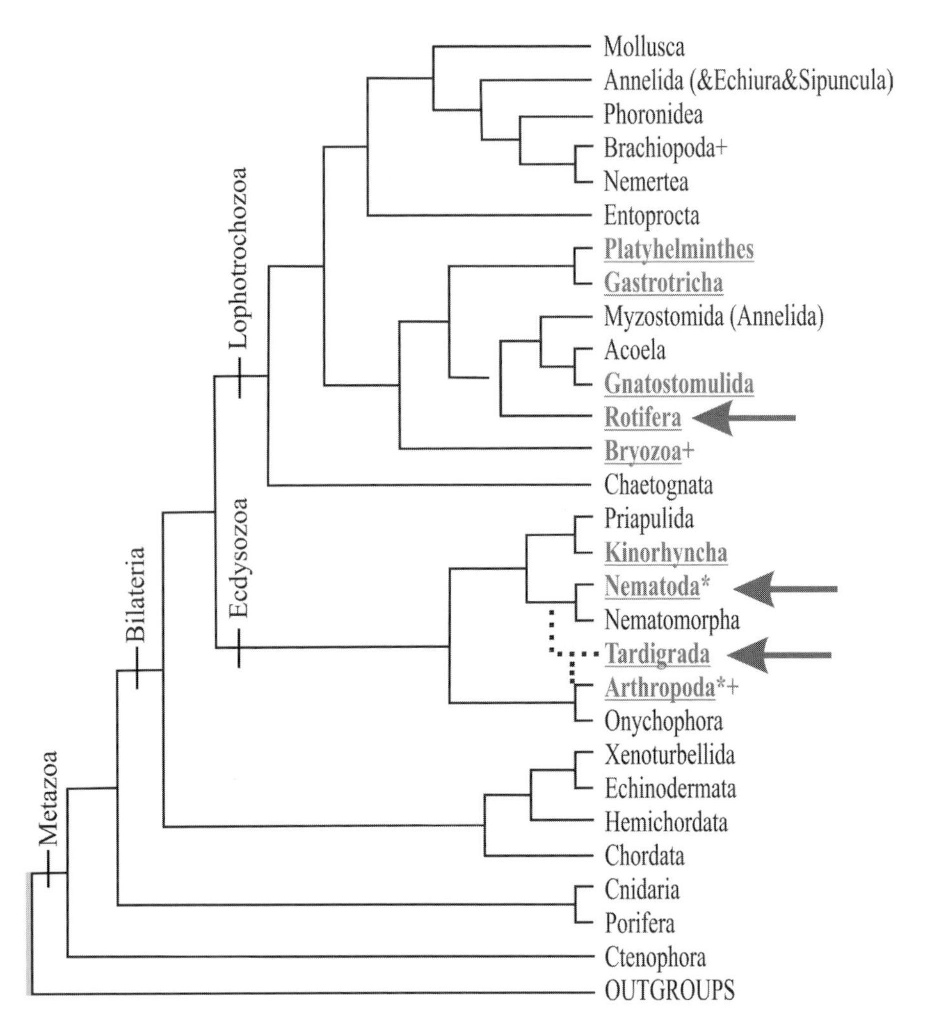

Fig 14.1 Resume of phylogenetic hypotheses proposed by Dunn et al. (2008) among animal phyla. Metazoans considered microscopic (< 2 mm) are underlined. Invertebrate phyla with a wide range of sizes among species, genera or families, including microscopic but also macroscopic sizes are marked with an asterisk (*). Metazoans with cryptobiotic episodes only in some stages of their life cycles are indicated with a plus character (+). Micrometazoans with cryptobiosis during their whole life cycle are signed with arrows. Tardigrada showed an uncertain position within Ecdysozoa: with Nematoida (Nematoda + Nematomorpha) or with Panarthropoda (Arthropoda + Onycophora).

and Soluk, 2002; Bohonak and Jenkins, 2003; Cohen and Shurin, 2003). However, one study collecting organisms dispersed by wind and rainfall found very few dormant stages of rotifer (Jenkins and Underwood, 1998), suggesting that these animals may not be dispersed as well as assumed by these vectors. Passive dispersal

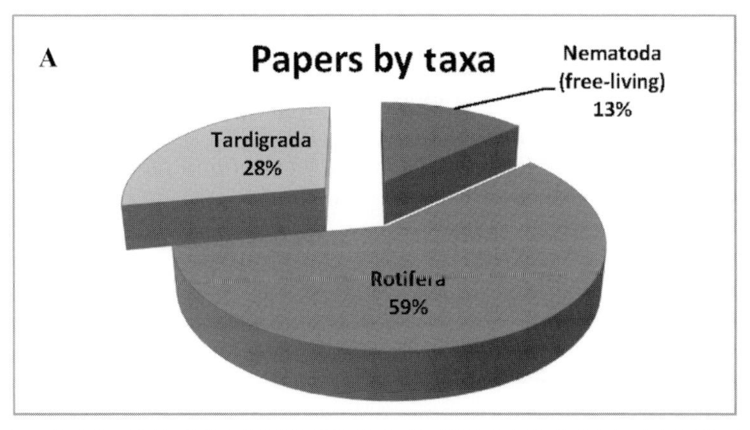

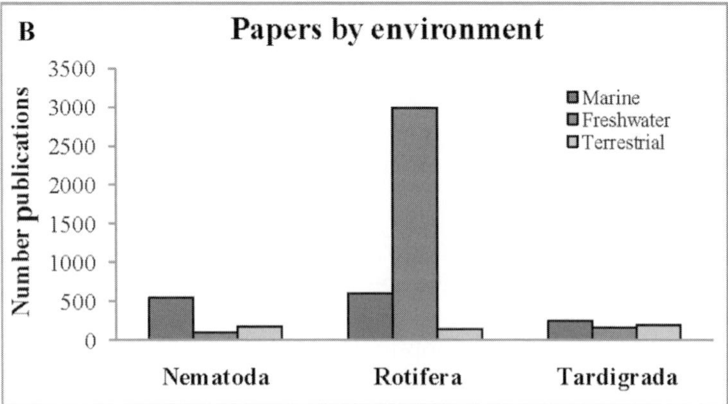

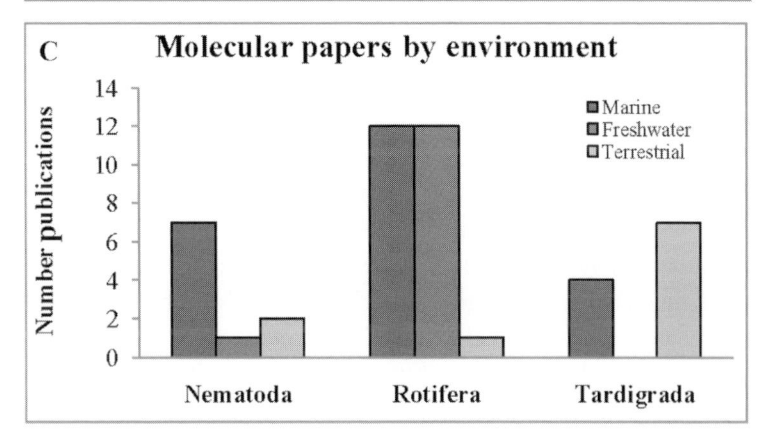

Fig 14.2 Information about papers published from the ISI Web of Knowledge (March 2010). (A) Percentage of papers published by phylum among the three micrometazoans studied. (B) Absolute number of papers published by environment and taxa (Nematoda, Rotifera, Tardigrada). (C) Absolute number of papers published dealing with molecular analyses by environment and taxa (Nematoda, Rotifera, Tardigrada).

by wind for tardigrades for long distances has been assumed to be frequent for a long time, but only two experiments actually measuring it have been published. They both deal with local scales, one under laboratory conditions (Sudzuki, 1972) and the other in a mountain slope (Janiec, 1996). On the other hand, transcontinental dispersal for propagules of lichens, mosses, liverwort and pteridophytes have variously been demonstrated (Muñóz et al., 2004; Buschbom, 2007; see also Chapters 10–12); it is thus plausible also for other microorganisms such as nematodes, rotifers and tardigrades which inhabit the same environments and have dormant stages of similar sizes and resistance capabilities.

14.2.2 Indirect evidence of dispersal

If dispersal does occur regularly and the dispersing micrometazoans successfully colonise new areas, high gene flow between populations would be expected. Provided there was a very high dispersal rate, no patterns in the genetic structure of populations would be expected, neither at local nor regional scales. Gene flow is usually measured with the fixation index (F_{ST}): this quantifies the inbreeding effect of population substructure by comparing the least inclusive to the most inclusive levels of the population hierarchy (Wright, 1921); higher F_{ST} values indicate lower gene flow and higher genetic differentiation between populations (Bohonak, 1999). However, dispersal may not translate into gene flow (Katz et al., 2005) for several reasons (e.g. failure of immigrants to breed in the new environment, unequal migration rates between populations, or artefacts due to small sample size or sampling at the wrong scale; Bohonak, 1999). In short, successful dispersal is a necessary but not sufficient requirement for gene flow and genetic homogenisation of populations within a species.

Some indirect measurements of effective dispersal through genetic information of gene flow (low F_{ST} values) have been obtained for tardigrades (global scales, Jørgensen et al., 2007; local scales, Guil and Giribet, 2009). In contrast, monogonont rotifers, which show cyclical parthenogenesis with dormant stages represented by resting eggs, showed low gene flow and thus high F_{ST} values despite their potentially high dispersal rates at local and regional scales (Gómez et al., 2002b; Gómez, 2005). This strong genetic structure for rotifers has been explained by the 'serial founder effect' hypothesis (De Meester et al., 2002; Gómez et al., 2002b; Mills et al., 2007). The serial founder effect, also known as the 'monopolisation hypothesis' (De Meester et al., 2002), assumes continuous waves of individuals reaching an area (immigrants) and not being able to colonise it because of competition with the local, better-adapted residents. Such a sequence of events results in low gene flow between resident and new immigrant individuals and in genetic differentiation and isolation-by-distance patterns at larger spatial scales (Gómez et al., 2002b). This process would start with the former colonisers rapidly increasing in numbers and establishing their genotypes in the newly colonised area. New immigrant genotypes would be in low numbers in comparison to resident ones since they would arrive by chance after passive dispersal. Thus, immigrant genotypes will not establish populations in the given area (De Meester et al., 2002), because no gene flow between resident and immigrant genotypes will occur, or not in significant numbers to be fixed in the population. Immigrant genotypes will establish populations in available, empty areas located further away (with no competition with resident genotypes), such that the areas further away would be colonised by the progeny of the first colonisation wave (as in a stepping stone model). A persistent founder effect will be even stronger in taxa with dormant stages, which could act as a genotype bank, buffering against the effect of immigrant genotypes (De Meester et al., 2002). Population structure such as isolation-by-distance patterns could be explained by a persistent founder effect (neutral processes) and local adaptation (selective processes), particularly effective in aquatic organisms, either in combination with or separately from palaeogeographic effects such as glacial refugia (Gómez et al., 2007). Results confirming this hypothesis have been obtained for rotifers (Gómez et al., 2002b, 2007; Mills et al., 2007, Derycke et al., 2008b; Fontaneto et al., 2008a), some crustaceans such as *Daphnia* (Palsson, 2000) and *Artemia* (Naihong et al., 2000), and nematodes (Derycke et al., 2005).

14.2.3 Survival of dormant stages

Dormancy appears in many phyla (Wright, 2001; Watanabe et al., 2004; Dunn et al., 2008) and has been acquired independently many times (Fig 14.1). Micrometazoans may survive in the dormant stage for a very long time: at least 28 years for nematodes (Fielding, 1951), nine years for rotifers (Guidetti and Jönsson, 2002) and 20 years for tardigrades (Jørgensen et al., 2007). Factors such as humidity levels, speed of the desiccation and rehydration processes, substrate texture and oxygen concentration can be important for survival in dormant animals (Guidetti and Jönsson, 2002; McSorley, 2003; Horikawa and Higashi, 2004; Ricci et al., 2008; Faurby et al., 2008). These processes are important for oxidation which seems to be involved in the biological degradation during dormancy (Guidetti and Jönsson, 2002; Neumann et al., 2009). High survival rates after dormancy periods found in tardigrades would facilitate their dispersal and colonisation (Ramløv and Westh, 2001; Li and Wang, 2005; Rebecchi et al., 2007, 2009; Hengherr et al., 2008). Considerable within-species variability in survival rates after dormancy was documented in terrestrial tardigrades; it can be associated with different genotypes (Faurby et al., 2008), biological aspects (Rebecchi et al., 2006), geographic distribution (Jönsson et al., 2001), habitat conditions (Wright, 1991), or a mix of habitat and geographic characteristics (Horikawa and Higashi, 2004). Both in rotifers and tardigrades, higher fecundity and longevity could be observed in individuals and populations which underwent dormancy than in individuals and populations kept under constantly hydrated conditions (Ricci and Caprioli, 2005; Rebecchi et al., 2009). So, dormancy may also increase fitness; the proposed mechanisms has been suggested to be a DNA repairing system activated during the recovery steps (Schill et al., 2008; Neumann et al., 2009); nevertheless, no clear hypotheses on any process have been tested yet.

Apart from dispersal and survival ability, reproduction mode is important for establishing a population after dispersal. Micrometazoans reproduce both sexually and asexually, and even alternate between the two (De Meester et al., 2002). Asexual species seem to be better adapted to colonisation by passive dispersal because they can create a population from a single individual arriving in a new area. In contrast, sexual specimens would need at least two individuals of both genders. This sexual vs. asexual issue (and the possibility to change within the same species) is important within the EiE framework since it could determine success or failure in the colonisation of a new area. The main problem for parthenogenesis would be the loss of genetic variability, which is one of the main explanations for the overwhelming presence of sexual recombination in living organisms (Gómez, 2005). However, the recent finding of horizontal gene transfer in parthenogenetic bdelloid rotifers with bacterial, fungi and plant origins has opened the question of whether DNA segments are released from related individuals, as then rotifers could have genetic exchange resembling that in sexual populations (Gladyshev et al., 2008). However, evolutionary influence of the reproductive modes is a complex topic which is not the objective of the present chapter; for overviews see Crow (1994), Hillis (2007) and Schurko et al. (2008).

14.3 Cosmopolitan distributions: a matter of overlooked diversity?

Diversity, taxonomy and distribution of species are main topics within the EiE because based on this information we would determine if organisms are or are not widely distributed, and so if micrometazoans can be everywhere.

14.3.1 Diversity

Certainly, some microorganisms tend to have wider distributions than macroscopic animals, but this could also be explained by their older phylogenetic ages which provided them with more time to disperse (Foissner, 2006). Based on the EiE, high local diversity (since broad distributions of micrometazoan species) and high ratios of local:global diversity would be expected; but little can be concluded when local, regional and global diversity is widely unknown, as is the case with nematodes, rotifers and tardigrades. Under the EiE, given one habitat type, species that are present in this habitat in one geographic area should in principle be present in the same habitat worldwide and so their local:global diversity would be high. However, it is impossible to make habitats comparable in different geographic areas, finding exactly the same environmental characteristics and biotic interrelationships within habitats, consequently, it is impossible to test this assumption (Foissner, 2006). We thus have to cope with two important limitations: first, a lack of knowledge on species distribution at all scales; and second, incomplete and unreliable taxonomy. The two issues are related because biodiversity assessments can be distorted if based on unreliable taxonomy. Information is lacking for basic aspects of micrometazoan taxonomy, biogeography and biology (e.g. Gastrotricha and Rotifera, Ricci and Balsamo, 2000; Nematoda, Eyualem Abebe et al., 2008; Gastrotricha, Balsamo et al., 2008; Tardigrada, Guil and Cabrero-Sañudo, 2007); this could be due to their lack of economic and/or health interests. Some attempts to widen the economic interest in these organisms have been undertaken, for example in applications of dormancy to ageing, tumorigenesis and cancer (Huang and Tunnacliffe, 2006). Less effort is, however, invested on basic research lines. Few taxonomists working on morphology exist and less basic information is available for any further research. Without the basic taxonomic information, biodiversity analyses both at the local and global scale are bound to fail.

14.3.2 Cryptic species and DNA taxonomy

Due to the small size of micrometazoans and homogeneous morphologies among species and genera, some diversity could be hidden. Cryptic species are frequently discovered in a wide range of micrometazoans as soon as molecular information is included (e.g. nematodes, Derycke et al., 2005, 2006, 2007, 2008a; Fonseca et al., 2008; Nieberding et al., 2005; rotifers, Gómez et al., 2002a; Gómez, 2005; Birky, 2007; Fontaneto et al., 2008a, 2008b, 2009; tardigrades, Jørgensen et al., 2007; Guil and Giribet, 2009). Moreover, during their collection and isolation, many rare species can be overlooked, especially if they are particularly small and/or have a cryptic appearance within the environment. This leads to a massive underestimation of the actual diversity, since these rare species can comprise more than 80% of the species in a community (Schwerdtfeger, 1975; Foissner et al., 2002). The introduction of molecular techniques in systematic studies could help in solving taxonomic problems and discover some of the cryptic diversity. Measures of diversity based on traditional versus molecular taxonomy may then give inconsistent estimates; as an example, DNA taxonomy resulted in 2–2.5 times higher estimates at the community level for moss-dwelling rotifers (Kaya et al., 2009). However, protocols to delimit species using DNA taxonomy are still under development (summarised in Moritz and Cicero, 2004; Birky, 2007; Chang et al., 2009; Valentini et al., 2009). Cytochrome oxidase I (COI) is the most widely used molecular marker for species barcoding (Hebert et al., 2003). In addition, other genetic markers have been proposed for micrometazoans, either because problems appeared with COI in certain taxa, or to complement the information provided by COI (e.g. 18S RNA,

Bhadury et al., 2006; COb, Birky, 2007; ITS1, Gómez, 2005; ITS2, Jørgensen et al., 2007). The main problem is to define the level of molecular divergence at which species should be distinguished. A general genetic divergence threshold of 3% was proposed for COI barcoding, and it identified 196 out of 200 species from five phyla (Hebert et al., 2003). In micrometazoans, thresholds of 3% have been suggested for tardigrades (Cesari et al., 2009), 4% for some rotifers (Derry et al., 2003) and over 5% for some nematodes (Nieberding et al., 2005). More elaborate approaches to species delimitation based on phylogenetic and/or population analyses have been proposed (see for example, Pons et al., 2006; Wiens, 2007), but critiques and discussions still remain open, and a clear conclusion has yet to be reached (e.g. Lohse, 2009; Papadopoulou et al., 2009).

14.3.3 Cosmopolitan distributions

Given the mentioned considerations concerning micrometazoan distributions and unreliable taxonomy, what can we say about cosmopolitan distributions among micrometazoans? Unfortunately, very little. From morphology, some species have been considered cosmopolitans, but doubts have arisen because some morphological differences among populations of these species have been found. As a consequence, these 'cosmopolitan' species have started to be considered complexes of species (common in many taxa, for example: polychaetes, Westheide and Schmidt, 2003; harpacticoids, Gómez et al., 2004; rotifers, Suatoni et al., 2006, Ricci and Fontaneto, 2009; cycliophorans, Baker and Giribet, 2007, Baker et al., 2007; gastrotrichs, Leasi and Todaro, 2009). Recently, it has been discovered that some micrometazoan species previously considered cosmopolitans are not supported as units of diversity by morphological and/or molecular information (i.e. Fontaneto et al., 2009). And in many cases, molecular information has supported the existence of more than one independent phylogenetic lineage (*sensu* Guil and Giribet, 2009) within a complex previously considered as a single cosmopolitan species. In those cases, either the rate of molecular evolution might be comparatively rapid, or the rate of morphological divergence might be slow (Todaro et al., 1996), and some taxa identified by molecular taxonomy might not be differentiated morphologically. Contrary to these results, a complex of morphologically distinct tardigrade species was found to be only one independent phylogenetic lineage using COI sequences for a local geographic scale (*Echinicus blumi-canadensis* series in a Spanish mountain range, Guil and Giribet, 2009).

But are there micrometazoan species which could be truly cosmopolitan? Truly cosmopolitan species and haplotypes have been suggested within micrometazoans (Gómez, 2005; Fontaneto et al., 2007), and probably, the whole range of alternatives (from cosmopolitanism to local endemism) can be possible (Segers and De Smet, 2008). We are really limited in what we can conclude about micrometazoan diversity and distribution, which makes it impossible to really test the EiE hypothesis for micrometazoans at the current state of knowledge.

14.4 Patterns and processes

According to the EiE hypothesis, populations of microorganisms would be isolated only by ecological specialisation and not by geographic limitations because of their highly efficient dispersal capabilities; historical events would thus be irrelevant (Fenchel and Finlay, 2004). Micrometazoans often show ecological and/or biotic gradients in their patterns of biodiversity. However, among the few papers that have studied genetic structure in micrometazoans, the majority analysed environmental variables (e.g. nematodes, Derycke et al., 2006; tardigrades, Guil and Giribet, 2009), and only a few papers analysed phylogeography (e.g. rotifers, Gómez, 2005). Ecological patterns for different haplotypes (gradients of pollution and salinity, habitat characteristics and soil type) have been found in micrometazoans (e.g. nematodes, Derycke et al., 2005, 2006, 2007; tardigrades, Guil and Giribet, 2009) relating them more to what the EiE hypothesis postulates than to the situation seen in macroscopic animals, where mixed effects of historical and ecological drivers are observed. However, only low levels of genetic variance were explained by those ecological variables. This could indicate both that the actually important environmental variables have not been considered and/or that historical events have also influenced the genetic structure of the micrometazoans, but have not been tested. Exclusive influence of either ecological or historical/geographic events on population structure is not biologically intuitive: occasional migrants cross even the most extreme barriers (e.g. colonisation of oceanic islands), and limited dispersal and patchy environments of most organisms do restrict gene flow at some level (Butlin et al., 2008).

Isolation-by-distance patterns have been found in certain micrometazoans but not in others. This pattern reflects structure in the geographic distribution (if existing) through correlation between genetic differences (measured by F_{ST} values) and geographic distances. If isolation-by-distance is occurring, genetic differences are expected to increase exponentially with geographic distance because gene flow is restricted, and then we observe genetic structure. So, we would expect no isolation-by-distance for microorganisms under the EiE hypothesis because significant gene flow does happen (caused by high efficiency in dispersal and colonisation), and so there is a homogenisation effect on the gene pool within the population. However, some micrometazoans with potentially high dispersal efficiency show an isolation-by-distance pattern explained by a serial founder effect, as was used originally to explain human dispersion and distribution in modern history (Cavalli-Sforza and Feldman, 2003), as I have explained in section

14.2.2. In other micrometazoans, either an isolation-by-distance pattern has not been tested or insignificant genetic differences have been found when correlated with geographic distances (for a more detailed analysis of this processes, refer to Chapter 16).

Phylogeography studies spatial relationships using gene genealogies to deduce evolutionary histories; more than one genetic marker is usually used since each genomic region has its own genealogy; species phylogenetic history is the combination of these (Avise, 2000; Emerson and Hewitt, 2005; Edwards, 2009). The few studies on micrometazoan phylogeography at regional and larger scales have shown that the observed geographic patterns are probably the result of complex interactions of both historical and ecological events (rotifers, Gómez, 2005, Gómez et al., 2007; nematodes, Nieberding et al., 2005; Derycke et al., 2008b). Gómez (2005) showed that the strong genetic structure found in some aquatic rotifers reflected the impact of Pleistocene glaciations, but also that this genetic structure was linked to ecological characteristics such as water salinity and temperature (Gómez et al., 1997; Serra et al. 1998) or food preferences (Ciros-Perez et al., 2001). The low number of phylogeographic studies in other micrometazoans prevents us from neglecting the influence of historical events on genetic structure, and does not allow estimation of the importance of environmental variables. So, a great effort in multiple disciplines is needed to improve our knowledge about the influence of different factors on diversity patterns found in micrometazoans.

14.5 Obstacles to molecular approaches in micrometazoans

This review on molecular studies on micrometazoans has highlighted a lack of published works on said topic. There are specific difficulties involved in molecular analyses of micrometazoans which is probably the reason for the scarcity of molecular information for this animal groups. Their size is the main issue, creating several practical difficulties: (1) for collecting, since rare morphospecies could be overlooked (Guil and Cabrero-Sañudo, 2007) and many species have been found only in remote and unapproachable locations; (2) for morphological identification to species level before DNA extraction, because temporal microscopic slide preparations are done and later dismantled to recover the animal, and the whole animal has finally to be used for DNA extraction; and (3) because of the low amount of DNA obtained per individual for sequencing analyses. Moreover, the published studies focus on very few taxa which are easy to find, widely distributed, etc. Information is thus very limited and extrapolations are very probably biased. In addition, the scarcity of GenBank (or any other molecular database available) information makes these groups hard to work with. For example, the commonly used Folmer universal primers for COI do not seem to amplify the locus in some species of Nematoda (Bhadury et al., 2006) and Tardigrada (Guil and Giribet, 2009). Moreover, a lack of primary information does not allow designing a priori specific primers, slowing down research within this topic. Finally, similar to what happens with other organisms, rare species appear in low numbers of individuals per site, which means that intraspecific variability is difficult to estimate (e.g. tardigrades, Guil, 2008; Guil and Giribet, 2009).

14.6 Tardigrade case

These tiny animals are a very interesting case of micrometazoans within the EiE framework; despite this, they have been completely neglected. Tardigrades are microscopic (the largest animals are around 2 mm, Guil, 2008), and many species are capable of entering dormancy at any moment of their life cycle, when unfavourable conditions are present in the environment. Tardigrade survival capacity of dormant and active forms, together with their capability to adapt in very different environments (terrestrial and extraterrestrial) fulfils the requirements for dispersal and survival of the EiE hypothesis. Tardigrade dispersal has been studied with direct (Sudzuki, 1972; Janiec, 1996), but also with indirect measurements through molecular information (Jørgensen et al., 2007; Guil and Giribet, 2009). No other information about dispersal, such as vectors (wind, rainfall and animal) is available for tardigrades. On the other hand, tardigrades survive both dormancy episodes and extreme conditions in dormant (but also in active) forms (Rebecchi et al., 2009), and have DNA repairing systems involved in the wake-up step after dormancy (Schill et al., 2008; Neumann et al., 2009). Moreover, higher fitness is found in some species when cyclically desiccated (Rebecchi et al., 2009). High survival rates have been found in tardigrades after exposing them to extreme environmental conditions (e.g. low – almost absolute zero – and high temperatures, Ramløv and Westh, 2001; Li and Wang, 2005; Rebecchi et al., 2007, 2009; Hengherr et al., 2008; immersion in organic solvents, Ramløv and Westh, 2001; exposure to ionising radiation, Horikawa et al., 2006; Jönsson and Schill, 2007; Jönsson et al., 2008), mainly in dormant forms but also in active forms (Jönsson et al., 2005; Horikawa et al., 2006, 2008), in terrestrial but also under space vacuum, microgravity and cosmic radiation in extraterrestrial environments (Jönsson et al., 2008; Rebecchi et al., 2009). Their capability to adapt to microgravity under extraterrestrial conditions (Rebecchi et al., 2009) gives us a clue about their plasticity in terms of adaptation to new environmental conditions, even when species' ecological requirements, ranges and plasticity are widely unknown. Sexual but also asexual species can be found within Tardigrada (Ramazzotti and Maucci, 1983), thus colonisation of new areas could be achieved by these two strategies.

For example, species for which population genetic structure have been studied (*E. testudo*, Jørgensen et al., 2007; *E. blumi-canadensis*, Guil and Giribet, 2009) (Figs 14.3A and 14.3B) can reproduce either asexually (Dastych, 1987; Bertolani et al., 1990; Jørgensen et al., 2007) or sexually (Claxton, 1996) with variable percentages of males within the populations: from 2.6% in Spain's *E. blumi-canadensis* (Guil, 2008; Guil and Giribet, 2009) and 7% in North American *E. mauccii* (Mitchell and Romano, 2007) to 28–53% in eight species of *Echiniscus* in Australia (Claxton, 1996). Furthermore, within the same species different sexual reproductive modes have being observed in populations inhabiting different habitats (R. Guidetti, personal communication).

Little can be concluded about local diversities and local:global diversity ratios in tardigrades, so we cannot support or reject the hypothesis that tardigrades tend to have high or low local, regional or global diversities. Few revisions concerning the global distribution of Tardigrada species have been published (all species, Ramazzotti and Maucci, 1983; only limnoterrestrial species, McInnes, 1994), thus, little updated information is available. These revisions showed species with broad but also species with narrow distributions (McInnes, 1994). However, taxonomic studies in tardigrades have been focused on selected geographic areas and habitats, potentially producing biased results and with limited extrapolation in species distributions (Guil, 2002; Guil and Cabrero-Sañudo, 2007). As in many other micrometazoans, tardigrade taxonomy is difficult due to the low number of morphological characters available (Guidetti and Bertolani, 2005; Cesari et al., 2009), distorting diversity at all scales. Little information is available in tardigrades for species delimitation from DNA taxonomy and the majority comes from phylogenetic analyses at a higher taxonomic level (Garey et al., 1996, 1999; Giribet et al., 1996; Regier et al., 2003; Jørgensen and Kristensen, 2004; Guidetti et al., 2005; Nichols et al., 2006; Møbjerg et al., 2007; Schill and Steinbruck, 2007; Sands et al., 2008b), which have focused on marine and terrestrial environments (Fig 14.2C). Only six out of 46 papers using molecular information dealt with molecular differentiation within species and focused on cryptic species (Jørgensen et al., 2007; Faurby et al., 2008; Sands et al., 2008a; Cesari et al., 2009; Guidetti et al., 2009; Guil and Giribet, 2009; Fig 14.1C). Cryptic species have been discovered with recent molecular analyses in both Heterotardigrada (*Echiniscus testudo*, Jørgensen et al., 2007; Fig 14.3A) and Eutardigrada (Guidetti et al., 2009); but the opposite has also been found in heterotardigrades: a complex of species (at least five morphological species) of the *Echiniscus blumi-canadensis* series (Fig 14.3B) with no genetic differentiation based on COI sequences (Guil and Giribet, 2009). Besides, some species considered cosmopolitans (such as *Milnesium tardigradum*; Fig 14.3C) are now known to have morphological and molecular differences within the same nominal species (personal observations, and R. Guidetti and R. M. Kristensen, personal communication). A preliminary threshold of 3% genetic divergence (Cesari et al.,

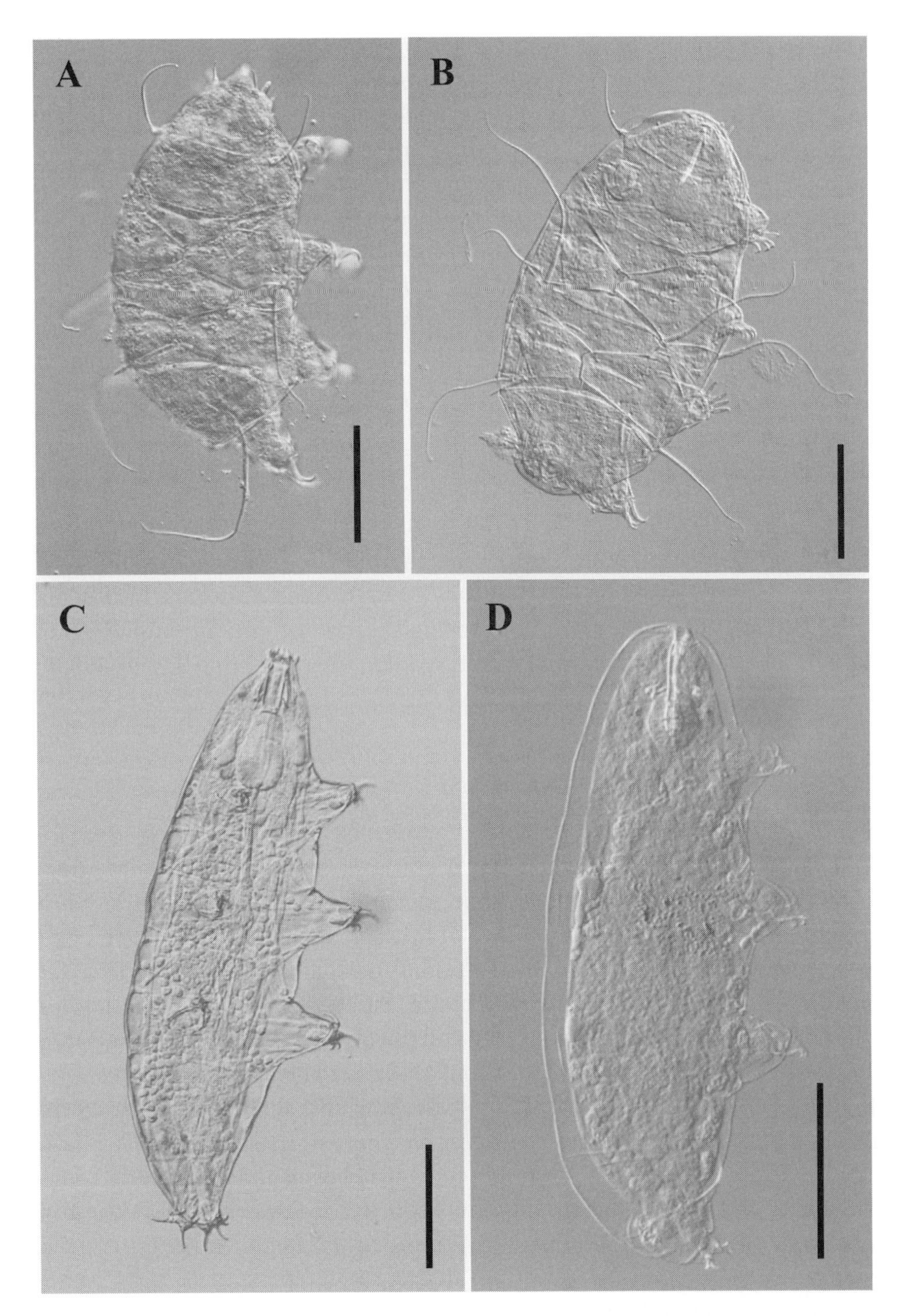

Fig 14.3 Tardigrada species. (A) *Echiniscus testudo* (Echiniscidae, Echiniscoidea, Heterotardigrada). (B) *Echiniscus trisetosus* (from the *Echinicus blumi-canadensis* series) (Echiniscidae, Echiniscoidea, Heterotardigrada). (C) *Milnesium tardigradum* (Milnesiidae, Apochela, Eutardigrada). (D) *Macrobiotus hufelandi* (Macrobiotidae, Parachela, Eutardigrada). Scale bar 100 μm.

2009) has been proposed to determine cryptic species from molecular information, but based only on sequences from *Macrobiotus* species (Fig 14.3D).

Finally, no clear conclusions can be reached about the influence of ecological or geographic/historical factors over tardigrade patterns and phylogeography, although preliminary results are promising and informative. Molecular spatial patterns in tardigrades have been hardly studied in relation either with ecological or historical/geographic factors. From the two published papers on tardigrade genetic structure, one at global (Jørgensen et al., 2007) and another at local (Guil and Giribet, 2009) scales, some clues can be deduced. On the one hand, tardigrades have not shown an isolation-by-distance pattern so far; this is perhaps due to the small size of the area studied (Guil and Giribet, 2009), or because DNA extracted from pooled specimens per species (Jørgensen et al., 2007) mixed up into a consensus sequence per site, and any geographic pattern was overlooked. On the other hand, *E. testudo* did not show any phylogeographic pattern, either with COI and ITS2, even when global scale was used, involving Europe, Asia, North America and Greenland (Jørgensen et al., 2007). The *E. blumi-canadensis* series did show an ecological pattern based on soil type for COI even when it was at a local scale (Guil and Giribet, 2009). However, this ecological pattern only explained 8% of genetic variance; it is unknown what factors explain the remaining 92% of genetic variance.

14.7 Summary

Many, if not all questions remain open. Little information is available about micrometazoans and, in particular, about nematodes, rotifers and tardigrades; this paucity of knowledge does not allow for any conclusion on the patterns and processes involved in their present distribution. The spatial patterns found so far explain low percentages of genetic variance, and the few phylogeographic analyses show signs that both historical/geographic and ecological events are involved in shaping the distribution of micrometazoans similarly to what is known in macroscopic animals. The whole range of alternatives (from endemism to cosmopolitan) is possible within these microinvertebrates, but some of these alternatives have yet to be discovered. Information on micrometazoan species, geography and genetics is lacking and perhaps biased, and thus conclusions drawn are unclear and extrapolations premature. Consequently, no hypothesis, including the EiE, can be accepted or rejected so far.

Nematodes, rotifers and tardigrades are a model to study evolutionary hypotheses, e.g. the influence of environment vs. historical events on their distribution, since they have a common evolutionary history with all other animals, while sharing environmental and biological characteristics such as cryptobiosis with unicellular organisms. Special characteristics of micrometazoans allow study of the influence of factors such as survival in extreme conditions or dormancy on adaptation, colonisation and evolution, which cannot be tested in macroscopic animals. However, the most basic taxonomical, distributional, biological and ecological information in these micrometazoans is far from complete and biased towards few species and environments. It is evident that much more effort is needed at all levels and in all topics to increase our knowledge about microinvertebrates, especially where recent molecular studies have been illustrative. More questions and new perspectives will arise as our knowledge on these creatures increases.

Acknowledgements

The author would like to thank Annie Machordom Barbe and Sara Sánchez Moreno for their valuable comments and suggestions. I would like to thank Diego Fontaneto for his invitation to write a chapter in this book, and subsequent discussions about the topic. In addition, I thank helpful comments and suggestions from an anonymous reviewer. This research was supported by: a Marie Curie Intra European Fellowship within the 7th European Community Framework Programme, the CSIC (*Consejo Superior de Investigaciones Científicas*) through the JAE-DOC programme and a scientific project funded by the Ministry of Education and Science (number CTM2008-00496).

References

Avise, J. C. (2000). *Phylogeography: The History and Formation of Species.* Cambridge, MA: Harvard University Press.

Baker, J.M., Giribet, G. (2007). A molecular phylogenetic approach to the phylum Cycliophora provides further evidence for cryptic speciation in *Symbion americanus. Zoologica Scripta* **36**, 353–359.

Baker, J.M., Funch, P., Giribet, G. (2007). Cryptic speciation in the recently discovered American cycliophoran *Symbion americanus*; genetic structure and population expansion. *Marine Biology* **151**, 2183–2193.

Balsamo, M., D´Hondt, J.L., Kisielewski, J., Pierboni, L. (2008). Global diversity of gastrotrichs (gastrotricha) in fresh waters. *Hydrobiologia* **595**, 85–91.

Beijerinck, M.W. (1913). *De infusies en de ontdekking der backteriën*. Jaarboek van de Koninklijke Akademie v. Wetenschappen. Amsterdam: Müller.

Bertolani, R., Rebecchi, L., Beccaccioli, G. (1990). Dispersal of *Ramazzottius* and other tardigrades in relation to type of reproduction. *Invertebrate Reproduction and Development* **18**, 153–157.

Bhadury, P., Austen, M.C., Bilton, D.T. et al. (2006). Development and evaluation of a DNA-barcoding approach for the rapid identification of nematodes. *Marine Ecology Progress Series* **320**, 1–9.

Birky, C.W. (2007). Workshop on barcoded DNA: application to rotifer phylogeny, evolution, and systematics. *Hydrobiologia* **593**, 175–183.

Bohonak, A.J. (1999). Dispersal, gene flow, and population structure. *Quarterly Review of Biology* **74**, 21–45.

Bohonak, A.J., Jenkins, D.G. (2003). Ecological and evolutionary significance of dispersal by freshwater invertebrates. *Ecology Letters* **6**, 783–796.

Buschbom, J. (2007). Migration between continents: geographical structure and long-distance gene flow in *Porpidia flavicunda* (lichen-forming Ascomycota). *Molecular Ecology* **16**, 1835–1846.

Butlin, R.K., Galindo, J., Grahame, J.W. (2008). Sympatric, parapatric or allopatric: the most important way to classify speciation? *Philosophical Transactions of the Royal Society Series B* **363**, 2997–3007.

Cáceres, C.E., Soluk, D.A. (2002). Blowing in the wind: a field test of overland dispersal and colonization by aquatic invertebrates. *Oecologia* **131**, 402–408.

Cavalli-Sforza, L.L., Feldman, M.W. (2003). The application of molecular genetic approaches to the study of human evolution. *Nature Genetics* **33**, 266–275.

Cesari, M., Bertolani, R., Rebecchi, L, Guidetti, R. (2009). DNA barcoding in Tardigrada: the first case study on *Macrobiotus macrocalix* Bertolani & Rebecchi 1993 (Eutardigrada, Macrobiotidae). *Molecular Ecology Resources* **9**, 699–706.

Chang, C.H., Rougerie, R., Chen, J.H. (2009). Identifying earthworms through DNA barcodes: pitfalls and promise. *Pedobiologia* **52**, 171–180.

Ciros-Pérez, J., Carmona M.J., Serra, M. (2001). Resource competition between sympatric sibling rotifer species. *Limnology and Oceanography* **46**, 1511–1523.

Claxton, S.K. (1996). Sexual dimorphisms in Australian *Echiniscus* (Tardigrada, Echiniscidae) with description of three new species. *Zoological Journal of the Linnean Society* **116**, 13–33.

Cohen, G.M., Shurin, J.B. (2003). Scale-dependence and mechanisms of dispersal in freshwater zooplankton. *Oikos* **103**, 603–617.

Crow, J.F. (1994). Advantages of sexual reproduction. *Developmental Genetics* **15**, 205–213.

Dastych, H. (1987). Two new species of Tardigrada from the Canadian Subarctic with some notes on sexual dimorphism in the family Echiniscidae. *Entomologische Mitteilungen aus dem Zoologischen Museum Hamburg* **8**, 319–334.

De Meester, L., Gómez, A., Okamura, B., Schwenk, K. (2002). The Monopolization Hypothesis and the dispersal–gene flow paradox in aquatic organisms. *Acta Oecologica* **23**, 121–135.

Derry, A.M., Hebert, P.D., Prepas, E.E. (2003). Evolution of rotifers in saline and subsaline lakes: a molecular phylogenetic approach. *Limnology and Oceanography* **48**, 675–685.

Derycke, S., Remerie, T., Vierstraete, A. et al. (2005). Mitochondrial DNA variation and cryptic speciation within the free-living marine nematode *Pellioditis marina*. *Marine Ecology Progress Series* **300**, 91–103.

Derycke, S., Backeljau, T., Vlaeminck, C. et al. (2006). Seasonal dynamics of population genetic structure in cryptic taxa of the *Pellioditis marina* complex (Nematoda: Rhabditida). *Genetica* **128**, 307–321.

Derycke, S., Backeljau, T., Vlaeminck, C. et al. (2007). Spatiotemporal analysis of population genetic structure in *Geomonhystera disjuncta* (Nematoda, Monhysteridae) reveals high levels of molecular diversity. *Marine Biology* **151**, 1799–1812.

Derycke, S., Fonseca, G., Vierstraete, A. et al. (2008a). Disentangling taxonomy within the *Rhabditis* (*Pellioditis*) *marina* (Nematoda, Rhabditidae) species complex using molecular and morphological tools. *Zoological Journal of the Linnean Society* **152**, 1–15.

Derycke, S., Remerie, T., Backeljau, T. et al. (2008b). Phylogeography of the *Rhabditis* (*Pellioditis*) *marina* species complex: evidence for long-distance dispersal, and for range expansions and restricted gene flow in the northeast Atlantic. *Molecular Ecology* **17**, 3306–3322.

Dunn, C.W., Hejnol, A., Matus, D.Q. et al. (2008). Broad phylogenomic sampling improves resolution of the animal tree of life. *Nature* **452**, 745–750.

Edwards, S.V. (2009). Is a new and general theory of molecular systematics emerging? *Evolution* **63**, 1–19.

Emerson, B.C., Hewitt, G.M. (2005). Phylogeography. *Current Biology* **15**, R367–R371.

Eyualem Abebe, Decraemer, W., De Ley, P. (2008). Global diversity of nematodes (Nematoda) in freshwater. *Hydrobiologia* **595**, 67–78.

Faurby, S., Jönsson, K.I., Rebecchi, L., Funch, P. (2008). Variation and anhydrobiotic survival of two eutardigrade morphospecies: a story of cryptic species and their dispersal. *Journal of Zoology* **275**, 139–145.

Fenchel, T., Finlay, B.J. (2004). The ubiquity of small species: patterns of local and global diversity. *BioScience* **54**, 777–784.

Fielding, M.J. (1951). Observations on the length of dormancy in certain plant infecting nematodes. *Proceedings of the Helminthological Society of Washington DC* **18**, 110–112.

Figuerola, J., Green, A.J., Michot, T.C. (2005). Invertebrate eggs can fly: evidence of waterflow-mediated gene flow in aquatic invertebrates. *American Naturalist* **165**, 274–280.

Foissner, W. (2006). Biogeography and dispersal of micro-organisms: a review emphasizing protists. *Acta Protozoologica* **45**, 111–136.

Foissner, W., Agatha, S., Berger, H. (2002). Soil ciliates (Protozoa,Ciliophora) from Namibia (Southwest Africa), with emphasis on two contrasting environments, the Etosha Region and the Namib Desert. *Denisia* **5**, 1–1459.

Fonseca, G., Derycke, S., Moens, T. (2008). Integrative taxonomy in two free-living nematode species complexes. *Biological Journal of the Linnean Society* **94**, 737–753.

Fontaneto, D., Herniou, E.A., Barraclough, T.G., Ricci, C. (2007). On the global distribution of microscopic animals: new worldwide data on bdelloid rotifers. *Zoological Studies* **46**, 336–346.

Fontaneto, D., Barraclough, T.G., Kimberly, C., Ricci, C., Herniou, E.A. (2008a). Molecular evidence for broad-scale distributions in bdelloid rotifers: everything is not everywhere but most things are very widespread. *Molecular Ecology* **17**, 3136–3146.

Fontaneto, D., Boschetti, C., Ricci, C. (2008b). Cryptic diversification in ancient asexuals: evidence from the bdelloid rotifer *Philodina flaviceps*. *Journal of Evolutionary Biology* **21**, 580–587.

Fontaneto, D., Kaya, M., Herniou, E.A., Barraclough, T.G. (2009). Extreme levels of hidden diversity in microscopic animals (Rotifera) revealed by DNA taxonomy. *Molecular Phylogenetics and Evolution* **53**, 182–189.

Frisch, D., Green, A.J., Figuerola, J. (2007). High dispersal capacity of a broad spectrum of aquatic invertebrates via waterbirds. *Aquatic Sciences* **69**, 568–574.

Garey, J.R., Krotec, M., Nelson, D.R., Brooks, J. (1996). Molecular analysis supports a tardigrade-arthropod association. *Invertebrate Biology* **115**, 79–88.

Garey, J.R., Nelson, D.R., Mackey, L.J., Li, J. (1999). Tardigrade phylogeny: congruency of morphological and molecular evidence. *Zoological Anzeiger* **238**, 205–210.

Giribet, G., Carranza, S., Baguna, J., Rintot, M., Ribera, C. (1996). First molecular evidence for the existence of a Tardigrada + Arthropoda clade. *Molecular Biology and Evolution* **13**, 76–84.

Gladyshev, E.A., Meselson, M., Arkhipova, I.R. (2008). Massive horizontal gene transfer in bdelloid rotifers. *Science* **320**, 1210–1213.

Gómez, A., Carmona, M.J., Serra, M. (1997). Ecological factors affecting gene flow in the *Brachionus plicatilis* complex (Rotifera). *Oecologia* **111**, 350–356.

Gómez, A., Serra, M., Carvalho, G.R., Lunt, D.H. (2002a). Speciation in ancient cryptic species complexes: evidence from the molecular phylogeny of *Brachionus plicatilis* (Rotifera). *Evolution* **56**, 1431–1444.

Gómez, A., Adcock, G.J., Lunt, D.H., Carvalho, G.R. (2002b). The interplay between colonisation history and gene flow in passively dispersing zooplankton: microsatellite analysis of rotifer resting egg banks. *Journal of Evolutionary Biology* **15**, 158–171.

Gómez, S. (2005). Molecular ecology of rotifers: from population differentiation to speciation. *Hydrobiologia* **546**, 83–99.

Gómez, S., Fleeger, J.W., Rocha, O.A., Foltz, D. (2004). Four new species of *Cletocamptus* Schmankewitsch, 1875, closely related to *Cletocamptus deitersi* (Richard, 1897) (Copepoda: Harpacticoida). *Journal of Natural History* **38**, 2669–2732.

Gómez, S., Montero-Pau, J., Lunt, D.H., Serra, M., Campillo, S. (2007). Persistent genetic signatures of colonization in *Brachionus manjavacas* rotifers in the Iberian Peninsula. *Molecular Ecology* **16**, 3228–3240.

Guidetti, R., Bertolani, R. (2005). Tardigrade taxonomy: an updated check list of the taxa and a list of characters for their identification. *Zootaxa* **845**, 1–46.

Guidetti, R., Jönsson, K.I. (2002). Long-term anhydrobiotic survival in semi-terrestrial micrometazoans. *Journal of Zoology* **257**, 181–187.

Guidetti, R., Gandolfi, A., Rossi, V., Bertolani, R. (2005). Phylogenetic analysis of Macrobiotidae (Eutardigrada, Parachela): a combined morphological and molecular approach. *Zoologica Scripta* **34**, 235–244.

Guidetti, R., Schill, R.O., Bertolani, R., Dankekar, T., Wolf, M. (2009). New molecular data for tardigrade phylogeny, with the erection of

Paramacrobiotus gen. nov. *Journal of Zoological Systematics and Evolutionary Research* **47**, 315–321.

Guil, N. (2002). Diversity and distribution of tardigrades (Bilateria, Tardigrada) from the Iberian Peninsula, Balearic Islands and Chafarinas Islands. *Graellsia* **58**, 75–94.

Guil, N. (2008). New records and within-species variability of Iberian tardigrades (Tardigrada), with comments on the species from the *Echiniscus blumi-canadensis* series. *Zootaxa* **1757**, 1–30.

Guil, N., Cabrero-Sañudo, F. (2007). Analysis of the species description process for a little known invertebrate group: the limnoterrestrial tardigrades (Bilateria, Tardigrada). *Biodiversity and Conservation* **16**, 1063–1086.

Guil, N., Giribet, G. (2009). Fine scale population structure in the *Echiniscus blumi-canadensis* series (Heterotardigrada, Tardigrada) in an Iberian mountain range – When morphology fails to explain genetic structure. *Molecular Phylogenetics and Evolution* **51**, 606–613.

Hebert, P.D.N., Cywinska, A., Ball, S.L., De Ward, J.R. (2003). Biological identifications through DNA barcodes. *Proceedings of the Royal Society of London B* **270**, 313–321.

Hengherr, S., Brümmer, F., Schill, R.O. (2008). Anhydrobiosis in tardigrades and its effects on longevity traits. *Journal of Zoology* **275**, 216–220.

Hillis, D.M. (2007). Asexual evolution: can species exist without sex? *Current Biology* **17**, R543–R544.

Horikawa, D.D., Higashi, S. (2004). Desiccation tolerance of the tardigrade *Milnesium tardigradum* collected in Sapporo, Japan, and Bogor, Indonesia. *Zoological Sciences* **21**, 813–816.

Horikawa, D.D., Sakashita, T., Katagiri, C. et al. (2006). Radiation tolerance in the tardigrade *Milnesium tardigradum*. *International Journal of Radiation Biology* **82**, 843–848.

Horikawa, D.D., Kunieda, T., Abe, W. et al. (2008). Establishment of a rearing system of the extremotolerant tardigrade *Ramazzottius varieornatus*: a new model animal for astrobiology. *Astrobiology* **8**, 549–556.

Huang, Z., Tunnacliffe, A. (2006). Cryptobiosis, aging, and cancer: Yin-Yang balancing of signaling networks. *Rejuvenation Research* **9**, 292–296.

Janiec, K. (1996). Short distance wind transport of microfauna in maritime Antarctic (King George Island, South Shetland Islands). *Polish Polar Research* **17**, 203–211.

Jenkins, D.G., Underwood, M.O. (1998). Zooplankton may not disperse readily in wind, rain, or waterfowl. *Hydrobiologia* **387**/388, 15–21.

Jönsson, K.I., Schill, R.O. (2007). Induction of Hsp70 by desiccation, ionising radiation and heat-shock in the eutardigrade *Richtersius coronifer*. *Comparative Biochemistry and Physiology Part B* **146**, 456–460.

Jönsson, K.I., Borsari, S., Rebecchi, L. (2001). Anydrobiotic survival in populations of the tardigrades *Richtersius coronifer* and *Ramazzottius oberhaeuseri* from Italy and Sweden. *Zoologischer Anzeiger* **240**, 419–423.

Jönsson, K.I., Harms-Ringdhal, M., Torudd, J. (2005). Radiation tolerance in the eutardigrade *Richtersius coronifer*. *International Journal of Radiation Biology* **81**, 649–656.

Jönsson, K.I., Rabbow, E., Schill, R.O., Brümmer, F. (2008). Tardigrades survive exposure to space in low Earth orbit. *Current Biology* **18**, R729–R731.

Jørgensen, A., Kristensen, R.M. (2004). Molecular phylogeny of Tardigrada – investigation of the monophyly of Heterotardigrada. *Molecular Phylogenetics and Evolution* **32**, 666–670.

Jørgensen, A., Møbjerg, N., Kristensen, R.M. (2007). A molecular study of the tardigrade *Echiniscus testudo* (Echiniscidae) reveals low DNA sequence diversity over a large geographical area. *Journal of Limnology* **66**, 77–83.

Katz, L.A., McManus, G.B., Snoeyenbos-West, O.L.O. et al. (2005). Reframing the 'Everything is everywhere' debate: evidence for high gene flow and diversity in ciliate morphospecies. *Aquatic Microbial Ecology* **41**, 55–65.

Kaya, M., Herniou, E.A., Barraclough, T.G., Fontaneto, D. (2009). Inconsistent estimates of diversity between traditional and DNA taxonomy in bdelloid rotifers. *Organisms, Diversity and Evolution* **9**, 3–12.

Leasi, F., Todaro, M.A. (2009). Meiofaunal cryptic species revealed by confocal microscopy: the case of *Xenotrichula intermedia* (Gastrotricha). *Marine Biology* **156**, 1335–1346.

Li, X., Wang, L. (2005). Effect of thermal acclimation on preferred temperature, avoidance temperature and lethal thermal maximum of *Macrobiotus harmsworthi* Murray (Tardigrada, Macrobiotidae). *Journal of Thermal Biology* **30**, 443–448.

Lohse, K. (2009). Can mtDNA barcodes be used to delimit species? A response to Pons et al. (2006). *Systematics Biology* **58**, 439–442.

McInnes, S.J. (1994). Zoogeographic distribution of terrestrial/freshwater tardigrades from current literature. *Journal of Natural History* **28**, 257–352.

McSorley, R. (2003). Adaptations of nematodes to environmental extremes. *Florida Entomologist* **86**, 138–142.

Mills, S., Lunt, D.H., Gómez, A. (2007). Global isolation by distance despite strong regional phylogeography in a small metazoan. *BMC Evolutionary Biology* **7**, 225.

Mitchell, C.R., Romano, F.A. (2007). Sexual dimorphism, population dynamics and some aspects of life history of *Echiniscus mauccii* (Tardigrada; Heterotardigrada). *Journal of Limnology* **66**, 126–131.

Møbjerg, N., Jørgensen, A.,Eibye-Jacobsen, J. et al. (2007). New records on cyclomorphosis in the marine eutardigrade *Halobiotus crispae* (Eutardigrada: Hypsibiidae). *Journal of Limnology* **66**, 132–140.

Moritz, C., Cicero, C. (2004). DNA barcoding: promise and pitfalls. *PLoS Biology* **2**, 1529–1531.

Muñóz, J., Felicisimo, A.M., Cabezas, F., Burgaz, A.R., Martinez, I. (2004). Wind as a long-distance dispersal vehicle in the southern hemisphere. *Science* **304**, 1144–1147.

Naihong, X., Audenaert, E., Vanoverbeke, J. et al. (2000). Low among-population genetic differentiation in Chinese bisexual *Artemia* populations. *Heredity* **84**, 238–243.

Neumann, S., Reuner, A., Brümmer, F., Schill, R.O. (2009). DNA damage in storage cells of anhydrobiotic tardigrades. *Comparative Biochemistry and Physiology Part A Molecular and Integrative Physiology* **153**, 425–429.

Nichols, P.B., Nelson, D.R., Garey, J.R. (2006). A family level analysis of tardigrade phylogeny. *Hydrobiologia* **558**, 53–60.

Nieberding, C., Libois, R., Douady, C.J., Morand, S., Michaux, R. (2005). Phylogeography of a nematode (*Heligmosomoides polygyrus*) in the western Palearctic region: persistence of northern cryptic populations during ice ages? *Molecular Ecology* **14**, 765–779.

Palsson, S. (2000). Microsatellite variation in *Daphnia pulex* from both sides of the Baltic sea. *Molecular Ecology* **9**, 1075–1088.

Papadopoulou, A., Monaghan, M.T., Barraclough, T.G., Vogler, A.P. (2009). Sampling error does not invalidate the Yule-Coalescent Model for species delimitation. A response to Lohse (2009). *Systematic Biology* **58**, 442–444.

Pons, J., Barraclough, T.G., Gomez-Zurita, J. et al. (2006). Sequence-based species delimitation for the DNA taxonomy of undescribed insects. *Systematic Biology* **55**, 595–610.

Ramazzotti, G., Maucci, W. (1983). Il phylum Tardigrada. III Edizione riveduta e aggiornata. *Memorie dell'Istituto Italiano di Idrobiologia Dott. Marco de Marchi* **41**, 1-1012.

Ramløv, H., Westh, P. (2001). Cryptobiosis in the eutardigrade *Adorybiotus* (*Richtersius*) *coronifer*: tolerance to alcohols, temperature and de novo protein synthesis. *Zoologischer Anzeiger* **240**, 517–523.

Rebecchi, L., Altiero, T., Guidetti, R. (2007). Anhydrobiosis: the extreme limit of desiccation tolerance. *Invertebrate Survival Journal* **4**, 65–81.

Rebecchi, L., Guidetti, R., Borsari, S., Altiero, T., Bertolani, R. (2006). Dynamics of longterm anhydrobiotic survival of lichen-dwelling tardigrades. *Hydrobiologia* **558**, 23–30.

Rebecchi, L., Altiero, T., Guidetti, R. et al. (2009). Tardigrade resistance space effect: first results of experiments on the LIFE-TARSE Mission on FOTON-M3 (September 2007). *Astrobiology* **9**, 581–591.

Regier, J.C., Shultz, J.W., Kambic, R.E., Nelson, D.R. (2003). Robust support for tardigrade clades and their ages from three protein-coding nuclear genes. *Invertebrate Biology* **123**, 93–100.

Ricci, C., Balsamo, M. (2000). The biology and ecology of lotic rotifers and gastrotrichs. *Freshwater Biology* **44**, 15–28.

Ricci, C., Caprioli, M. (2005). Anhydrobiosis in bdelloid species, populations and individuals. *Integrative and Comparative Biology* **45**, 759–763.

Ricci, C., Fontaneto, D. (2009). The importance of being a bdelloid: ecological and evolutionary consequences of dormancy. *Italian Journal of Zoology* **76**, 240–249.

Ricci, C., Melone, G., Santo, D., Caprioli, M. (2008). Morphological response of a bdelloid rotifer to desiccation. *Journal of Morphology* **257**, 246–253.

Sands, C.J., McInnes, S.J., Marley, N.J. et al . (2008a). Phylum Tardigrada: an "individual" approach. *Cladistics* **24**, 861–871.

Sands, C.J., Convey, P., Linse, K., McInnes, S.J. (2008b). Assessing meiofaunal variation among individuals utilising morphological and molecular approaches: an example using the Tardigrada. *BMC Ecology* **8**, 7.

Schill, R.O., Steinbruck, G. (2007). Identification and differentiation of Heterotardigrada and Eutardigrada species by riboprinting. *Journal of Zoological Systematics and Evolutionary Research* **45**, 184–190.

Schill, R., Neumann, S., Reuner, A., Brümmer, F. (2008). Detection of DNA damage with single-cell gel electrophoresis in anhydrobiotic tardigrades. *Comparative Biochemistry and Physiology Part A Molecular and Integrative Physiology* **151**, 32.

Schurko, A.M., Neiman, M., Logsdon, J.M. (2008). Signs of sex: what we know and how we know it. *Trends in Ecology and Evolution* **24**, 208–217.

Schwerdtfeger, F. (1975). *Synökologie*. Hamburg: P. Parey.

Segers, H., De Smet, W.H. (2008). Diversity and endemism in Rotifera: a review, and *Kerarella* Bory de St Vincent. *Biodiversity and Conservation* **17**, 303–316.

Serra, M., Gómez, A., Carmona, M.J. (1998). Ecological genetics of *Brachionus* sympatric sibling species. *Hydrobiologia* **388**, 373–384.

Suatoni, E., Vicario, S., Rice, S., Snell, T., Caccone, A. (2006). An analysis of species boundaries and biogeographic patterns in a cryptic species complex: the rotifer–*Brachionus plicatilis*. *Molecular Phylogenetics and Evolution* **41**, 86–98.

Sudzuki, M. (1972). An analysis of colonization in freshwater micro-organisms. II. Two simple experiments on the dispersal by wind. *Japanese Journal of Ecology* **22**, 222–225.

Todaro, M.A., Fleeger, J.W., Hu, Y.P., Hrincevich, A.W., Foltz, D.W. (1996). Are meiofaunal species cosmopolitan? Morphological and molecular analyses of Xenotrichula intermedia (Gastrotricha: Chaetonotida). *Marine Biology* **125**, 735–742.

Valentini, A., Pompanon, F., Taberlet, P. (2009). DNA barcoding for ecologists. *Trends in Ecology and Evolution* **24**, 110–117.

Watanabe, M., Kikawada, T., Fujita, A. et al. (2004). Physiological traits of invertebrates entering cryptobiosis in a post-embryonic stage. *European Journal of Entomology* **101**, 439–444.

Westheide, W., Schmidt, H. (2003). Cosmopolitan versus cryptic meiofaunal polychaete species. An approach to a molecular taxonomy. *Helgoland Marine Research* **57**, 1–6.

Wiens, J. (2007). Species delimitation: new approaches for discovering diversity. *Systematics Biology* **56**, 875–879.

Wright, J.C. (1991). The significance of four xeric parameters in the ecology of terrestrial Tardigrada. *Journal of Zoology* **224**, 59–77.

Wright, J.C. (2001). Cryptobiosis 300 years on from van Leuwenhoek: What have we learned about tardigrades? *Zoologischer Anzeiger* **240**, 563–582.

Wright, J.C., Westh, P., Ramløv, H. (1992). Cryptobiosis in Tardigrada. *Biological Reviews* **67**, 1–29.

Wright, S. (1921). Systems of mating. *Genetics* **6**, 111–178.

Part V

Processes

15

Microbes as a test of biogeographic principles

David G. Jenkins[1], Kim A. Medley[1] and Rima B. Franklin[2]

[1] *Department of Biology, University of Central Florida, Orlando, USA*
[2] *Department of Biology, Virginia Commonwealth University, Richmond, USA*

15.1 Introduction

In the hierarchy of scientific knowledge, a principle, rule or law describes consistent observations and precedes hypothesis and theory. Given consistent observations, other information or insight may suggest mechanisms, and a hypothesis can be formed. For example, the first principle of biogeography, Buffon's law, states that disjunct regions have distinct species assemblages despite similar environments. Buffon proposed a mechanism to explain biogeographic patterns: that species 'improve' or 'degenerate' according to their environment. Given generality and often incorporating multiple facets, a theory may emerge that explains the patterns well (e.g. evolutionary theory).

As in ecology, biogeographic principles may include speculations that 'have often been elevated to laws merely by the passing of time' (Loehle, 1987). Tests of

Biogeography of Microscopic Organisms: Is Everything Small Everywhere?, ed. Diego Fontaneto. Published by Cambridge University Press.

biogeographic laws/principles/rules are thus valuable for biogeography in general and for understanding the tested system.

In that context, the statement for microbes that 'Everything is everywhere, but the environment selects' (Finlay, 2002; de Wit and Bouvier, 2006; hereafter abbreviated as EiE) is valuable to test the generality of biogeography's principles and their hypothesised mechanisms. Generality is tested best by extremes, and microbes (defined here as < ~1–2 mm; Finlay, 2002) certainly represent the lower margin of body size for most biogeographic evidence because most biogeography research has been conducted with macrobes (defined here as larger than 1–2 mm; Finlay, 2002). According to EiE, microbes have no biogeographic pattern due to their enormous population sizes and high probability of ubiquitous dispersal (Finlay, 2002). If so, then biogeographic principles derived from macrobes are not general, and subsequent hypotheses and theory must be also be constrained. In addition, the EiE claim tests biogeographic principles because EiE argues that macrobes have biogeographies (Finlay, 2002). The EiE claim is thus double-edged because it also expects definitive patterns (laws, principles or rules) for macrobes.

In this chapter we evaluate the evidence for biogeographic principles of macrobes and the extension of those principles to microbes. We do not claim to have found all literature on this rather broad topic, though we conducted a thorough literature search. Specifically, we evaluate the evidence that:

(1) Abundance, body size and distribution are inter-related for both macrobes and microbes.
(2) Niche affects spatial distribution for both macrobes and microbes.
(3) Microbes and small macrobes have phylogeographies (i.e. geographic pattern in phylogenetic structure).

Topics 1 and 2 address mechanisms (e.g. high abundance causes a large range), while topic 3 is about biogeographic patterns that may result from multiple mechanisms. These topics are important to biogeography (Lomolino et al., 2006) and have not been explored for microbial biogeography, while other related topics have been explored. For example, Green and Bohannan (2006) focused on questions of spatial scale (greater community dissimilarity with greater distance, taxa–area relationships and the ratio of local:global taxa richness). Martiny et al. (2006) considered non-random spatial distributions of microbes and general approaches to examine contemporary and/or historical processes acting on microbial community structure. Others have considered speciation and extinction rates (e.g. Horner-Devine et al., 2004; Ramette and Tiedje, 2007) but concluded that too few data exist, especially for extinction rates.

15.2 Abundance, body size and distribution

Abundance is important to ecological, biogeographic and macroecological concepts. Here we focus on three abundance relationships: abundant-centre, abundance-range and size-abundance.

15.2.1 Abundant–centre

According to the abundant-centre principle, a species reaches its greatest local abundance near its range centre, related to increasingly detrimental conditions toward its range edge (Andrewartha and Birch, 1954; Whittaker, 1956; Westman, 1980; Hengeveld and Haeck, 1982; Brown, 1984; Brown et al., 1995; Thomas and Kunin, 1999; Gaston, 2003). This relationship has been influential in ecology and biogeography (Sagarin et al., 2006) and assumes that a species' range is determined by environmental conditions, that the species' range has an edge, and that the range is roughly equilibrial. These assumptions are most likely true for native species inhabiting a relatively stable landscape, but may not be expected for native species during climate change, for an invasive species still expanding its non-native range, or in the case of invasional ratcheting, in which an invasive species adapts to a new range and then is re-introduced to its native region and expands that native range (Medley, 2010).

Evidence for the abundant-centre relationship was reviewed by Sagarin and Gaines (2002). They found only 39% of studies support the relationship and concluded that 'more exploration of species' abundance distributions is necessary', including more sampling near range edges. The abundant-centre principle is better characterised as an assumption than as a principle for macrobes (Sagarin and Gaines, 2002; Sagarin et al., 2006).

The EiE claim for cosmopolitan distributions and 'astronomical' abundances of microbes (Finlay, 2002) translates to an expectation that microbes do not decline in abundance from range centre to range edge (no range edge exists for cosmopolitan species). Most biogeographic information has been collected for macrobes, so it should be no surprise that less is known about the distribution of abundance across microbial species ranges. The best example we could find for microbes was that of Krasnov et al. (2008), in which fleas and mites on Palearctic small mammals tended to correspond to the expected abundant-centre pattern for macrobes. However, parasitic organisms have been excluded from the EiE claim (Finlay, 2002; Finlay and Fenchel, 2004) because patterns should mirror host patterns, plus Krasnov et al. (2008) demonstrated that the patterns are likely affected by other factors. We conclude that the abundant-centre 'principle' can hardly be considered definitive for macrobes, and is far less understood for microbes.

15.2.2 Abundance–range and size–abundance

The abundance-range principle holds that species with greater local abundance have greater distributional ranges, and has been considered a generality among diverse macrobes (e.g. Andrewartha and Birch, 1954; Blackburn et al., 1997; Gaston et al., 1997; Hubbell, 2001; Harte et al., 2001). The EiE claim is a corollary of this principle because microbial species can attain 'astronomical' local abundance and thus are argued to have very large (i.e. cosmopolitan) distributions (Finlay, 2002). As described above for the abundant–centre principle, the EiE claim essentially states that the abundance-range principle is saturated for microbes. Likewise, a negative relationship between body size and local abundance is regarded as well-supported for macrobes (Damuth, 1987; Brown et al., 1995) and is consistent with EiE (Finlay, 2002). This principle has the advantage that it is intuitive, in that many microbes can be visualised as fitting into the space occupied by one macrobe.

Given that abundance appears to be positively related to range area and that body size is logically and negatively related to abundance, then body size should be negatively related to range area (smaller organisms should have larger ranges; Fig 15.1). In addition, this relationship should apply to macrobes and microbes. However, this does not seem to be the case. Most (80%) of macrobial studies reviewed by Gaston (1996) observed a positive relationship between body size and range, rather than a negative relationship as predicted by the combination of the abundance–range and size–abundance principles. We know of no comparable data to evaluate the size–range relationship among microbes, but a random pattern may be expected (Martiny et al., 2006; Jenkins et al., 2007).

What may reconcile the contrast between individual well-founded principles and observations of their combination? A negative size–range relationship requires only simple diffusive (random) dispersal because no factors are needed to explain the pattern other than a density-dependent probability of dispersal from a local population into the surrounding landscape. This relationship should be most appropriate for passive dispersers, including free-living microbes that are the focus of the EiE claim (Finlay, 2002). On the other hand, actively dispersing organisms (typically macrobes) have a positive size–range relationship (Gaston, 1996). As evidence to support this difference between passive and active dispersers, maximal observed dispersal distance is a random function of body size for passive dispersers, while dispersal distance increases with body size for active dispersers (Jenkins et al., 2007). Maximal observed dispersal distance is relevant to range area but should be more proximal to dispersal-based differences among organisms because many other

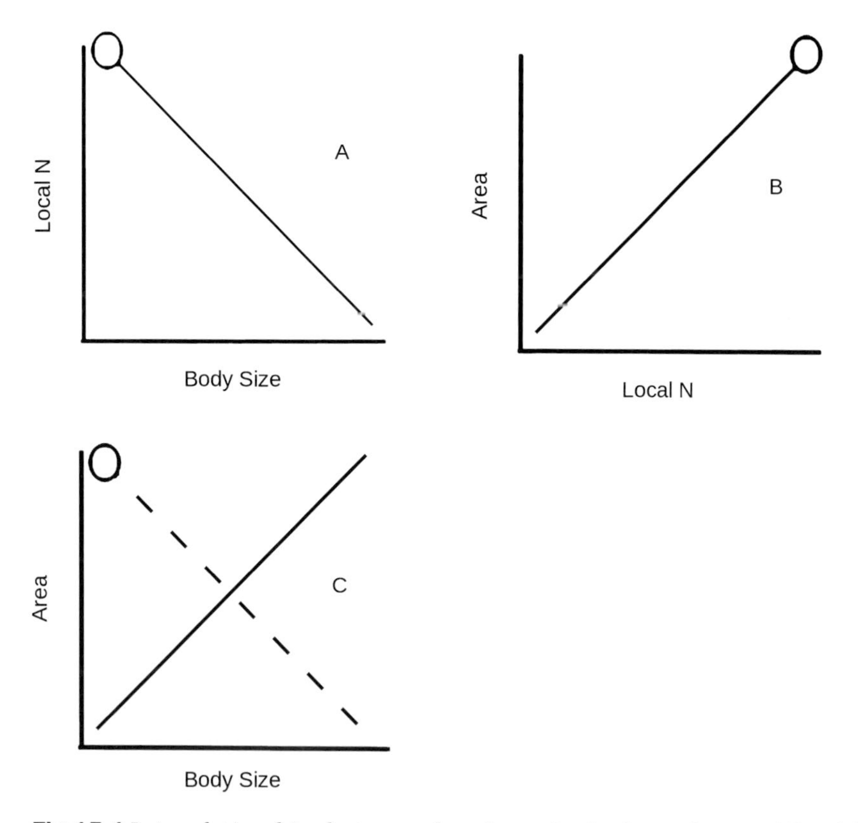

Fig 15.1 Interrelationships between abundance, body size and range. Microbial organisms are indicated with the open circle on each plot. (A) Logic and evidence support the negative relationship between body size and local abundance (Damuth, 1987; Brown et al., 1995). (B) The positive relationship between local abundance and range size is also well documented for macrobes (Gaston et al., 1997). (C) Given A and B, then smaller organisms that have larger local abundance should also have larger range sizes, whereas larger organisms with less abundance should have smaller range sizes (dashed line). In fact, the opposite (solid line) is well documented for macrobes (Gaston, 1996).

factors (e.g. landscape heterogeneity, climate, biological interactions) also may affect range area.

Our brief evaluation of abundance, body size and distribution for microbes and macrobes suggests that dispersal mode (passive vs. active) actually causes observed patterns, rather than simple body size per EiE. Overall, the macrobial and microbial evidence for abundance, body size and distribution do not support the EiE claim because the principles for macrobes are not definitive and because the evidence for microbes is grossly inadequate at this time.

15.3 Niche and distribution

The ecological niche has been conceptually related to organismal distributions for nearly a century (Grinnell, 1917) and niche-based distribution models continue to be important for predicting biogeographic distributions (Wiley et al., 2003; Reed et al., 2008; Kearney and Porter, 2009; Medley, 2010). Much has been written about the niche concept (see reviews by Pulliam, 2002; Chase and Leibold, 2003; Colwell and Rangel, 2009; Soberón and Nakamura, 2009). The niche is classically related to distribution in terms of the fundamental niche, defined as the multidimensional space within which a species can attain positive population growth. When the fundamental niche is projected onto geographic space, species occupy that subset of the fundamental niche that is actually available at a given space and time (potential niche, Jackson and Overpeck, 2000; Soberón and Nakamura, 2009). Finally, additional constraints by biotic interactions yield the realised niche (Hutchinson, 1957; Pulliam, 2002; Colwell and Rangel, 2009; Soberón and Nakamura, 2009). These niche concepts do not incorporate other processes (e.g. source–sink dynamics, dispersal limitation) that appear to also affect distributions (Fig 15.2; after Pulliam, 2002). The EiE claim ('... but the environment selects'; Fig 15.2A) is consistent with the Grinnelian niche concept, or the Hutchinsonian niche concept if biotic interactions further limit distributions (Fig 15.2B). However, alternative mechanisms of source–sink dynamics (Fig 15.2C) or dispersal limitation (Fig 15.2D) are inconsistent with EiE because microbial species are presumed to be uniformly abundant and cosmopolitan (Finlay, 2002).

What evidence exists that the niche affects microbial distributions? We surveyed the literature for studies examining either niche or distribution for organisms with propagules < 1–2 mm. While many studies report ecological differences between species, we focused our search on those studies of quantitative niche characteristics that cause spatial segregation between species or result in apparent distributional boundaries at some scale. All studies we found consistently reported niche differences or local adaptation at intra- or interspecific levels, consistent with the fundamental niche in all cases and potentially related to the realised niche in a few cases (Table 15.1). Given that niche constraints on local persistence/occurrence have been observed for microbes, it is reasonable to expect that niche affects distribution of multiple microbial species, consistent with the 'environment selects' portion of the EiE claim (and with much of evolutionary ecology). Tests for source–sink dynamics or dispersal limitation as alternative explanations of microbial niche-distribution relationships will require that the fundamental niche for a species is already

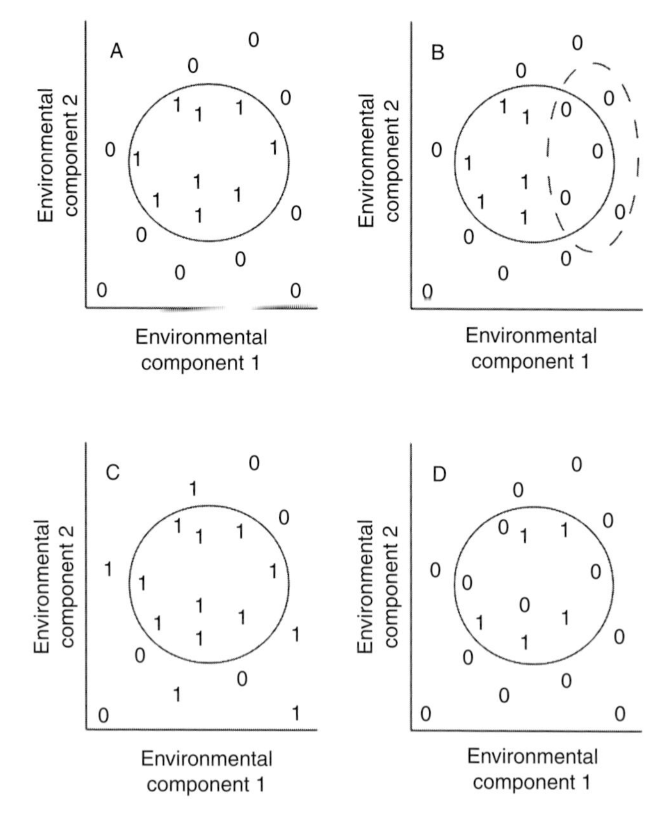

Fig 15.2 Niche-distribution relationships, based on Pulliam (2002). Zeros represent absence, and ones represent presence in niche space (e.g. two ordination axes). (A) The fundamental (Grinnellian) niche (or potential niche, Jackson and Overpeck, 2000; Soberón and Nakamura, 2009) is related to abiotic interactions. (B) The realised niche (*sensu* Hutchinson) is due to the combined influence of abiotic and biotic interactions, where the dashed line represents niche space of a second species. (C) Source-sink dynamics represent one alternative to (A) and (B), where sink populations outside the fundamental (or realised) niche exist due to immigration from source populations. (D) Dispersal limitation is a second alternative, where some combinations of environmental components have not been colonised, even within the fundamental niche space.

well characterised and that multiple sites within and beyond that niche space are thoroughly sampled for microbes and environmental conditions (Fig 15.2). Such data do not yet exist, but may soon be approached for marine microbes in the form of the International Census of Marine Microbes (ICoMM; http://icomm.mbl.edu/microbis/).

Table 15.1 Example evidence of niche differences resulting in spatial discontinuity for microbial species or genera with body sizes < ~1–2 mm. Niche differences have been demonstrated for diverse microbes at multiple scales and using both experimental and observational evidence.

Taxon	Habitat [1]	Approach [2]	Niche-related traits measured	Approximate spatial scale (km)	Conclusions	Source
Euchaeta norvegica, *Calanus finmarchicus*, *Pleuromamma robusta*, *Metrida lucens* (copepods)	M	O	Abundance	North Atlantic Ocean	Species partition niches in horizontal and vertical space	Williams (1988)
Balanion planctonicum, *Urotricha farcta*, *U. furcata* (ciliates)	F	E	Temperature and food (growth rate)	700	Niche differentiation within and between competing species and genera	Weisse et al. (2001)
Cephalodella hoodi, *Elosa worallii* (rotifers)	F	E	Temperature, food, and predation	0.008 (vertical samples)	Vertical niche separation related to temperature and food	Weithoff (2004)
Actinobacteria (bacteria)	F	E	Temperature	13 000	Local thermal adaptation; identical 16S sequences but some genetic variation at other loci	Hahn and Pöckl (2005)
Spumella sp. (chrysophyte flagellate)	F and T	E	Temperature (growth rate)	17 000	Local thermal adaptation among strains	Boenigk et al. (2007)

Sellaphora pupula, *S. bacillum*, *S. laevissima* (diatoms)	F	O	Trophic status of local habitat	600	*Sellaphora* demes (putative species) differ in environmental tolerances	Poulíčková et al. (2008)
Micromonas pusilla (chlorophyte flagellate)	M	O	Genetic differentiation	12500	Niche partitioning evident for this widely distributed morphospecies	Foulon et al. (2008)
Daphnia magna (cladoceran)	F	E and O	Ca++, pH requirements	600	Experimental niche accurately predicted 56 of 58 occurrences in Europe	Hooper et al. (2008)
diatoms, rotifers, crustacean zooplankton, aquatic insects	F	O	multiple physical-chemical and biotic variables	3000	Local habitat variables and regional location determine community structure.	Kernan et al. (2009)

[1] Habitats: F: Freshwater, M: Marine, T: Terrestrial.
[2] Study Approach: E: Experimental, O: Observational.

15.4 Microbial phylogeographies

Phylogeography is pattern analysis that indicates evolutionary processes in space and time, and thus enables phylogenetic and geographic history to be evaluated as a potential mechanism of microbial biogeography. In contrast to the large body of knowledge on macrobe biogeography (e.g. Lomolino et al., 2006), EiE argues that the high dispersal rates and frequent dispersal events of microbes swamp any spatial structure that may otherwise arise through vicariance, historical dispersal and local adaptation. Given the repeated reshuffle of microbial populations predicted by EiE, phylogeographic patterns concordant with geological processes of plate tectonics, glaciations, geographic barriers, etc. should not apply because phylogeography should be swamped by contemporary dispersal.

According to the EiE claim, microbes do not have biogeographies while macrobes do. Finlay (2002) presented the 1 mm cutoff between microbes and macrobes as two mirror-image, logistic curves (Fig 15.3A); the proportion of species that are ubiquitous purportedly decreases abruptly at ~1 mm (dashed line, Fig 15.3A), while the proportion of species that have biogeographies increases abruptly at ~1 mm (solid line, Fig 15.3B). Because these two curves are mirror images, we can focus here on the curve for species with biogeographies, with the understanding that evidence for one curve necessarily provides evidence for the other. In addition, Finlay (2002) stated that ubiquity–biogeography transition should be in the 1–10 mm size range.

We tested Finlay's clear and specific prediction (Fig 15.3A) for the presence of a logistic function in the proportion of species with biogeographies and a transition in the 1–10 mm size range. Phylogeography studies focus on closely related lineages and provide specific tests of the EiE claim that microbes do not have biogeographies. We collected 51 phylogeographic studies published in the peer-reviewed literature (1998–2009) of organisms for which the dispersive life stage is < 10 mm. All studies applied molecular phylogeographic approaches at regional to global spatial scales and included Archaea, Bacteria, Protista, fungi, bryophytes, Rotifera, Annelida, Mollusca, Copepoda and Cladocera.

We evaluated the evidence by recording whether or not the authors concluded that the subject species had phylogeographies (1 = yes, 0 = no). We then computed a logistic regression of those binary conclusions as a function of body size to estimate the probability of a biogeography for a given body size. If Finlay's prediction is correct, a significant logistic function with a transition ~1–10 mm should be observed. The alternative null model (i.e. biogeography is not a function of body size) is a linear fit that has no significant slope but a significant intercept.

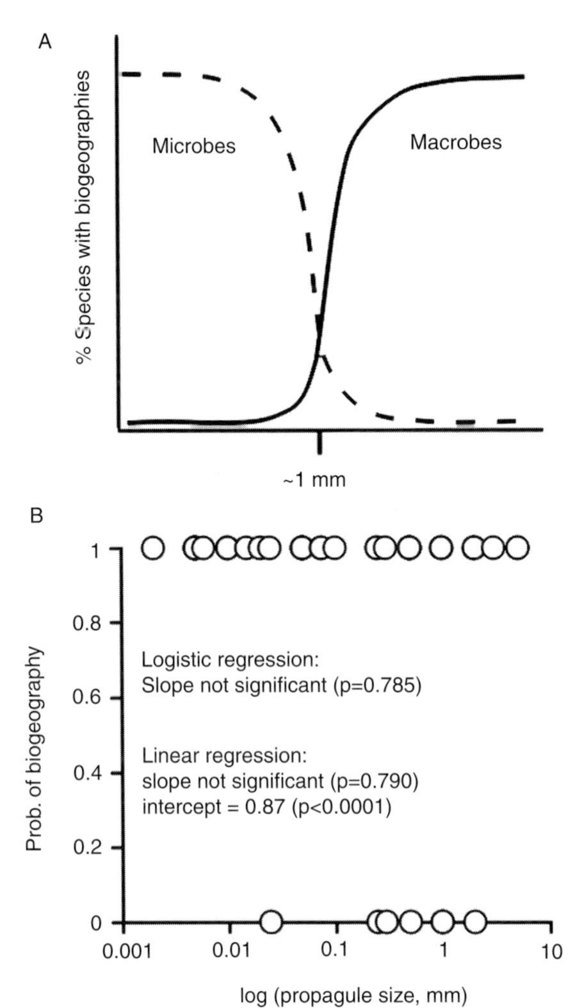

Fig 15.3 Size-based expectations and empirical results from the 'Everything is everywhere' claim (EiE; Finlay, 2002). (A) Microbes and macrobes present mirror-image trends in the predicted proportion of species that have biogeographies (from Finlay, 2002). We tested evidence for the macrobe curve (solid line) in empirical phylogeographies. (B) Empirical patterns, where circles represent conclusions by phylogeography study authors for the study organism's propagule size (0: no biogeography observed; 1: biogeography observed; N = 51).

Forty-four of the 51 papers concluded that studied organisms had biogeographies, while only seven of 51 found no evidence for biogeographic structure (Fig 15.3B). A logistic regression did not significantly fit the data ($p = 0.785$), nor did a linear regression have a significant slope ($p = 0.790$), though the linear regression did have a significant intercept ($\beta_0 = 0.87$, $p < 0.0001$). Thus, evidence we found indicates that microbes (< ~1 mm) are just as likely to exhibit biogeographies as macrobes, and that there is no support for the logistic, mirror-imaged distinction between ubiquitous microbes vs. macrobes with biogeographies.

Phylogeographies may arise by multiple mechanisms, but the fact that they are repeatedly observed for microbes is strong evidence that the same biogeographic mechanisms (e.g. vicariance, dispersal, speciation, adaptation, extinction) that affect macrobes also affect microbes. A more interesting challenge is to learn why some microbial species are widespread while others are not. To begin to address this challenge we will need to move beyond simple size-based distinctions and take account of life-history traits more likely to be related to dispersal (e.g. active or passive mode, dormancy, adaptations for phoretic transport) *and* success upon arrival (abiotic tolerance limits, nutrient requirements, trophic interactions, etc.).

In summary, we conclude that:

- Too few data exist to evaluate relationships between abundance, body size and distribution for microbes, and remain unclear (in part) for macrobes. Thus, the EiE claim is not supported for these basic components of biogeography. However, the EiE claim has been useful for biogeographic principles because it led to consideration of relationships for macrobes and microbes and revealed potential new research directions.
- Evidence exists for fundamental niche constraints in microbes, plus some evidence for realised niche constraints. Niche-distribution relationships that are consistent with the EiE claim await more extensive and intensive sampling to fully characterise the role of niche in affecting microbial distributions. As for macrobes, we expect niche-distribution relationships will be found to constrain some microbes to distributions that are less than cosmopolitan.
- Most (86%) of phylogeographic analyses do not support the EiE claim that microbes have no biogeography. Contrary to the EiE prediction that the proportion of species with biogeographies declines logistically ~1–10 mm in body size, no such trend was observed among empirical data sets.
- The EiE claim has helped turn biogeographic research attention to small organisms, especially in its recent revival during the era of molecular systematics. We expect that the stark contrasts in the EiE claim will be replaced over time with more sophisticated understanding of patterns and processes that more fully reflect Nature's complexity. The clear and simple EiE claim will likely give way to a more nuanced but representative understanding of microbial biogeography that is based on more salient metrics than body size alone.

References

Andrewartha, H.G., Birch, L.C. (1954). *The Distribution and Abundance of Animals*. Chicago, IL: University of Chicago Press.

Blackburn, T.M., Gaston, K.J., Quinn, R.M., Arnold, H., Gregory, R.D. (1997). Of mice and wrens: the relation between abundance and geographic range size in British mammals and birds. *Philosophical Transactions of the Royal Society of London B* **352**, 419–427.

Boenigk, J., Jost, S., Stoeck, T., Garstecki, T. (2007). Differential thermal adaptation of clonal strains of a protist morphospecies originating from different climatic zones. *Environmental Microbiology* **9**, 593–602.

Brown, J. H. (1984). On the relationship between abundance and distribution of species. *American Naturalist* **124**, 255–279.

Brown, J.H., Mehlman, D.W., Stevens, G.C. (1995). Spatial variation in abundance. *Ecology* **76**, 2028–2043.

Chase, J.M., Leibold, M.A. (2003). *Ecological Niches: Linking Classical and Contemporary Approaches*. Chicago, IL: University of Chicago Press.

Colwell, R. K., Rangel, T.F. (2009). Hutchinson's duality: The once and future niche. *Proceedings of the National Academy of Sciences USA* **106**, 19651–19658.

Damuth, J. (1987). Interspecific allometry of population density in mammals and other animals: the independence of body mass and population energy use. *Biological Journal of the Linnean Society* **31**, 193–246.

de Wit, R., Bouvier, T. (2006). 'Everything is everywhere, but the environment selects'; what did Baas Becking and Beijerinck really say? *Environmental Microbiology* **8**, 755–758.

Finlay, B.J. (2002). Global dispersal of free-living microbial eukaryote species. *Science* **296**, 1061–1063.

Finlay, B.J., Fenchel, T. (2004) Cosmopolitan metapopulations of free-living microbial eukaryotes. *Protist* **155**, 237–244.

Foulon, E., Not, F., Jalabert, F., Cariou, T. et al. (2008). Ecological niche partitioning in the picoplanktonic green alga *Micromonas pusilla*: evidence from environmental surveys using phylogenetic probes. *Environmental Microbiology* **10**, 2433–2443.

Gaston, K.J. (1996). Species-range-size distributions: patterns, mechanisms and implications. *Trends in Ecology and Evolution* **11**, 197–201.

Gaston, K.J. (2003). *The Structure and Dynamics of Geographic Ranges*. New York, NY: Oxford University Press.

Gaston, K.J., Blackburn, T.M., Lawton, J.H. (1997). Interspecific abundance-range size relationships: an appraisal of mechanisms. *Journal of Animal Ecology* **66**, 579–601.

Green, J., Bohannan, B.J.M. (2006). Spatial scaling in microbial biodiversity. *Trends in Ecology and Evolution* **21**, 501–507.

Grinnell, J. (1917). The niche-relationships of the California Thrasher. *The Auk* **34**, 427–433.

Hahn, M.W., Pöckl, M . (2005). Ecotypes of planktonic actinobacteria with identical 16s rrna genes adapted to thermal niches in temperate, subtropical, and tropical freshwater habitats. *Applied and Environmental Microbiology* **71**, 766–773.

Harte, J., Blackburn, T., Ostling, A. (2001). Self-similarity and the relationship between abundance and range size. *American Naturalist* **157**, 374–386.

Hengeveld, R., Haeck, J. (1982). The distribution of abundance. I. Measurements. *Journal of Biogeography* **9**, 303–316.

Hooper, H.L., Connon, R., Callaghan, A., Fryer, G. et al. (2008). The ecological niche of *Daphnia magna* characterized using population growth rate. *Ecology* **89**, 1015–1022.

Horner-Devine, M.C., Lage, M., Hughes, J.B., Bohannan, B.J. (2004). A taxa-area relationship for bacteria. *Nature* **432**, 750–753.

Hubbell, S.P. (2001). *The Unified Neutral Theory of Biodiversity and Biogeography*. Princeton, NJ: Princeton University Press.

Hutchinson, G.E. (1957). Concluding remarks. *Cold Spring Harbor Symposium on Quantitative Biology* **22**, 415–427.

Jackson, S.T., Overpeck, J.T. (2000). Responses of plant populations and communities to environmental changes of the late Quaternary. *Paleobiology* **26 (Suppl)**, 194–220.

Jenkins, D.G., Brescacin, C.R., Duxbury, C.V. et al. (2007). Does size matter for dispersal distance? *Global Ecology and Biogeography* **16**, 415–425.

Kearney, M., Porter, W. (2009). Mechanistic niche modelling: combining physiological and spatial data to predict species ranges. *Ecology Letters* **12**, 334–350.

Kernan, M., Ventura, M., Bitušík, P., Brancelj, A. et al. (2009). Regionalisation of remote European mountain lake ecosystems according to their biota: environmental versus geographical patterns. *Freshwater Biology* **54**, 2470–2493.

Krasnov, B.R., Shenbrot, G.I., Khokhlova, I.S., Vinarski, M. et al. (2008). Geographical patterns of abundance: testing expectations of the ‘abundance optimum’ model in two taxa of ectoparasitic arthropods. *Journal of Biogeography* **35**, 2187–2194.

Loehle, C. (1987). Hypothesis testing in ecology: psychological aspects and the importance of theory maturation. *Quarterly Review of Biology* **62**, 397–409.

Lomolino, M.V., Riddle, B.R., Brown, J.H. (2006). *Biogeography*, 3rd Edition. Sunderland, MA: Sinauer Associates.

Martiny, J.B.H., Bohannan, B., Brown, J. et al. (2006) Microbial biogeography: putting microorganisms on the map. *Nature Reviews Microbiology* **4**, 102–112.

Medley, K.A. (2010). Niche shifts during the global invasion of the Asian tiger mosquito, *Aedes albopictus* Skuse (Culicidae), revealed by reciprocal distribution models. *Global Ecology and Biogeography* **19**, 122–133.

Poulíčková, A., Špačková, J., Kelly, M.G., Duchoslav, M., Mann, D.G . (2008). Ecological variation within *Sellaphora* species complexes (Bacillariophyceae): specialists or generalists? *Hydrobiologia* **614**, 373–386.

Pulliam, H.R. (2002). On the relationship between niche and distribution. *Ecology Letters* **3**, 349–361.

Ramette, A., Tiedje, J.M. (2007). Biogeography: an emerging cornerstone for understanding prokaryotic diversity, ecology, and evolution. *Microbial Ecology* **53**, 197–207.

Reed, K.D., Meece, J.K., Archer, J.R., Peterson, A.T. (2008). Ecologic niche modeling of *Blastomyces dermatitidis* in Wisconsin. *PLoS ONE* **3**, 1–7.

Sagarin, R.D., Gaines, S.D. (2002). The 'abundant centre' distribution: to what extent is it a biogeographical rule? *Ecology Letters* **5**, 137–147.

Sagarin, R. D, Gaines, S.D., Gaylord, B. (2006). Moving beyond assumptions to understand abundance distributions across the ranges of species. *Trends in Ecology and Evolution* **21**, 526–530.

Soberón, J., Nakamura, M. (2009). Niches and distributional areas: concepts, methods, and assumptions. *Proceedings of the National Academy of Sciences USA* **106**, 19644–10650.

Thomas, C.D., Kunin, W.E. (1999). The spatial structure of populations. *Journal of Animal Ecology* **68**, 647–657.

Weisse, T., Karstens, N., Meyer, V.C.L. et al. (2001). Niche separation in common prostome freshwater ciliates: the effect of food and temperature. *Aquatic Microbial Ecology* **26**, 167–179.

Weithoff, G. (2004). Vertical niche separation of two consumers (Rotatoria) in an extreme habitat. *Oecologia* **139**, 594–603.

Westman, W.E. (1980). Gaussian analysis: identifying environmental factors influencing bell-shaped species distributions. *Ecology* **61**, 733–739.

Whittaker, R.H. (1956). Vegetation of the Great Smoky Mountains. *Ecological Monographs* **26**, 2–80.

Wiley, E.O., McNyset, K.M., Peterson, A.T., Robins, C.R., Stewart, A.M. (2003). Niche modeling and geographic range predictions in the marine environment using a machine-learning algorithm. *Oceanography* **16**, 120–127.

Williams, R. (1988). Spatial heterogeneity and niche differentiation in oceanic zooplankton. *Hydrobiologia* **167**-168, 151–159.

16

A metacommunity perspective on the phylo- and biogeography of small organisms

LUC DE MEESTER

Laboratory of Aquatic Ecology and Evolutionary Biology, Katholieke Universiteit Leuven, Leuven, Belgium

16.1 Dispersal in small organisms

Small organisms rely on passive dispersal for colonising new habitats. Especially when they form resistant stages, passive dispersal does not translate into weak or limited dispersal (Bilton et al., 2001; Havel and Shurin, 2004). The main cost of passive dispersal is that the organism has no control over the trajectory and destination. By having adaptations for specific vectors (e.g. animals instead of wind), directionality and destination can to a certain extent be influenced. In aquatic organisms and plants, there is increasing evidence of widespread and potentially long-distance dispersal by a multitude of vectors, ranging from wind (Vanschoenwinkel et al., 2008a) and birds (Green et al., 2002; Figuerola et al., 2005) to insects (Van de Meutter et al., 2008), mammals (Vanschoenwinkel et al., 2008b) and humans and their transportation means (Havel et al., 2002). This translates into relatively high dispersal rates, as is shown by rapid colonisation rates of new habitats and rapid spread of exotic species (e.g. Louette and De Meester, 2005;

Biogeography of Microscopic Organisms: Is Everything Small Everywhere?, ed. Diego Fontaneto.

Havel and Shurin, 2004). Specific characteristics may make some species better dispersers than others, and dispersal rates in practice will also largely depend on abundance (i.e. sources of individuals). Effective dispersal, i.e. dispersal followed by establishment success, will in addition depend on the occurrence of habitats and their suitability for the focal species, and thus also on ecological specialisation and habitat preference of these species. In addition, potentially suitable habitats may become practically unsuitable because of the presence of other species, implying that effective dispersal may also depend on the dispersal rates and ecological specialisation of other species. The discrepancy between dispersal rates quantified as movement of organisms from habitat to habitat and effective dispersal rates, i.e. the establishment of new populations in target habitats following immigration, can be very large. This makes quantifying dispersal rates very difficult: quantifying moving individuals or estimating their number by extrapolation through modelling is a very difficult and time-consuming activity, yet does not suffice to obtain a reliable estimate of effective dispersal between communities or populations. Although it does provide valuable data on colonisation potential of organisms, it should not be carelessly translated into patterns of gene flow. The reverse is also true, as gene flow estimated from population genetic analyses using neutral markers should not be used as a substitute of dispersal rates. In many organisms including human populations (Ramachandran et al., 2005), population genetic structure and phylogeographic patterns reflect colonisation dynamics rather than ongoing gene flow, and show a pattern of serial colonisation.

16.2 Metacommunity structure of small organisms

Metacommunities are defined as collections of local communities that interact through the exchange of individuals, while local dynamics still dominate over regional impact. Leibold et al. (2004) introduced four paradigms (species sorting, mass effects, patch dynamics and the neutral model) as guidelines to grasp the relative importance of processes that structure natural metacommunities. In the species sorting paradigm, patches differ ecologically and dispersal is high enough so that the right species reaches the right patches. However, dispersal is not so high to result in homogenisation of species composition across ecologically different patches. Mass effects occur when dispersal is so high that it results in source-sink dynamics across patches, so that the match between species composition and local environmental conditions is reduced as species are found in habitats in which they would normally not occur. In the patch dynamics and neutral model paradigms, dispersal limitation is more important. The patch dynamics paradigm implies a trade-off between dispersal capacity and competitive strength among species, while species are ecologically equivalent in the neutral model.

If small organisms show relatively high dispersal rates, overall metacommunity structure in these organisms is expected to be dominated by species sorting and mass effects rather than by dispersal limitation and patch dynamics. I here refer to the overall metacommunity structure, because one has to consider that the likelihood of dispersal limitation will become higher for rare species in the metacommunity, as rareness translates in fewer sources of individuals and thus overall lower dispersal. Most data sets on small organisms so far tend to conform to the above expectation: in most cases metacommunities of small organisms (microbial organisms, protists, small metazoans like zooplankton) conform to the species sorting paradigm (Cottenie et al., 2003; Leibold et al., 2004; Cottenie, 2005; Van der Gucht et al., 2007; Vanormelingen et al., 2008), although some studies report an impact of dispersal limitation (Whitaker et al., 2003) or mass effects (Lindström et al., 2006). In metacommunity analysis, spatial scale and landscape connectivity are important. In studies in which several organism groups are compared for their metacommunity structure in the same set of habitats, an overall increase in the importance of species sorting as a force structuring local metacommunities is expected with decreasing size of the organism. This is illustrated by the study of Beisner et al. (2006) on metacommunity structure of fish, zooplankton, phytoplankton and bacteria in 18 Canadian lakes. Similarly, in a number of parallel studies on the same set of interconnected ponds at 'De Maten' in Belgium, there is an increase in the degree to which environmental gradients (i.e. species sorting) determine species composition of local communities as one moves from macro-invertebrates (Van de Meutter et al., 2007) through zooplankton (Cottenie et al., 2003) to phyto- and bacterioplankton (Van der Gucht et al., 2007; Vanormelingen et al., 2008). The reason why few cases of mass effects have been reported is probably related to the high population growth rates of many small organisms, which increases the impact of local as compared with regional dynamics (Van der Gucht et al., 2007). In organisms that have resistant dormant stages or can remain viable in an inactive state (like many prokaryotes), the impact of species sorting is enhanced because dispersal can act in a cumulative way thus reducing dispersal limitation (Fenchel, 2003). Dormant stages may reach an unsuitable habitat and reside there, remaining sensitive to emerge when environmental conditions would change so that the habitat becomes suitable. In this way, dispersal must not continuously be high to overcome dispersal limitation, as a low dispersal rate combined with a capacity to remain viable during unfavourable conditions during a reasonably long time may lead to the capacity of local communities to effectively respond to environmental gradients by species and lineage sorting.

In summary, the general pattern is that species (lineage) sorting becomes more important as organisms become smaller, which increases their capacity for passive long-distance dispersal and is associated with higher population growth rates. It remains that in habitat specialists favouring rare habitats and/or rare species, dispersal limitation may still occur and result in a decrease of the matching of species (lineage) composition to environmental gradients and an associated increase of the spatial signal in a metacommunity analysis.

16.3 Extending the spatial scale: phylo- and biogeography of small organisms

Where does metacommunity ecology stop and biogeography begin? As one increases spatial scales, interactions among communities through exchange of individuals becomes weaker. In principle, a set of communities that are so distant from each other that an exchange of individuals is extremely rare could still be viewed as an example of a metacommunity at the low dispersal end of the continuum. One can thus view metacommunity dynamics as one of the major structuring forces of biogeographic patterns. Entirely in parallel, at the population level, metapopulation dynamics can be viewed as one of the major structuring forces of phylogeographic patterns (Roderick, 1996; Avise, 2000; Hanski and Gaggiotti, 2004). Which large-scale patterns does one expect for small organisms? Does one expect similar patterns to the typical biogeography seen for macroorganisms, reflecting geographic dispersal barriers (e.g. at the continental scale) and past geological events? The observation that dispersal rates in space and time (through dormant stages) can be high would lead one to suggest that the biogeographic patterns observed for macroorganisms are unlikely to be common for microorganisms. There is a vivid debate on the degree to which microorganisms have cosmopolitan distributions (e.g. Fenchel, 2003; Martiny et al., 2006; Green et al., 2008). There are numerous reports suggesting that the biogeographic patterns observed for macroorganisms are absent among microorganisms (Fierer and Jackson, 2006; Van der Gucht et al., 2007). Also, haplotypes for specific genes (e.g. 16SrRNA gene) are often shared by individuals inhabiting different latitudes and even continents (Zwart et al., 2003). These data imply that a substantial number of microorganisms may have much broader distributions than most macroorganisms (Finlay, 2002; Katz et al., 2005). Yet, there are also studies showing biogeographic signals in microorganisms that seem reminiscent of those observed for macroorganisms (Whitaker et al., 2003). Indeed, many studies have reported striking provincialism in the phylogeography of small organisms such as zooplankton and protists (e.g. Hebert and Wilson, 1994). The controversy is strongly enhanced by the problematic nature of identifying prokaryotes and protists to species level (Heger et al., 2009). A major argument of authors claiming against cosmopolitanism in protists and prokaryotes is that taxonomic resolution is too low or unreliable, so that what is being claimed to be a cosmopolitan species is actually a conglomerate of related species or a species complex. There are clear cases in which formerly cosmopolitan species were recognised to belong to different taxa, meaning that what are currently considered to be cosmopolitan species may instead be cosmopolitan genera (zooplankton: Frey, 1982). Yet, even if one considers those lineage clusters as genera rather than species, the emerging pattern seems that cosmopolitanism at genus level in prokaryotes and protists is much higher than that observed for macroorganisms.

16.4 Towards a reconciliation of these conflicting patterns and expectations?

There are a number of processes that interfere with the extrapolation of potentially high dispersal rates to widespread distributions and an absence of spatial signal. In the following, I argue that these modifying processes result in predictions that may match with observed phylo- and biogeographic patterns in microorganisms.

The first process to interfere with global distributions of organisms is habitat specialisation and rareness, two features that are often associated with each other. If organisms are specialised to rare habitat conditions such as hot water springs (e.g. Whitaker et al., 2003), it is more likely that dispersal limitation will impact their distribution. As organisms become more rare or adapted to more rare habitats, populations and communities tend to become more isolated. The more habitats are isolated islands in an unfavourable matrix, the more dispersal limitation may become important.

The second process that interferes with global distributions of microorganisms is priority effects and monopolisation (*sensu* De Meester et al., 2002; Urban and De Meester, 2009). Microorganisms are not only characterised by high dispersal rates, but also by high population growth rates, short generation times and, often, the capacity to produce dormant stage banks. Rapid population growth combined with the formation of dormant stage banks may result in strong numerical advantage of first colonisers, as they rapidly reach carrying capacity in the newly colonised habitat. Once a habitat has been successfully colonised, the bank of dormant individuals effectively prevents population sizes from reaching very low numbers. For microorganisms, soon after colonisation of even a small habitat it may very soon be inhabited by millions to billions of individuals, providing the resident population with a powerful advantage over secondary immigrants (De Meester et al., 2002). If rapid local adaptation, fostered by short generation times, also increases the match between the fitness profile of the resident population and local habitat conditions, then the priority effect may be strongly enhanced and may become permanent (Urban and De Meester, 2009). This process, by which local adaptation may result in first colonisers monopolising the habitat to such an extent that the subsequent immigrants fail to successfully establish, will only be effective for species that occupy a similar niche and therefore may compete with each other. This process is thus most likely to influence intraspecific phylogeographic patterns, and patterns of occurrence of ecologically similar species, which may or may not be phylogenetically strongly related.

If the above processes are important in microorganisms, what phylo- and biogeographic patterns would one expect to see? The predictions are broad, ranging from cosmopolitanism to extreme provincialism. First, for some very common species, dispersal rates may be high enough at a global scale such that they are truly cosmopolitan. For more rare species or species that are highly specialised, a more typical biogeographic pattern reminiscent of most macroorganisms may appear because of dispersal limitation. In cosmopolitan species too, however, one may expect structure at the phylogeographic level, reflecting monopolisation events. First, serial colonisation is likely, yielding a pattern of isolation by distance not driven by a gradual reduction in gene flow but because of historical colonisation events. Serial colonisation receives increasing attention, also in macroorganisms including humans (Ramachandran et al., 2005). In microorganisms more than in macroorganisms, however, one may also expect a more complicated mosaic pattern of haplotype distribution, and sometimes 'enclaves' in which specific haplotypes dominate in a certain region that is surrounded by other haplotypes. These mosaic and enclave patterns result because of a combination of long-distance dispersal and monopolisation. In most cases, new habitats will be colonised by lineages from nearby habitats, resulting in continuous distribution patterns. But in organisms that engage in long-distance dispersal and show a high capacity of monopolisation (rapid population growth combined with rapid local adaptation), from time to time new habitats may get colonised through long-distance dispersal, and, if these early arrivers grow with sufficient rapidity to enforce a priority effect over secondary immigrants from nearby habitats, this may result in discontinuous distribution patterns of haplotypes (Fig 16.1).

I predict that phylogeographic patterns like mosaic and enclave distributions are much more common in microorganisms than in macroorganisms, because of their different capacity for monopolisation (cf. capacity for high population growth rates and rapid local adaptation; De Meester et al., 2002). Long-distance dispersal may be quite common in many microorganisms, e.g. when dispersal is mediated by migrating birds. In a phylogeographic study of the water flea *Daphnia magna*, De Gelas and De Meester (2005) revealed such discontinuous distribution of a number of haplotypes (e.g. Scandinavia and Israel). Mills

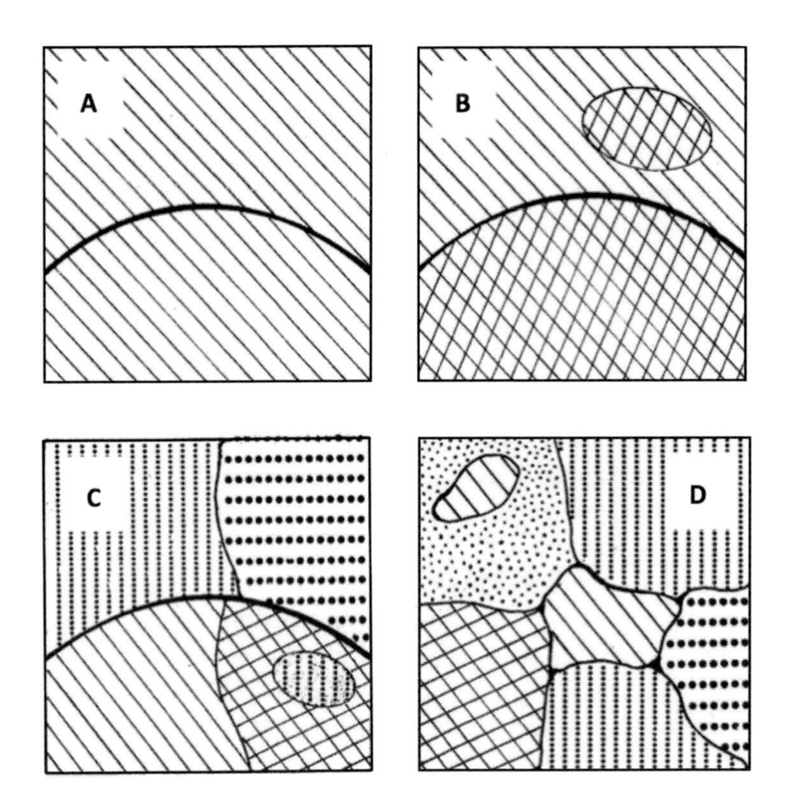

Fig 16.1 Hypothetical distributions of species (biogeography) or haplotypes (phylogeography) in a landscape. Different hatching patterns refer to different species or genetic lineages. The thick line in A–C depicts a geographic barrier; there is no geographic barrier in panel D. I assume no environmental gradients, so that the distributions of species and genetic lineages are independent of habitat preferences. (A) Ubiquitous species that colonised the whole area ('cosmopolitan' if this pattern occurs at a global scale). (B) Distribution that matches a geographic barrier, here also combined with an 'enclave' distribution. This pattern results from priority effects, partly modulated by a geographic barrier, and partly by a chance colonisation event (enclave distribution). (C) A more complex pattern partly mediated by a geographic barrier, partly by priority effects not associated with a geographic barrier. (D) A complex 'mosaic' pattern that is independent from true geographic barriers.

et al. (2007) also interpret isolation by distance in the globally distributed rotifer *Brachionus plicatilis* as reflecting persistent founder events. In essence, the monopolisation process, i.e. founder events being reinforced by local genetic adaptation, is quite similar to insular settings and isolated habitats that get colonised and undergo independent specialisation, leading to incipient or true speciation (Schluter, 2000; Hendry et al., 2002; Gillespie, 2004). I expect monopolisation to occur more frequently in micro- than in macroorganisms, because of their capacity for rapid local adaptation. Intriguingly, the elevated capacity of microorganisms to become locally adapted leads to a prediction that they should often show provincialism, thus running in the counter-direction to the prediction that they should be widespread due to their dispersal abilities. Here, spatio-temporal dynamics may be important, one prediction being that taxa may initially appear widespread but become increasingly provincial during the course of time, because of adaptation to local conditions. Although this process is expected to be more apparent across greater spatial scales, the relationship with space is difficult to predict, as monopolisation may already reduce effective gene flow at small spatial scales. The key message here is that more studies are needed, and that the contrast between high potential for long-distance dispersal combined with a strong capacity for monopolisation (evolution-driven priority effects) provides an interesting framework for interpreting phylogeographic patterns in small organisms.

The above suggests that there should be differences between the typical biogeographic signal in macroorganisms as compared with microorganisms. I expect many more mosaic distributions and disjunct distributions in microorganisms because of their capacity for long-distance dispersal combined with rapid population development fuelled by asexual reproduction during the parthenogenetic/asexual phase of their reproduction cycle. The biogeographic signal is thus 'disturbed' by unexpected occurrence of haplotypes or species that have locally colonised and occupied habitat. To the extent that species strongly overlap in niches and are ecologically similar, the same mosaic patterns may also be found at the community level, in which, depending on the region, one or the other of ecologically similar species dominates. Although there may often be a clear spatial signal, it may not be a straightforward increase in genetic differentiation with increasing geographic distance.

16.5 Summary

If the above ideas bear out, biogeographic patterns among macro- and microorganisms would be caused by quite different mechanisms: more dispersal limitation driven in macroorganisms, while more monopolisation driven in microorganisms. Isolation by distance in our view is often not driven by ongoing dispersal and gene flow, but may rather reflect historical colonisation events. And especially in microorganisms, occasional long-distance dispersal may disturb this isolation by distance and generate mosaic and enclave distributions, both at the among (biogeography) and within species (phylogeography) level.

References

Avise, J.C. (2000). *Phylogeography: The History and Formation of Species.* Cambridge, MA: Harvard University Press.

Beisner, B.E., Peres Neto, P.R., Lindström, E.S., Barnett A., Longhi, M.L. (2006). The role of environmental and spatial processes in structuring lake communities from bacteria to fish. *Ecology* **87**, 2985–2991.

Bilton, D.T., Freeland, J.R., Okamura, B. (2001). Dispersal in freshwater invertebrates. *Annual Reviews of Ecology and Systematics* **32**, 159–181.

Cottenie, K. (2005). Integrating environmental and spatial processes in ecological community dynamics. *Ecology Letters* **8**, 1175–1182.

Cottenie, K., Michels, E., Nuytten N., De Meester, L. (2003). Zooplankton metacommunity structure: regional versus local processes in highly interconnected ponds. *Ecology* **84**, 991–1000.

De Gelas, K., De Meester, L. (2005). Phylogeography of *Daphnia magna* in Europe. *Molecular Ecology* **14**, 754–763.

De Meester, L., Gómez, Q., Okamura, B., Schwenk, K. (2002). The Monopolization Hypothesis and the dispersal-gene flow paradox in aquatic organisms. *Acta Oecologica* **23**, 121–135.

Fenchel, T. (2003). Biogeography for bacteria. *Science* **301**, 925–926.

Fierer, N., Jackson, R.B. (2006). The diversity and biogeography of soil bacterial communities. *Proceedings of the National Academy of Sciences USA* **103**, 626–631.

Figuerola, J., Green A.J., Michot, T.C. (2005). Invertebrate eggs can fly: evidence of waterfowl-mediated gene flow in aquatic invertebrates. *American Naturalist* **165**, 274–280.

Finlay, B.J. (2002). Global dispersal of free-living microbial eukaryote species. *Science* **296**, 1061–1063.

Frey, D.G. (1982). Questions concerning cosmopolitanism in Cladocera. *Archiv für Hydrobiologie* **93**, 484–502.

Gillespie, R. (2004). Community assembly through adaptive radiation in Hawaiian spiders. *Science* **303**, 356–359.

Green, A.J., Figuerola, J., Sanchez, M.T. (2002). Implications of waterbird ecology for the dispersal of aquatic organisms. *Acta Oecologica* **23**, 177–189.

Green, J.L., Bohannan, B.J.M., Whitaker, R.J. (2008). Microbial biogeography: from taxonomy to traits. *Science* **320**, 1039–1043.

Hanski, I., Gaggiotti, O.E. (2004). *Ecology, Genetics, and Evolution of Metapopulations.* Amsterdam: Elsevier.

Havel, J.E., Shurin, J.B. (2004). Mechanisms, effects, and scale of dispersal in zooplankton. *Limnology and Oceanography* **49**, 1229–1238.

Havel, J.E., Shurin, J.B., Jones, J.R. (2002). Estimating dispersal from patterns of spread: spatial and local control of lake invasions. *Ecology* **83**, 3306–3318.

Hebert, P.D.N., Wilson, C.C. (1994). Provincialism in plankton: endemism and allopatric speciation in Australian *Daphnia. Evolution* **48**, 1333–1349.

Heger, T.J., Mitchell, E.A.D., Ledeganck, P. et al. (2009). The curse of taxonomic uncertainty in biogeographical studies of free-living terrestrial protists: a case study of testate amoebae from Amsterdam Island. *Journal of Biogeography* **36**, 1551–1560.

Hendry, A.P., Taylor, E.B., McPhail, J.D . (2002). Adaptive divergence and the balance between selection and gene flow: lake and stream stickleback in the Misty system. *Evolution* **56**, 1199–1216.

Katz, L.A., McManus, G.B., Snoeycnbos-West, O.L.O. et al. (2005). Reframing the 'Everything is everywhere' debate: evidence for high gene flow and diversity in ciliate morphospecies. *Aquatic Microbial Ecology* **41**, 55–65.

Leibold, M.A., Holyoak, M., Mouquet, N. et al. (2004). The metacommunity concept: a framework for multiple-scale community ecology. *Ecology Letters* **7**, 601–613.

Lindström, E.S., Forlsund, M., Algesten, G., Bergström, A.-K . (2006). External control of bacterial community structure in lakes. *Limnology and Oceanography* **51**, 339–342.

Logue, J.B., Lindström, E.S . (2008). Biogeography of bacterioplankton in inland waters. *Freshwater Reviews* **1**, 99–114.

Louette, G., De Meester, L. (2005). High dispersal capacity in aquatic organisms: species richness in cladoceran communities colonizing newly created habitats. *Ecology* **86**, 353–359.

Martiny, J.B.H., Bohannan, B.J.M., Brown, J.H. et al. (2006). Microbial biogeography: putting micro-organisms on the map. *Nature Reviews Microbiology* **4**, 102–112.

Mills, S., Lunt, D.H., Gomez, A. (2007). Global isolation by distance despite strong regional phylogeography in a small metazoan. *BMC Evolutionary Biology* **7**, 225.

Ramachandran, S., Deshpande, O., Roseman, C. et al. (2005). Support from the relationship of genetic and geographic distance in human populations for a serial founder effect originating in Africa. *Proceedings of the National Academy of Sciences USA* **102**, 15942–15947.

Roderick, G.K. (1996). Geographic structure of insect populations: gene flow, phylogeography, and their uses. *Annual Review on Entomology* **41**, 325–352.

Schluter, D. (2000). *The Ecology of Adaptive Radiation*. Oxford: Oxford University Press.

Urban, M., De Meester, L. (2009). Community monopolization: local adaptation enhances priority effects in an evolving metacommunity. *Proceedings of the Royal Society of London B* **276**, 4129–4138.

Van De Meutter, F., De Meester, L., Stoks, R. (2007). Metacommunity structure of pond macro invertebrates: effects of dispersal mode and generation time. *Ecology* **88**, 1687–1695.

Van de Meutter, F., Stoks, R., De Meester, L. (2008). Size-selective dispersal of *Daphnia* resting eggs by backswimmers (*Notonecta maculata*). *Biology Letters* **4**, 494–496.

Van der Gucht, K., Cottenie, K., Muylaert, K. et al. (2007). The power of species sorting: local factors drive bacterial community composition over a wide range of spatial scale. *Proceedings of the National Academy of Sciences USA* **104**, 20404–20409.

Vanormelingen, P., Cottenie, K., Michels, E. et al. (2008). The relative importance of dispersal and local processes in structuring phytoplankton communities in a set of highly interconnected ponds. *Freshwater Biology* **53**, 2170–2183.

Vanschoenwinkel, B., Gielen, S., Seaman, M., Brendonck, L. (2008a). Any way the wind blows – frequent wind dispersal drives species sorting in ephemeral aquatic communities. *Oikos* **117**, 125–134.

Vanschoenwinkel, B., Waterkeyn, A., Vandecaetsbeek, T. et al. (2008b). Dispersal of freshwater invertebrates by large terrestrial mammals: a case study with wild boar (*Sus scrofa*) in Mediterranean wetlands. *Freshwater Biology* **53**, 2264–2273.

Whitaker, R.J., Grogan, D.W., Taylor, J.W. (2003). Geographic barriers isolate endemic populations of hyperthermophilic archaea. *Science* **301**, 976–978.

Zwart, G., van Hannen, E.J., Kamst-van Agterveld, M.P. et al. (2003). Rapid screening for freshwater bacterial groups by using reverse line blot hybridization. *Applied and Environmental Microbiology* **69**, 5875–5883.

17

Geographic variation in the diversity of microbial communities: research directions and prospects for experimental biogeography

JOAQUÍN HORTAL

Departamento de Biodiversidad y Biología Evolutiva, Museo Nacional de Ciencias Naturales (CSIC), Madrid, Spain; and Azorean Biodiversity Group - CITA A, Department of Agricultural Sciences, University of the Azores, Angra do Heroísmo, Terceira, Açores, Portugal

17.1 Introduction

Traditionally, most ecologists understand the world from a human scale. Ecosystems are often understood as large visible units of the landscape,[1] usually homogeneous land patches or a series of adjacent patches with intense flows of individuals, energy or biomass and nutrients. However, there is more in a landscape than meets the eye. An arguably homogeneous land patch within a

[1] Commonly 'all the *visible* features of an area of land' (*Compact Oxford English Dictionary*, revised edition 2008; my italics). In ecology, a series of spatial units occupied by an heterogeneous species assemblage (e.g. Polis et al., 2003).

Biogeography of Microscopic Organisms: Is Everything Small Everywhere?, ed. Diego Fontaneto. Published by Cambridge University Press.

landscape hosts many small ecosystems, or microhabitat patches, where many different communities of microbes[2] dwell and interact. For example, imagine you are standing in a clearing of an open forest in a temperate region. A terrestrial ecologist studying macroscopic organisms would think he is looking at part of a single ecosystem. On the contrary, a microbial ecologist will identify a plethora of different ecosystems, including leaf litter of different degrees of humidity, the bark of each different tree and shrub species, treeholes, temporary puddles and pools, moss cushions of different life forms growing over different substrates, etc. Not to mention soil communities. In other words, a 1 ha clearing within a forest could be considered a whole landscape for many groups of microbes.

A key question in microbial ecology is thus whether the patterns and organisation of microbial communities differ from those of macroscopic organisms just in terms of scale or they are so radically different that the rules affecting macrobes cannot be extrapolated to microbes. The debate on this question extends to the biogeography of microorganisms. Strikingly, it has been argued that most microorganisms do not have biogeography; that is, that contrary to macroorganisms, the distributions of microorganism species are just limited by local environmental conditions (e.g. Fenchel and Finlay, 2003, 2004 and below). But, are microbes so different from their larger relatives than they follow different ecological and biogeographic rules?

Here I will argue that when it comes to the spatial distribution (and basic ecology) of their communities, many microbes (especially multicellular ones) are just smaller than large organisms, rather than radically different in their ecological organisation and biogeographic responses. More precisely, I will first argue that not everything small is everywhere, and then provide a brief account of current evidence on the spatial distribution of microbe diversity at the community level. Given the purpose of this chapter, this review will be argumentative rather than exhaustive. Based on such review, I will propose the study of microbes as a way of advancing current biogeographic and macroecological theory,[3] under the hypothesis that some biogeographic principles can be evaluated with success on microorganisms, controlling for many of the confounding factors acting at large scales, or even allowing to develop experiments in biogeography.

[2] By microbes I refer to all microscopic and small-sized organisms. I will use this term (and eventually microorganisms for style reasons) throughout. Microbes are often defined as organisms less than 1–2 mm in size (in opposition to macrobes, i.e. those larger than 2 mm; see Finlay, 2002), but here I will use a relaxed definition of microorganisms, and follow the common use of including also bryophytes, ferns and fungi, whose spores are smaller than 2 mm.

[3] Note that I refer to patterns at the community level (i.e. species richness, species replacement or functional diversity); complementarily, Jenkins and colleagues (Chapter 15) propose using microbes to study a series of biogeographical principles at the species level, namely the relationships between species distributions and abundance, body size and niche characteristics, and the phylogenetic structure of species across space.

17.2 Spatial variations in the diversity of microscopic organisms

17.2.1 Is everything small everywhere?

Perhaps the most striking difference between the known spatial distributions of macrobe and microbe species is that while restricted distributions are the rule for the former, it has been argued that they may be exceptions for the latter (Fenchel and Finlay, 2003, 2004; Kellogg and Griffin, 2006 Fontaneto and Hortal, 2008). However, the level of knowledge about the spatial distribution of microbial diversity is not comparable to that of macroscopic organisms. Despite the causes of the spatial distribution of diversity still being under debate (see section 17.4 below), the current degree of knowledge on macrobes is rather good. Although most of the groups with well-known diversity patterns at the global scale are vertebrates (Grenyer et al., 2006; Schipper et al., 2008), the variations in the numbers of species of plants (Kreft and Jetz, 2007) or insects (Dunn et al., 2009) throughout the globe are also starting to be well-known, and at least partly understood (Lomolino et al., 2006). In contrast, the level of knowledge on the spatial distribution of most (if not all) groups of microorganisms is quite limited (Foissner, 2008; Fontaneto and Hortal, 2008). Whether such deficit in knowledge is the cause of the apparent lack of biogeography of most microorganisms or not is perhaps the hottest debate in current microbial ecology.

The realisation that, apparently, many microbial species are found in quite distant localities led to the proposition of the 'Everything is everywhere' (EiE) hypothesis at the beginning of the twentieth century (Beijerinck, 1913; Baas Becking, 1934). This hypothesis is further supported by the high dispersal potential (*sensu* Weisse, 2008) of most microbes (Finlay, 2002). Their small size, large population numbers and, especially, the ability to either enter dormant states or produce small spores allow many microorganisms to produce vast numbers of propagules that are easily dispersed in a passive way (i.e. 'ubiquitous dispersal': Fenchel, 1993; Finlay et al., 1996a, 2006; Cáceres, 1997; Wilkinson, 2001; Fenchel and Finlay, 2004). Arguably, this would permit many microbes to maintain cosmopolitan distributions. Although the EiE hypothesis has been the dominant paradigm for microbial biogeography until relatively recently (O'Malley, 2007, 2008), it has been hotly debated during the last decade, dividing microbial ecologists into two factions (Whitfield, 2005). Some argue that the rule for microorganisms is 'Everything is everywhere, but the environment selects' (Finlay, 2002; de Wit and Bouvier, 2006; Fenchel and Finlay, 2006). Others counter that many apparently cosmopolitan ranges are actually artifacts of the deficient taxonomy of microbes, which does not permit distinguishing between morphologically similar but spatially and genetically isolated lineages (Coleman, 2002; Foissner, 2006, 2008; Taylor et al., 2006).

In my opinion there is now enough information to develop a theoretical (and analytical) framework that will resolve the EiE debate and lay the foundations for a general theory of microbial biogeography. However, this is beyond the intended scope of the chapter; more information can be found in several chapters of this book, or by consulting the references above (see also Martiny et al., 2006; Green and Bohannan, 2006; Telford et al., 2006; Green et al., 2008). Having said this, any study on the spatial distribution of microorganism communities shall necessarily address the question of whether everything small is everywhere. Should microbes be locally abundant and extremely widespread, their local diversity would be the result of random colonisation processes, as argued by, for example, Finlay et al. (1999, 2001). Here, differences among communities would be determined only by local environmental conditions. However, such cosmopolitanism seems to be far from universal. Many microbe species have been found to have restricted distributions (Mann and Droop, 1996; Smith and Wilkinson, 2007; Frahm, 2008; Segers and De Smet, 2008; Vanormelingen et al., 2008; Spribille et al., 2009). Hence, the dependence of range size on body size hypothesised by Finlay and colleagues (e.g. Finlay et al., 1996b; Finlay, 2002; Finlay and Fenchel, 2004) is not as general as they argue (Valdecasas et al., 2006; Pawlowski and Holzman, 2008; but see Martiny et al., 2006). More importantly, the EiE hypothesis is challenged in its assumption that the large dispersal potential of microbes necessarily results in high rates of *effective dispersal* (i.e. successful dispersal events, see Weisse, 2008). Rather, the propagules of many (but not all) microorganisms are not 'universally successful' in maintaining significant levels of gene flow between geographically remote populations (Jenkins, 1995; Jenkins and Underwood, 1998; Bohonak and Jenkins, 2003; Foissner, 2006, 2008; Jenkins et al., 2007; Weisse, 2008; Frahm, 2008).

As a direct consequence of the total or partial isolation of populations in relation to distance, phylogeographic variations (i.e. geographically structured genetic differences) have been found for many microbial taxa. Increasing spatial distance between populations results in genetic divergence and isolation even for prokaryotes (Whitaker et al., 2003; Prosser et al., 2007; Vos and Velicer, 2008). This may emerge as the common rule for many protists and multicellular microbes, once traditional approaches to their taxonomy are complemented with more detailed molecular studies (Foissner, 2008; Pawlowski and Holzman, 2008; Weisse, 2008). In fact, many recent studies finding significant hidden genetic divergence within morphologically based microbial species find also that these genetically different populations or species occupy geographically distant areas (Gómez et al., 2007; Mills et al., 2007; Fontaneto et al., 2008a; Weisse, 2008; Xu et al., 2009). Therefore, it could be expected that as knowledge of the phylogeny of microorganisms and their taxonomy improves, the number of microbes with restricted distributions will increase as well. The actual proportion of microbe species with reduced range sizes remains as a mystery, although some estimates indicate that at least one third of protist species will show restricted distributions (according to the moderate endemicity model; see Foissner, 2006, 2008). Nevertheless, such hidden microbial diversity will have an impact on the patterns of diversity observed at the community level.

17.2.2 Spatial variations in microbe communities

If the distributions of microorganisms are not cosmopolitan, microbe communities in similar substrates situated in geographically distant areas ought to show significant differences in their diversity and species composition. The spatial replacement of species in macrobial communities is typically the result of both environmental differences and geographic distance, no matter whether they are lake fishes (Genner et al., 2004), mammals (Hortal et al., 2005), birds or land snails (Steinitz et al., 2006). Microbes are to some extent similar to macrobes in this particular aspect. Although environmental heterogeneity is the main driver of the decay of compositional similarity with distance in microorganisms (see Green and Bohannan, 2006; Martiny et al., 2006), it is not the only source of spatial variation. Using an array of studies on lake diatoms encompassing several continents, Verleyen et al. (2009) found that, although environment accounts for the larger part of the spatial replacement of species, connectivity[4] also explains a large proportion of the compositional similarities between lakes: all else being equal, the closer the lakes, the more similar their species composition. Similar patterns were found in the phytoplankton communities of the Swedish lakes studied by Jankowski and Weyhenmeyer (2006). Interestingly, the strength of such replacement may vary according to the kind of habitat for both microbes and macrobes. Macrobial communities often show different patterns of distance decay of similarity[5] in different kinds of habitats (e.g. palm trees, Bjorholm et al., 2008; birds and land snails, Steinitz et al., 2006). Similarly, the degree of compositional similarity of the communities of bdelloid rotifers in a valley of northern Italy varies from one ecological system to another: whereas stream communities were relatively similar to one another throughout the valley, the species integrating the communities from terrestrial habitats and lakes were highly variable (Fontaneto et al., 2006).

[4] Connectivity measures distance as perceived by the studied organisms; that is, how difficult it would be to move between two sites or colonise a given one taking into account the existence of barriers or facilitations to dispersal (e.g. mountains or wind currents in the direction of the dispersal, respectively). Therefore, it can be considered a biologically meaningful analogue to distance.

[5] The similarity between several ecological and evolutionary phenomena often decreases or decays as the distance between them increases; the distance decay relationship is thus defined as the negative relationship between distance and the similarity of biological communities (see Nekola & White, 1999; Soininen et al., 2007).

In contrast with these qualitative similarities between small- and large-sized organisms, the *magnitude* of the distance-driven changes in species composition across similar habitats may not be comparable. The slope of the taxa–area relationship in contiguous habitats can be used as a raw measure of the accumulation of species or other taxonomic units with area, and hence of compositional changes in space (Prosser et al,, 2007; Santos et al., 2010; see also Rosenzweig, 1995 and Whittaker and Fernández-Palacios, 2007 for extensive reviews on the species–area relationship). To date, the slopes recorded for microbes are typically smaller than those of macrobes; while the former may range from 0.043 to 0.114 in natural systems (in a power model; Finlay et al., 1998; Azovsky, 2002; Green et al., 2004; Bell et al., 2005; Smith et al., 2005; Green and Bohannan, 2006; Prosser et al., 2007) the latter are typically larger than 0.15. This indicates that in the absence of environmental differences, the spatial replacement in the composition of local communities occurs at much larger scales for microbes, varying in the range of a few hundreds to thousands of kilometres, instead of the hundreds or even tens of kilometres usually found for macrobes.

Strikingly, however, when taxa–area relationships are calculated for habitat islands (i.e. separate territories/habitat patches instead of contiguous habitats), the slopes may reach values well over 0.2 for bacteria, which are similar to those of large-sized organisms (Bell et al., 2005; van der Gast, 2005; see also Green and Bohannan, 2006; Prosser et al., 2007). This indicates that in spite of their minute size, habitat area plays a critical role in determining the number of bacterial species that can coexist in a given place, as it does for macroorganisms. In other words, although microbial communities change with distance at a lower pace than macrobes, the increase in the number of species with area is similar for both groups. Further study is required to determine whether this dependence on area is due to the carrying capacity of the locality (which increases with its area), the increase of habitat diversity with increasing area, or to passive sampling mechanisms (i.e. larger areas receive more propagules and therefore may be colonised by more species).

The similarities between the macroecological patterns of micro- and macroorganisms do not end with their shared dependence on area. As with their larger counterparts, the diversity of microbe communities varies along environmental gradients. One of the most studied is the altitudinal gradient: species richness varies with altitude, increasing or decreasing for localities closer to or farther from an optimal altitudinal band (Rahbek, 1995, 2005). These changes in richness are often accompanied by changes in species composition (e.g. the Alpine dung beetles studied by Jay-Robert et al., 1997). This macroecological pattern has been also observed in many microorganisms. A good example is the altitudinal variations in richness shown by rotifers in the Alps (Fontaneto and Ricci, 2006; Fontaneto et al., 2006; Obertegger et al., 2010). The diversity of tardigrades inhabiting moss and leaf litter at the Guadarrama mountain range also shows a typical hump-shaped relationship with elevation (Guil et al., 2009a). As with macrobes, altitudinal variations in microbe species richness are caused by the varying environmental conditions at different elevations. Sometimes the decrease in productivity with altitude limits local diversity, as occurs for phytoplankton richness in a series of Swedish lakes (Jankowski and Weyhenmeyer, 2006). In other cases, the climatic variations associated with elevational changes cause spatial gradients in local richness, as with epiphytic bryophyte communities in Guiana (Oliveira et al., 2009) or northwest Spain (N.G. Medina, B. Albertos, F. Lara, V. Mazimpaka, R. Garilleti, D. Draper and J. Hortal, unpublished manuscript). Such climate-driven diversity variations in microbes may arise simply because of habitat sorting (i.e. due to the differences in niche requirements of each of the species regionally available; see e.g. Whittaker, 1972). In fact, in the two examples given above many species of both bdelloid rotifers and tardigrades show strong habitat selection (i.e. habitat sorting; Fontaneto and Ricci, 2006 and Guil et al., 2009b, respectively). This is consistent with a high potential for dispersal and local environmental selection (although this process may involve an important degree of stochasticity; see Fontaneto et al., 2006).

However, geographic changes in the diversity of microbe species inhabiting similar microhabitats are not only the result of local productivity and/or carrying capacity and environmental variations. The pool of species that can colonise each microhabitat varies also in space, therefore limiting the number and identity of the colonising species. The regional differences produced by changes in the species pool are one of the major determinants of the geographic variations of assemblage diversity for macroorganisms (Ricklefs, 1987, 2004, 2007; Huston, 1999; Hawkins et al., 2003a; Hortal et al., 2008; Hawkins, 2010). Some environments or particular habitats may host fewer species simply because fewer of the available species have evolved adaptations to these environments, either because these environments are rare in nature, they are too recent in the region, or they were affected by important changes in the past, like glaciations. In fact, although the dispersal distances and the location of refugia may differ, the diversity of several microbial groups has been shaped by post-glacial dispersal (e.g. Gómez et al., 2007; Smith and Wilkinson, 2007; Smith et al., 2008). More importantly, the composition of the regionally available species pool varies among continents, at least for lake diatoms (Verleyen et al., 2009). As with macrobes, these differences in the species pool produce different responses to elevational or climatic gradients, as shown also for lake diatoms by Telford et al. (2006).

The ultimate consequence of the spatial variations in microbe communities is the existence of geographic differences in ecosystem functioning. Community richness, composition, assembly and functional dissimilarity affect ecosystem productivity and functioning (Laakso and Setälä, 1999; Fukami and Morin, 2003; Heemsbergen et al., 2004; Sánchez-Moreno et al., 2008). Given that microbes perform many ecosystem services, geographic variations in community composition are likely to have important functional consequences (Naeslund and Norberg, 2006; Green et al., 2008). However, the impact of these geographic differences in the functions provided by microbial communities in ecological processes at regional and global scales is, to date, poorly known.

17.3 A frontier of biogeography

An overall insight from the short review above is that the biogeography and macroecology of microbial assemblages present both similarities and differences with those of macroorganisms. To synthesise, microbes and macrobes both show distance decay in community similarity. It follows that not everything is everywhere. Hence, although the decay of similarity and the associated compositional changes are mainly caused by environmental gradients, they are also driven by geographic variations in the composition of the species pool and the degree of connectivity or the presence of barriers to dispersal between localities. The geographic variations in microbe assemblages typically occur at larger distances than for macrobes, and thus they show a shallower increment of species with increasing area. In spite of this, the relationships between local richness and area and environmental gradients are comparable to the ones found in macrobes. The differences between micro- and macroorganisms seem thus limited to their differences in size and dispersal power, which link microbes to smaller microhabitats and make their distributions typically larger.

Some of the challenges awaiting biogeography are right in front of our eyes, rather than in distant places. Whether the similarities and differences between the diversity of macrobe and microbe assemblages arise from fundamentally similar or different processes needs further investigation. Here I outline a research agenda to help explore this particular frontier of biogeography (see also Martiny et al., 2006; Green and Bohannan, 2006; Prosser et al., 2007; Green et al., 2008). To understand the spatial variations in the diversity of microbial assemblages, research in four main areas is needed: microbial taxonomy; description of biogeographic and macroecological patterns; community ecology and assembly; and the study of functional diversity and ecosystem functioning.

17.3.1 Microbial taxonomy

One key question is to determine the extent to which the patterns observed for microbes so far are due to their deficient taxonomy, as argued by, for example, Foissner (2008). Strikingly, in a recent study on moss-dwelling bdelloid rotifers from the UK and Turkey, the pattern of variations in species richness based on traditional taxonomy varies when DNA-based taxonomy is applied, not only in the numbers of species identified, but also in the relative differences in richness between localities, and their similarity in composition (Kaya et al., 2009). In fact, many recent detailed studies reveal large amounts of hidden genetic, physiological and ecological diversity in microorganism communities (Mann and Droop, 1996; Weisse, 2008; Guil, 2008; Guil and Giribet, 2009; Fontaneto et al., 2009). Therefore, the establishment of a solid taxonomy based on the identification of ecological and evolutionarily meaningful units through DNA and ecophysiological analyses is a necessary prerequisite to the study of the diversity of microbes. Importantly, current genetic techniques allow the identification of operational taxonomic units (OTUs) just from the divergence in a reduced number of DNA sequences with relatively small costs. Thus, a final definition of the species concept (which remains elusive for some microbial groups) may not be necessary for describing the geographic (and ecological) variations in microorganism diversity.

17.3.2 Description of macroecological and biogeographic patterns

Once a good taxonomy is established, it will be possible to describe the patterns of variation of the diversity of any group of microorganisms. We know relatively little about how the richness and composition of microbes vary around the world. In particular, the relationships and compositional similarities between different regions, islands and continents are typically unknown, as well as whether the latitudinal diversity gradients observed for most groups of macroorganisms hold out for microbes. In the same way, the relationships between many microorganisms and the environmental conditions (temperature, nutrients, etc.) are often known in lab conditions, but how these responses translate to the real world is unknown. This prevents an accurate determination of the influence of climate change on the distribution of microbial species, one of the needs identified by Prosser et al. (2007).

Given current knowledge of the patterns of microbial diversity, three main lines of research need development within this particular area. Some of this research must first be conducted at the species (or species-group) level, since a degree of basic knowledge is necessary before questions can be addressed at the level of the community or assemblage.

(i) *Descriptive biogeography*: the pure description of geographic patterns of variation at the species level or higher (see e.g. Lomolino et al., 2006). This includes the identification of biogeographic regions and the inventory of their faunas or floras, as well as the study of the geographic gradients of diversity. Achieving this particular objective for any microorganism group may need a large amount of fieldwork, but the establishment of standardised survey protocols, the focus on one or a set of particular microhabitats and the cooperation between experts from different parts of the world may facilitate progress within relatively short periods of time. A good example of this kind of cooperation is the recent analysis of the global variations in ant richness performed by Dunn et al. (2009), who were able to compile over 1000 ant assemblages from all over the world coming from standardised surveys within a few years (N.J. Sanders, personal communication). Such kinds of data can be used to study the relationships of diversity with climate, to describe latitudinal gradients, to determine or define biogeographic regions, and for other purposes (see also the *Macroecology* section below).

(ii) *Phylogeography*: assessing the impact of geographic structure on the evolutionary relationships among populations of a particular taxon (Avise, 2009). This particular research line is already under development for a few microbial groups (e.g. Lowe et al., 2005; Gómez et al., 2007; Mills et al., 2007; Mikheyev et al., 2008; Fontaneto et al., 2008b). However, these efforts are still dispersed: a goal during the next few years or decades must be to develop a coordinated long-term research programme to facilitate a systematic characterisation of the recent evolution and geographic relationships between microorganism populations in nature. Thus, as before, some little discussion and coordination effort in this area, including the identification of specific research goals and certain geographic areas (e.g. islands, archipelagos or the borders of continents or biogeographic regions) will certainly enhance current knowledge of the phylogeographic patterns of microorganisms. Of particular importance is the potential role of humans as dispersal vectors for free-living and parasitic terrestrial microorganisms. As Wilkinson (2010) recently stated, the imprint of human-facilitated dispersion on the current biogeography of many microorganisms may have been underestimated. Determining the impact of increasing biotic homogenisation at the global scale needs large-scale phylogeographic studies on a number of cosmopolitan microbe species.

(iii) *Macroecology*. Once the basic biogeographic patterns are described for a microorganism group, macroecological analyses can determine the impact of climate, productivity and historical effects on its distribution and diversity. Martiny et al. (2006) review some particular topics that need further research in this area, and provide an analytical framework to study them. In particular, they advocate the study of the effects of environment and history, as well as the processes shaping microbial biogeography, including dispersal and colonisation, diversification and extinction, or the relationship between body size and distribution range. Apart from these research topics, the scaling of the relationship between microorganisms and climate needs further study. On the one hand, it is necessary to determine what is the relationship between the responses to environmental factors observed in laboratory conditions, and the responses to the same factors that are observed in the field. On the other hand, we need to identify the correct scale to assess the effects of large-scale environmental gradients (such as climate) on the diversity of microbial communities inhabiting minute microhabitats. Perhaps surprisingly, climate variables measured with a scale of 1 km^2 (i.e. yearly or monthly precipitation and temperature) are good predictors of the spatial variations in the richness of both tardigrades inhabiting 9 cm^2 moss and leaf litter samples (Guil et al., 2009b) and bryophytes found over oak trunks using quadrats of 400 cm^2 (N.G. Medina, B. Albertos, F. Lara, V. Mazimpaka, R. Garilleti, D. Draper and J. Hortal, unpublished manuscript). Thus, some particular research on the extent to which coarse-scale climate and/or microclimatic conditions are affecting micro-organism diversity in microhabitats is needed (see also section 17.3.3 below).

All these lines of research, particularly the first two, will benefit considerably from the cooperation and coordination among the experts of each microbial group. Thus, specific effort should be devoted to scientific networking, particularly among the associations of microbiologists or experts in particular groups. An additional goal is to increase the participation of microbiologists in the International Biogeography Society (IBS; http://www.biogeography.org/): a multidisciplinary association that seeks the advancement of all studies of the geography of nature. The results of these efforts could be enhanced with help and interest from the editorial boards of journals in both biogeography and microbial ecology, the inclusion of chapters on microbial biogeography in the general books on both disciplines, as well as the organisation of specific symposia, such as the one entitled 'The importance of being small: does size matter in biogeography?' organised by Diego Fontaneto, David Roberts and Juliet Brodie (held within the 7th Systematics Association Biennial Conference, Leiden, the Netherlands, August 2009), which led to the compilation of this book.

17.3.3 Community ecology and assembly

In parallel with the description and analysis of patterns in nature, the ecology of microbial communities needs further research. Two aspects deserve particular attention: disentangling what determines the number of species that a given system can host; and assessing whether the presence of some species (or any other taxa) prevents or facilitates the entrance of other species in the local assemblage.

Apart from climate and habitat gradients, the richness of local communities is typically related with their area (MacArthur and Wilson, 1963, 1967; Rosenzweig, 1995), habitat diversity (Triantis et al., 2003; Hortal et al., 2009) and productivity (Wright, 1983; Srivastava and Lawton, 1998). As discussed above (section 17.2.2), the richness of microbial communities is related to their area. However, the origin of such dependence on area requires further study, and the same occurs with the relationship with productivity. Here, experiments under laboratory and, especially, natural conditions may help to clarify the relative importance of these factors. Field experiments may not be particularly difficult to set up (see, e.g. Srivastava and Lawton, 1998), by: (1) selecting a particular microhabitat (such as treeholes, temporary ponds, moss cushions or tank bromeliads) that is patchily distributed in a relative small territory (e.g. a valley, a forest patch or a segment of coastline); (2) measuring or perhaps altering the area, habitat diversity and productivity (in terms of raw provision of nutrients) of these microhabitats; (3) sterilising them (or at least washing away as much of the pre-existing communities as possible); (4) letting these communities develop again by dispersal and colonisation from other microhabitats in the territory; before (5) measuring the diversity of the resulting assemblages.

Determining whether the coexistence of some species (or taxa) is determined by facilitation or exclusion during the assembly process due to the presence of other species may prove more difficult. One of the most controversial and bitter debates in ecology during the late twentieth century has been elucidating whether there are some assembly rules determining the composition of local communities, and the eventual nature of these rules (for review and synthesis see Gotelli and Graves, 1996; Weiher and Keddy, 1999; Gotelli, 2001; Gotelli and McCabe, 2002; Jenkins, 2006; Sanders et al., 2007). To study their role in microbial communities requires knowledge of the composition of the assemblage before and after the arrival of any new species. This makes it extremely difficult to use field experiments to study the determinants of coexistence and community assembly, although they reveal the effects of dispersal (Oliveira et al., 2009). Rather, laboratory experiments in controlled conditions, where species are added to previously set up communities, will be a more effective approach here (see Liess and Diehl, 2006 for an example). One particular topic that can be studied using field experiments would be the influence of the richness and composition of the species pool on local communities. In this case, experiments placed in geographically distant territories of environmentally similar regions, together with exhaustive surveys that allow identification of the whole species pool in each territory, could be used to determine how and to what extent local communities are shaped by the species available in their pool.

17.3.4 Functional diversity and ecosystem functioning

Describing the functional role of species within an ecosystem might at first be viewed as an unrelated, 'purely ecological' research topic. However, the attention of biogeography and community ecology is shifting to consider the geographic distribution of functional traits, species interactions and the organisation of the ecosystems provided by the distributions of these traits and interactions (Naeem and Wright, 2003; McGill et al., 2006; Petchey and Gaston, 2006; Green et al., 2008; Schemske et al., 2009). The relationship between biodiversity, biotic interactions and ecosystem functioning is well studied in microbial systems, particularly soil communities (Allsopp et al., 1994; Kennedy and Smith, 1995; Heemsbergen et al., 2004; Bruno et al., 2006; Wardle, 2006; Srivastava and Bell, 2009). New techniques that facilitate the measurement of traits and functioning in microbial communities (in comparison to macrobes; e.g. Lowe et al., 2005) offer encouraging possibilities for the future pursuit of this particular research topic (Green et al., 2008). This could be achieved through fieldwork and analyses designed in a similar way to the ones argued for macroecology above (section 17.3.2), but measuring specific functional traits and/or facets of ecosystem functioning instead of just the diversity of the communities and the inputs of materials and energy.

17.4 The biogeographer's wish list: how microbial ecology could reinvigorate the development of biogeographic theory

Adding a biogeographic perspective to the study of the diversity of microorganism communities will certainly improve current knowledge of their ecology and evolution, their responses to anthropogenic environmental changes, and the effect of these responses on the ecosystem services they provide. However, the field of biogeography itself stands to gain at least as much from this association. One of the main limitations of biogeographic research compared with other fields of biology is that large temporal and spatial scales are not conducive to field experiments, and when manageable long-term experiments are designed, they are often flawed by the limited number (or lack) of independent replicates, thus compromising the generality of the final results. This tends to limit biogeography to conceptual experiments, the analysis of the predictions of null and neutral models, or extrapolating the patterns observed in a few natural experiments such as island groups.

In contrast, the study of microbes may allow measurement of the equivalent to large-scale biogeographic effects within relatively small extents (see above). Generation times are also typically shorter than for large-sized organisms, permitting the study of the assembly and development of a community within limited time spans. In addition, they can be (and are) used for closed or semi-open experiments (i.e. performed in controlled laboratory conditions or natural field conditions, respectively), thus allowing to work with multiple replicates. Further, their minute size and the analytical tools currently available make it possible to measure virtually all components of an ecosystem, from the inputs of energy and materials, to the species present in the assemblage, the functioning of the ecosystem itself and its output in terms of biomass or ecosystem services.

Here I advocate the use of microscopic organisms for the development of a purely experimental biogeography of the diversity of biological communities. I do it in parallel with Jenkins and colleagues (Chapter 15), who propose a similar approach to the study of the geographic responses and environmental requirements of single species (or similar taxa). This will permit assessing the generality of many aspects of current biogeographic and macroecological theory, from the relationship between diversity and productivity/energy, to community composition or metacommunity dynamics. A careful choice of study system is important in the design of successful field experiments, which may eventually permit the exhaustive description of variations of microorganism diversity in space. The perfect candidates for such small-scale biogeography are a number of types of microhabitats, such as moss cushions, tank bromeliads, treeholes or temporary ponds, among others, that often appear scattered across apparently homogeneous territories or land patches (e.g. Gonzalez, 2000). By assuming no dispersal limitations within a sensible area and amount of time (which are taxon dependent), any field experiment could assume that this aspect has been controlled for, and therefore that no effect of dispersal would remain in the results. The scale of the localities (i.e. microhabitats) to study can allow manipulating dispersal itself, by restricting the arrival of propagules or inoculating the selected ones.

Implementing microbial experiments for biogeographic research might need some consideration and preparation. Depending on the system studied, it would be necessary to determine how transferable the results would be to the macroecological world. Also, some preparatory work may be needed to find a proper way of taking samples and measure functioning without having spurious effects on the studied assemblages. Nevertheless, many groups of unicellular organisms, and in particular bacteria, may not be appropriate for this kind of research, if the goal is to allow extrapolating the results obtained to macrobes. Their different evolutionary modes can result in substantially different patterns of diversity from those of any large-sized organism (e.g. bacteria can rapidly interchange significant amounts of genetic material, impeding any direct comparison with multicellular organisms). However, as discussed above, many multicellular microorganisms can be directly comparable to large-sized organisms in these respects.

Many areas of research in biogeography may benefit from micro-biogeographic experiments with multicellular microbes. In particular, designing experiments on island biogeography would be straightforward, providing powerful tools for the research in many of the current debates and questions in this area. Good examples of currently debated questions that would benefit from these experiments are: How predictive are Hubbell's (2001) Neutral Theory or current models of metacommunity dynamics (e.g. Mouquet and Loureau, 2003)? Which are the influences of dispersal and connectivity on species assembly (and assemblage composition)? What is the exact relationship between habitat diversity and species richness, and to what extent does this depend on area, *versus* the niche width of the species available in the species pool (Hortal et al., 2009)? How well does the recently proposed general dynamic theory of oceanic island biogeography (Whittaker et al., 2008) predict the patterns of diversity and regional endemism in microhabitat islands? How can speciation occur in the absence of proper allopatric populations (i.e. speciation in sympatry; e.g. Phillimore et al., 2008)?

The current theory on the causes and consequences of geographic biodiversity gradients can also be largely improved by studying the variations in the diversity of microbes across relatively small territories. Despite the level of knowledge on the latitudinal diversity gradient reached to date, its causes are still under debate. The role and relative importance of several processes in shaping species richness gradients is still unknown (i.e. water-energy dynamics, energy and/or resource availability, regional constraints and other historical effects and temporal changes, among others; Hawkins et al., 2003a, 2003b, 2007; Willig et al., 2003; Currie et al., 2004; Wiens and Donoghue, 2004; Storch et al., 2006; Mittelbach et al., 2007; Ricklefs, 2007; Hawkins, 2008; Hortal et al., 2008; Qian and Ricklefs, 2008). Further, the study of microbial communities may help in the understanding of the geographic variations in functional diversity, including gradients in functional traits (e.g. Diniz-Filho et al., 2009), functional structure (e.g. Rodríguez et al., 2006), or the effects of these spatial variations on ecosystem functioning and resilience.

17.5 Conclusion

The biogeography of microbial communities is in large part not qualitatively different from the biogeography of macroorganisms. Under current levels of knowledge, most differences between macro- and micro-assemblages are due to the large dispersal potential of microbes and their affinity to microhabitats. This results in a different scaling of the distance decay in community similarity and a higher patchiness of microbe communities. However, the current lack of knowledge on the geographic variations of microorganisms is still a challenge for both microbiologists and biogeographers. In my opinion, further research on microbial biogeography will not find the processes underlying the assembly of microbe communities to be fundamentally different from the ones already described for macrobes. Rather, the similarity between macrobial and microbial communities allows for the examining of biogeographic theory with a completely different perspective using microbes. Thus, here I argue for the development of an experimental research programme in microbial biogeography. Regardless of which particular question is studied, I am certain that the biogeography of small things and the experimental approach it permits will generate significant advances in biogeographic theory.

Acknowledgements

I wish to thank Diego Fontaneto for the invitation to write this chapter and participate in the corresponding symposium, and to an anonymous referee for a thoughtful review of the previous version of this manuscript. I am also indebted to Noemi Guil, Sara Sánchez Moreno and Diego Fontaneto, for introducing me to the world of the very very small things, as well as for many insightful discussions throughout the last ten years. JH is funded by a Spanish CSIC JAE-Doc research grant, and obtained additional funding from a travel grant of the Azorean Biodiversity Group - CITA-A.

References

Allsopp, D., Hawksworth, D.L., Colwell, R.R. (1994). *Microbial biodiversity and ecosystem function.* CAB International, Wallington.

Avise, J.C. (2009). Phylogeography: retrospect and prospect. *Journal of Biogeography* **36**, 3–15.

Azovsky, A.I. (2002). Size-dependent species–area relationships in benthos: is the world more diverse for microbes? *Ecography* **25**, 273–282.

Baas Becking, L.G.M. (1934). *Geobiologie of inleiding tot de milieukunde.* The Hague: Van Stockum and Zoon.

Beijerinck, M.W. (1913). De infusies en de ontdekking der bakterien. In: *Jaarboek van de Koninklijke Akademie van Wetenschappen.* 1–28.

Bell, T., Ager, D., Song, J.-I., Newman, J.A., Thompson, I.P., Lilley, A.K., van der Gast, C.J. (2005). Larger islands house more bacterial taxa. *Science* **308**, 1884.

Bjorholm, S., Svenning, J.-C., Skov, F., Balslev, H. (2008). To what extent does Tobler's 1st law of geography apply to macroecology? A case study using American palms (Arecaceae). *BMC Ecology* **8**, 11.

Bohonak, A.J., Jenkins, D.G. (2003). Ecological and evolutionary significance of dispersal by freshwater invertebrates. *Ecology Letters* **6**, 783–796.

Bruno, J.F., Lee, S.C., Kertesz, J.S. et al. (2006). Partitioning the effects of algal species identity and richness on benthic marine primary production. *Oikos* **115**, 170–178.

Cáceres, C.E . (1997). Dormancy in invertebrates. *Invertebrate Biology* **116**, 371–383.

Coleman, A.W. (2002). Microbial eukaryote species. *Science* **297**, 337–337.

Currie, D.J., Mittelbach, G.G., Cornell, H.V. et al. (2004). Predictions and tests of climate-based hypotheses of broad-scale variation in taxonomic richness. *Ecology Letters* **7**, 1121–1134.

de Wit, R., Bouvier, T. (2006). 'Everything is everywhere, but the environment selects'; what did Baas Becking and Beijerinck really say? *Environmental Microbiology* **8**, 755–758.

Diniz-Filho, J.A.F., Rodríguez, M.Á., Bini, L.M. et al. (2009). Climate history, human impacts and global body size of carnivora at multiple evolutionary scales. *Journal of Biogeography* **36**, 2222–2236.

Dunn, R.R., Agosti, D., Andersen, A.N. et al. (2009). Climatic drivers of hemispheric asymmetry in global patterns of ant species richness. *Ecology Letters* **12**, 324–333.

Fenchel, T. (1993). There are more small than large species. *Oikos* **68**, 375–378.

Fenchel, T., Finlay, B.J. (2003). Is microbial diversity fundamentally different from biodiversity of larger animals and plants? *European Journal of Protistology* **39**, 486–490.

Fenchel, T., Finlay, B.J. (2004). The ubiquity of small species: Patterns of local and global diversity. *Bioscience* **54**, 777–784.

Fenchel, T., Finlay, B.J. (2006). The diversity of microbes: resurgence of the phenotype. *Philosophical Transactions of the Royal Society B* **361**, 1965–1973.

Finlay, B.J. (2002). Global dispersal of free-living microbial eukaryote species. *Science* **296**, 1061–1063.

Finlay, B.J., Fenchel, T. (2004). Cosmopolitan metapopulations of free-living microbial eukaryotes. *Protist* **155**, 237–244.

Finlay, B.J., Corliss, J.O., Esteban, G., Fenchel, T. (1996a). Biodiversity at the microbial level: The number of free-living ciliates in the biosphere. *Quarterly Review of Biology* **71**, 221–237.

Finlay, B.J., Esteban, G.F., Fenchel, T. (1996b). Global diversity and body size. *Nature* **383**, 132–133.

Finlay, B.J., Esteban, G.F., Fenchel, T. (1998). Protozoan diversity: converging estimates of the global number of free-living ciliate species. *Protist* **149**, 29–37.

Finlay, B.J., Esteban, G.F., Olmo, J.L., Tyler, P.A. (1999). Global distribution of free-living microbial species. *Ecography* **22**, 138–144.

Finlay, B.J., Esteban, G.F., Clarke, K.J., Olmo, J.L. (2001). Biodiversity of terrestrial protozoa appears homogeneous across local and global spatial scales. *Protist* **152**, 355–366.

Finlay, B.J., Esteban, G.F., Brown, S., Fenchel, T., Hoef- Emden, K. (2006). Multiple cosmopolitan ecotypes within a microbial eukaryote morphospecies. *Protist* **157**, 377–390.

Foissner, W. (2006). Biogeography and dispersal of micro-organisms: A review emphasizing protists. *Acta Protozoologica* **45**, 111–136.

Foissner, W. (2008). Protist diversity and distribution: some basic considerations. *Biodiversity and Conservation* **17**, 235–242.

Fontaneto, D., Ricci, C. (2006). Spatial gradients in species diversity of microscopic animals: the case of bdelloid rotifers at high altitude. *Journal of Biogeography* **33**, 1305–1313.

Fontaneto, D., Hortal, J. (2008). Do microorganisms have biogeography? *IBS Newsletter* **6**.2, 3–8.

Fontaneto, D., Ficetola, G.F., Ambrosini, R., Ricci, C. (2006). Patterns of diversity in microscopic animals: are they comparable to those in protists or in larger animals? *Global Ecology and Biogeography* **15**, 153–162.

Fontaneto, D., Barraclough, T.G., Chen, K., Ricci, C., Herniou, E.A. (2008a). Molecular evidence for broad-scale distributions in bdelloid rotifers: everything is not everywhere but most things are very widespread. *Molecular Ecology* **17**, 3136–3146.

Fontaneto, D., Boschetti, C., Ricci, C. (2008b). Cryptic diversification in ancient asexuals: evidence from the bdelloid rotifer *Philodina flaviceps*. *Journal of Evolutionary Biology* **21**, 580–587.

Fontaneto, D., Kaya, M., Herniou, E.A., Barraclough, T.G. (2009). Extreme levels of hidden diversity in microscopic animals (Rotifera) revealed by DNA taxonomy. *Molecular Phylogenetics and Evolution* **53**, 182–189.

Frahm, J.P. (2008). Diversity, dispersal and biogeography of bryophytes (mosses). *Biodiversity and Conservation* **17**, 277–284.

Fukami, T., Morin, P.J. (2003). Productivity-biodiversity relationships depend on the history of community assembly. *Nature* **424**, 423–426.

Genner, M.J., Taylor, M.I., Cleary, D.F.R. et al. (2004). Beta diversity of rock-restricted cichlid fishes in Lake Malawi: importance of environmental and spatial factors. *Ecography* **27**, 601–610.

Gómez, A., Montero-Pau, J., Lunt, D.H., Serra, M., Campillo, S. (2007). Persistent genetic signatures of colonization in *Brachionus manjavacas* rotifers in the Iberian Peninsula. *Molecular Ecology* **16**, 3228–3240.

Gonzalez, A. (2000). Community relaxation in fragmented landscapes: the relation between species richness, area and age. *Ecology Letters* **3**, 441–448.

Gotelli, N.J. (2001). Research frontiers in null model analysis. *Global Ecology and Biogeography* **10**, 337–343.

Gotelli, N.J., Graves, G.R. (1996). *Null models in ecology*. Smithsonian Institution Press, Washington, D. C.

Gotelli, N.J., McCabe, D.J . (2002). Species co-occurrence: a meta-analysis of J. M. Diamond's assembly rules model. *Ecology* **83**, 2091–2096.

Green, J., Bohannan, B.J.M. (2006). Spatial scaling of microbial biodiversity. *Trends in Ecology and Evolution* **21**, 501–507.

Green, J.L., Holmes, A.J., Westoby, M. et al. (2004). Spatial scaling of microbial eukaryote diversity. *Nature* **432**, 747–750.

Green, J.L., Bohannan, B.J.M., Whitaker, R.J. (2008). Microbial Biogeography: From Taxonomy to Traits. *Science* **320**, 1039–1043.

Grenyer, R., Orme, C.D.L., Jackson, S.F. et al. (2006). Global distribution and conservation of rare and threatened vertebrates. *Nature* **444**, 93–96.

Guil, N. (2008). New records and within-species variability of Iberian tardigrades (Tardigrada), with comments on the species from the *Echiniscus blumi-canadensis* series. *Zootaxa* **1757**, 1–30.

Guil, N., Giribet, G. (2009). Fine scale population structure in the *Echiniscus blumi-canadensis* series (Heterotardigrada, Tardigrada) in an Iberian mountain range—When morphology fails to explain genetic structure. *Molecular Phylogenetics and Evolution* **51**, 606–613.

Guil, N., Hortal, J., Sánchez-Moreno, S., Machordom, A. (2009a). Effects of macro and micro-environmental factors on the species richness of terrestrial tardigrade assemblages in an Iberian mountain environment. *Landscape Ecology* **24**, 375–390.

Guil, N., Sánchez-Moreno, S., Machordom, A. (2009b). Local biodiversity patterns in micrometazoans: Are tardigrades everywhere? *Systematics and Biodiversity* **7**, 259–268.

Hawkins, B.A. (2008). Recent progress toward understanding the global diversity gradient. *IBS Newsletter* **6**.1, 5–8.

Hawkins, B.A. (2010). Multiregional comparison of the ecological and phylogenetic structure of butterfly species richness gradients. *Journal of Biogeography* **37**, 647–656.

Hawkins, B.A., Porter, E.E., Diniz-Filho, J.A.F. (2003a). Productivity and history as predictors of the latitudinal diversity gradient of terrestrial birds. *Ecology* **84**, 1608–1623.

Hawkins, B.A., Field, R., Cornell, H.V. et al. (2003b). Energy, water, and broad-scale geographic patterns of species richness. *Ecology* **84**, 3105–3117.

Hawkins, B.A., Diniz-Filho, J.A.F., Jaramillo, C.A., Soeller, S.A. (2007).

Climate, niche conservatism, and the global bird diversity gradient. *The American Naturalist* **170**, S16-S27.

Heemsbergen, D.A., Berg, M.P., Loreau, M. et al. (2004). Biodiversity effects on soil processes explained by interspecific functional dissimilarity. *Science* **306**, 1019–1020.

Hortal, J., Nieto, M., Rodríguez, J., Lobo, J.M. (2005). Evaluating the roles of connectivity and environment on faunal turnover: patterns in recent and fossil Iberian mammals. In: Elewa, A.M.T. (ed.) *Migration in Organisms. Climate, Geography, Ecology*, Springer, Berlin. 301–327.

Hortal, J., Rodríguez, J., Nieto-Díaz, M., Lobo, J.M. (2008). Regional and environmental effects on the species richness of mammal assemblages. *Journal of Biogeography* **35**, 1202–1214.

Hortal, J., Triantis, K.A., Meiri, S., Thébault, E., Sfenthourakis, S. (2009). Island species richness increases with habitat diversity. *American Naturalist* **173**, E205-E217.

Hubbell, S.P. (2001). *The unified neutral theory of biodiversity and biogeography*. Princeton University Press, Princeton.

Huston, M.A. (1999). Local processes and regional patterns: appropriate scales for understanding variation in the diversity of plants and animals. *Oikos* **86**, 393–401.

Jankowski, T., Weyhenmeyer, G.A. (2006). The role of spatial scale and area in determining richness-altitude gradients in Swedish lake phytoplankton communities. *Oikos* **115**, 433–442.

Jay-Robert, P., Lobo, J.M., Lumaret, J.P. (1997). Altitudinal turnover and species richness variation in European montane dung beetle assemblages. *Arctic and Alpine Research* **29**, 196–205.

Jenkins, D.G. (1995). Dispersal-limited zooplankton distribution and community composition in new ponds. *Hydrobiologia* **313**, 15–20.

Jenkins, D.G. (2006). In search of quorum effects in metacommunity structure: Species co-occurrence analyses. *Ecology* **87**, 1523–1531.

Jenkins, D.G., Underwood, M.O. (1998). Zooplankton may not disperse readily in wind, rain, or waterfowl. *Hydrobiologia* **388**, 15–21.

Jenkins, D.G., Brescacin, C.R., Duxbury, C.V. et al. (2007). Does size matter for dispersal distance? *Global Ecology and Biogeography* **16**, 415–425.

Kaya, M., Herniou, E.A., Barraclough, T.G., Fontaneto, D. (2009). Inconsistent estimates of diversity between traditional and DNA taxonomy in bdelloid rotifers. *Organisms Diversity and Evolution* **9**, 3–12.

Kellogg, C.A., Griffin, D.W. (2006). Aerobiology and the global transport of desert dust. *Trends in Ecology and Evolution* **21**, 638–644.

Kennedy, A.C., Smith, K.L. (1995). Soil microbial diversity and the sustainability of agricultural soils. *Plant and Soil* **170**, 75–86.

Kreft, H., Jetz, W. (2007). Global patterns and determinants of vascular plant diversity. *Proceedings of the National Academy of Sciences USA* **104**, 5925–5930.

Laakso, J., Setälä, H. (1999). Sensitivity of primary production to changes in the architecture of belowground food webs. *Oikos* **87**, 57–64.

Liess, A., Diehl, S. (2006). Effects of enrichment on protist abundances and bacterial composition in simple microbial communities. *Oikos* **114**, 15–26.

Lomolino, M.V., Riddle, B.R., Brown, J.H. (2006). *Biogeography. Third Edition.* Sinauer Associates, Inc., Sunderland, Massachussets.

Lowe, C.D., Kemp, S.J., Montagnes, D.J.S. (2005). An interdisciplinary approach to assess the functional diversity of free-living microscopic eukaryotes. *Aquatic Microbial Ecology* **41**, 67–77.

MacArthur, R.H., Wilson, E.O. (1963). An equilibrium theory of insular zoogeography. *Evolution* **17**, 373–387.

MacArthur, R.H., Wilson, E.O. (1967). *The theory of island biogeography.* Princeton University Press, Princeton.

Mann, D.G., Droop, S.J.M. (1996). Biodiversity, biogeography and conservation of diatoms. *Hydrobiologia* **336**, 19–32.

Martiny, J.B.H., Bohannan, B.J.M., Brown, J.H. et al. (2006). Microbial biogeography: putting microorganisms on the map. *Nature Reviews Microbiology* **4**, 102–112.

McGill, B.J., Enquist, B.J., Weiher, E., Westoby, M. (2006). Rebuilding community ecology from functional traits. *Trends in Ecology and Evolution* **21**, 178–185.

Mikheyev, A.S., Vo, T., Mueller, U.G. (2008). Phylogeography of post-Pleistocene population expansion in a fungus-gardening ant and its microbial mutualists. *Molecular Ecology* **17**, 4480–4488.

Mills, S., Lunt, D.H., Gómez, A . (2007). Global isolation by distance despite strong regional phylogeography in a small metazoan. *BMC Evolutionary Biology* **7**, 225.

Mittelbach, G.G., Schemske, D.W., Cornell, H.V. et al. (2007). Evolution and the latitudinal diversity gradient: speciation, extinction and biogeography. *Ecology Letters* **10**, 315–331.

Mouquet, N., Loreau, M. (2003). Community patterns in source-sink metacommunities. *American Naturalist* **162**, 544–557.

Naeem, S., Wright, J.P. (2003). Disentangling biodiversity effects on ecosystem functioning: deriving solutions to a seemingly insurmountable problem. *Ecology Letters* **6**, 567–579.

Naeslund, B., Norberg, J. (2006). Ecosystem consequences of the regional species pool. *Oikos* **115**, 504–512.

Nekola, J.C., White, P.S. (1999). The distance decay of similarity in biogeography and ecology. *Journal of Biogeography* **26**, 867–878.

Obertegger, U., Thaler, B., Flaim, G. (2010). Rotifer species richness along an altitudinal gradient in the Alps. *Global Ecology and Biogeography* **79**, 895–904.

Oliveira, S.M., ter Steege, H., Cornelissen, J.H.C., Gradstein, S.R. (2009). Niche assembly of epiphytic bryophyte communities in the Guianas: a regional approach. *Journal of Biogeography* **36**, 2076–2084.

O'Malley, M.A . (2007). The nineteenth century roots of 'everything is everywhere'. *Nature Reviews Microbiology* **5**, 647–651.

O'Malley, M.A . (2008). *'Everything is everywhere*: but *the environment selects'*: ubiquitous distribution and ecological determinism in microbial biogeography. *Studies in History and Philosophy of Science C* **39**, 314–325.

Pawlowski, J., Holzman, M. (2008). Diversity and geographic distribution of benthic foraminifera: a molecular perspective. *Biodiversity and Conservation* **17**, 317–328.

Petchey, O.L., Gaston, K.J. (2006). Functional diversity: back to basics and looking forward. *Ecology Letters* **9**, 741–758.

Phillimore, A.B., Orme, C.D.L., Thomas, G.H. et al. (2008). Sympatric speciation in birds is rare: Insights from range data and simulations. *American Naturalist* **171**, 646–657.

Polis, G.A., Anderson, W.B., Holt, R.D. (2003). Toward an integration of landscape and food web ecology: The dynamics of spatially subsidized food webs. *Annual Review of Ecology and Systematics* **28**, 289–316.

Prosser, J.I., Bohannan, B.J.M., Curtis, T.P. et al. (2007). The role of ecological theory in microbial ecology. *Nature Reviews Microbiology* **5**, 384–392.

Qian, H., Ricklefs, R.E. (2008). Global concordance in diversity patterns of vascular plants and terrestrial vertebrates. *Ecology Letters* **11**, 547–553.

Rahbek, C. (1995). The elevational gradient of species richness – a uniform pattern. *Ecography* **18**, 200–205.

Rahbek, C. (2005). The role of spatial scale and the perception of large-scale species richness patterns. *Ecology Letters* **8**, 224–239.

Ricklefs, R.E. (1987). Community diversity: Relative roles of local and regional processes. *Science* **235**, 167–171.

Ricklefs, R.E. (2004). A comprehensive framework for global patterns in biodiversity. *Ecology Letters* **7**, 1–15.

Ricklefs, R.E. (2007). History and diversity: explorations at the intersection of ecology and evolution. *American Naturalist* **170**, S56-S70.

Rodríguez, J., Hortal, J., Nieto, M. (2006). An evaluation of the influence of environment and biogeography on community structure: the case of the Holarctic mammals. *Journal of Biogeography* **33**, 291–303.

Rosenzweig, M.L. (1995). *Species diversity in space and time*. Cambridge University Press, Cambridge.

Sánchez-Moreno, S., Ferris, H., Guil, N. (2008). Role of tardigrades in the suppressive service of a soil food web. *Agriculture, Ecosystems and Environment* **124**, 187–192.

Sanders, N.J., Gotelli, N.J., Wittman, S.E. et al. (2007). Assembly rules of ground-foraging ant assemblages are contingent on disturbance, habitat and spatial scale. *Journal of Biogeography* **34**, 1632–1641.

Santos, A.M.C., Whittaker, R.J., Triantis, K.A. et al. (2010). Are species–area relationships from entire archipelagos congruent with those of their constituent islands? *Global Ecology and Biogeography* **19**, 527–540.

Schemske, D.W., Mittelbach, G.G., Cornell, H.V., Sobel, J.M., Roy, K. (2009). Is there a latitudinal gradient in the importance of biotic interactions? *Annual Review of Ecology Evolution and Systematics* **40**, 245–269.

Schipper, J., Chanson, J.S., Chiozza, F. et al. (2008). The status of the world's land and marine mammals: diversity, threat, and knowledge. *Science* **322**, 225–230.

Segers, H., De Smet, W.H. (2008). Diversity and endemism in Rotifera: a review, and *Keratella* Bory de St Vincent. *Biodiversity and Conservation* **17**, 303–316.

Smith, V.H., Foster, B.L., Grover, J.P. et al. (2005). Phytoplankton species richness scales consistently from laboratory microcosms to the world's oceans. *Proceedings of the National Academy of Sciences USA* **102**, 4393–4396.

Smith, H.G., Wilkinson, D.M. (2007). Not all free-living microorganisms have cosmopolitan distributions – the case of *Nebela (Apodera) vas* Certes (Protozoa: Amoebozoa: Arcellinida). *Journal of Biogeography* **34**, 1822–1831.

Smith, H.G., Bobrov, A., Lara, E. (2008). Diversity and biogeography of testate amoebae. *Biodiversity and Conservation* **17**, 329–343.

Soininen, J., McDonald, R., Hillebrand, H. (2007). The distance decay of similarity in ecological communities. *Ecography* **30**, 3–12.

Spribille, T., Björk, C.R., Exman, S. et al. (2009). Contributions to an epiphytic lichen flora of northwest North America: I. Eight new species from British Columbia inland rain forests. *Bryologist* **112**, 109–137.

Srivastava, D.S., Lawton, J.H. (1998). Why more productive sites have more species: An experimental test of theory using tree-hole communities. *American Naturalist* **152**, 510–529.

Srivastava, D.S., Bell, T. (2009). Reducing horizontal and vertical diversity in a foodweb triggers extinctions and impacts functions. *Ecology Letters* **12**, 1016–1028.

Steinitz, O., Heller, J., Tsoar, A., Rotem, D., Kadmon, R. (2006). Environment, dispersal and patterns of species similarity. *Journal of Biogeography* **33**, 1044–1054.

Storch, D., Davies, R.G., Zajicek, S. et al. (2006). Energy, range dynamics and global species richness patterns: reconciling mid-domain effects and environmental determinants of avian diversity. *Ecology Letters* **9**, 1308–1320.

Taylor, J.W., Turner, E., Townsend, J.P., Dettman, J.R., Jacobson, D. (2006). Eukaryotic microbes, species recognition and the geographic limits of species: examples from the kingdom Fungi. *Philosophical Transactions of the Royal Society B* **361**, 1947–1963.

Telford, R.J., Vandvik, V., Birks, H.J.B. (2006). Dispersal limitations matter for microbial morphospecies. *Science* **312**, 1015.

Triantis, K.A., Mylonas, M., Lika, K., Vardinoyannis, K. (2003). A model for the species–area–habitat relationship. *Journal of Biogeography* **30**, 19–27.

Valdecasas, A.G., Camacho, A.I., Peláez, M.L . (2006). Do small animals have a biogeography? *Experimental and Applied Acarology* **40**, 133–144.

van der Gast, C.J., Lilley, Andrew K., Ager, D., Thompson, I.P. (2005). Island size and bacterial diversity in an archipelago of engineering machines. *Environmental Microbiology* **7**, 1220–1226.

Vanormelingen, P., Verleyen, E., Vyverman, W. (2008). The diversity and distribution of diatoms: from cosmopolitanism to narrow endemism. *Biodiversity and Conservation* **17**, 393–405.

Verleyen, E., Vyverman, W., Sterken, M. et al. (2009). The importance of dispersal related and local factors in shaping the taxonomic structure of diatom metacommunities. *Oikos* **118**, 1239–1249.

Vos, M., Velicer, G.J. (2008). Isolation by distance in the spore-forming soil bacterium *Myxococcus xanthus*. *Current Biology* **18**, 386–391.

Wardle, D.A. (2006). The influence of biotic interactions on soil biodiversity. *Ecology Letters* **9**, 870–886.

Weiher, E., Keddy, P. (1999). *Ecological assembly rules: Perspectives, advances, retreats*. Cambridge University Press, Cambridge.

Weisse, T. (2008). Distribution and diversity of aquatic protists: an evolutionary and ecological perspective. *Biodiversity and Conservation* **17**, 243–259.

Whitaker, R.J., Grogan, D.W., Taylor, J.W. (2003). Geographic barriers isolate endemic populations of hyperthermophilic archaea. *Science* **301**, 976–978.

Whitfield, J. (2005). Biogeography: is everything everywhere? *Science* **310**, 960–961.

Whittaker, R.H. (1972). Evolution and measurement of species diversity. *Taxon* **21**, 213–251.

Whittaker, R.J., Fernández-Palacios, J.M. (2007). *Island biogeography. Ecology, evolution, and conservation. Second edition*. Oxford University Press, Oxford.

Whittaker, R.J., Triantis, K.A., Ladle, R.J. (2008). A general dynamic theory of oceanic island biogeography. *Journal of Biogeography* **35**, 977–994.

Wiens, J.J., Donoghue, M.J. (2004). Historical biogeography, ecology and species richness. *Trends in Ecology and Evolution* **19**, 639–644.

Wilkinson, D.M. (2001). What is the upper size limit for cosmopolitan distribution in free-living microorganisms? *Journal of Biogeography* **28**, 285–291.

Wilkinson, D.M. (2010). Have we underestimated the importance of humans in the biogeography of free-living terrestrial microorganisms? *Journal of Biogeography* **37**, 393–397.

Willig, M.R., Kaufman, D.M., Stevens, R.D. (2003). Latitudinal gradients of biodiversity: Pattern, process, scale, and synthesis. *Annual Review of Ecology Evolution and Systematics* **34**, 273–309.

Wright, D.H. (1983). Species-energy theory — an extension of species-area theory. *Oikos* **41**, 496–506.

Xu, S., Hebert, P.D.N., Kotov, A.A., Cristescu, M.E. (2009). The noncosmopolitanism paradigm of freshwater zooplankton: insights from the global phylogeography of the predatory cladoceran *Polyphemus pediculus* (Linnaeus, **1761**) (Crustacea, Onychopoda). *Molecular Ecology* 18, 5161–5179.

Index

Systematics Association Publications

1. Bibliography of Key Works for the Identification of the British Fauna and Flora, 3rd edition (1967)[†]
   Edited by G.J. Kerrich, R.D. Meikie and N. Tebble
2. Function and Taxonomic Importance (1959)[†]
   Edited by A.J. Cain
3. The Species Concept in Palaeontology (1956)[†]
   Edited by P.C. Sylvester-Bradley
4. Taxonomy and Geography (1962)[†]
   Edited by D. Nichols
5. Speciation in the Sea (1963)[†]
   Edited by J.P. Harding and N. Tebble
6. Phenetic and Phylogenetic Classification (1964)[†]
   Edited by V.H. Heywood and J. McNeill
7. Aspects of Tethyan Biogeography (1967)[†]
   Edited by C.G. Adams and D.V. Ager
8. The Soil Ecosystem (1969)[†]
   Edited by H. Sheals
9. Organisms and Continents through Time (1973)[*]
   Edited by N.F. Hughes
10. Cladistics: A Practical Course in Systematics (1992)[‡]
    P.L. Forey, C.J. Humphries, I.J. Kitching, R.W. Scotland, D.J. Siebert and D.M. Williams
11. Cladistics: The Theory and Practice of Parsimony Analysis (2nd edition) (1998)[‡]
    I.J. Kitching, P.L. Forey, C.J. Humphries and *D.M. Williams*

---

† Published by the Systematics Association (out of print).

* Published by the Palaeontological Association in conjunction with the Systematics Association.

‡ Published by Oxford University Press for the Systematics Association.

Systematics Association Special Volumes

1. The New Systematics (1940)[a]
   Edited by J.S. Huxley (reprinted 1971)
2. Chemotaxonomy and Serotaxonomy (1968)*
   Edited by J.C. Hawkes
3. Data Processing in Biology and Geology (1971)*
   Edited by J.L. Cutbill
4. Scanning Electron Microscopy (1971)*
   Edited by V.H. Heywood
5. Taxonomy and Ecology (1973)*
   Edited by V.H. Heywood
6. The Changing Flora and Fauna of Britain (1974)*
   Edited by D.L. Hawksworth
7. Biological Identification with Computers (1975)*
   Edited by R.J. Pankhurst
8. Lichenology: Progress and Problems (1976)*
   Edited by D.H. Brown, D.L. Hawksworth and R.H. Bailey
9. Key Works to the Fauna and Flora of the British Isles and Northwestern Europe, 4th edition (1978)*
   Edited by G.J. Kerrich, D.L. Hawksworth and R.W. Sims
10. Modern Approaches to the Taxonomy of Red and Brown Algae (1978)*
    Edited by D.E.G. Irvine and J.H. Price
11. Biology and Systematics of Colonial Organisms (1979)*
    Edited by C. Larwood and B.R. Rosen
12. The Origin of Major Invertebrate Groups (1979)*
    Edited by M.R. House
13. Advances in Bryozoology (1979)*
    Edited by G.P. Larwood and M.B. Abbott
14. Bryophyte Systematics (1979)*
    Edited by G.C.S. Clarke and J.G. Duckett
15. The Terrestrial Environment and the Origin of Land Vertebrates (1980)*
    Edited by A.L. Panchen

16 Chemosystematics: Principles and Practice (1980)*
Edited by F.A. Bisby, J.G. Vaughan and C.A. Wright

17. The Shore Environment: Methods and Ecosystems (2 volumes) (1980)*
Edited by J.H. Price, D.E.C. Irvine and W.F. Farnham

18. The Ammonoidea (1981)*
Edited by M.R. House and J.R. Senior

19. Biosystematics of Social Insects (1981)*
Edited by P.E. House and J.-L. Clement

20. Genome Evolution (1982)*
Edited by G.A. Dover and R.B. Flavell

21. Problems of Phylogenetic Reconstruction (1982)*
Edited by K.A. Joysey and A.E. Friday

22. Concepts in Nematode Systematics (1983)*
Edited by A.R. Stone, H.M. Platt and L.F. Khalil

23. Evolution, Time and Space: The Emergence of the Biosphere (1983)*
Edited by R.W. Sims, J.H. Price and P.E.S. Whalley

24. Protein Polymorphism: Adaptive and Taxonomic Significance (1983)*
Edited by G.S. Oxford and D. Rollinson

25. Current Concepts in Plant Taxonomy (1983)*
Edited by V.H. Heywood and D.M. Moore

26. Databases in Systematics (1984)*
Edited by R. Allkin and F.A. Bisby

27. Systematics of the Green Algae (1984)*
Edited by D.E.G. Irvine and D.M. John

28. The Origins and Relationships of Lower Invertebrates (1985)[†]
Edited by S. Conway Morris, J.D. George, R. Gibson and H.M. Platt

29. Infraspecific Classification of Wild and Cultivated Plants (1986)[†]
Edited by B.T. Styles

30. Biomineralization in Lower Plants and Animals (1986)[†]
Edited by B.S.C. Leadbeater and R. Riding

31. Systematic and Taxonomic Approaches in Palaeobotany (1986)[†]
Edited by R.A. Spicer and B.A. Thomas

32. Coevolution and Systematics (1986)[†]
Edited by A.R. Stone and D.L. Hawksworth

33. Key Works to the Fauna and Flora of the British Isles and Northwestern Europe, 5th edition (1988)[†]
Edited by R.W. Sims, P. Freeman and D.L. Hawksworth

34. Extinction and Survival in the Fossil Record (1988)[†]
Edited by G.P. Larwood

35. The Phylogeny and Classification of the Tetrapods (2 volumes) (1988)[†]
Edited by M.J. Benton

36. Prospects in Systematics (1988)[‡]
    Edited by J.L. Hawksworth
37. Biosystematics of Haematophagous Insects (1988)[‡]
    Edited by M.W. Service
38. The Chromophyte Algae: Problems and Perspective (1989)[‡]
    Edited by J.C. Green, B.S.C. Leadbeater and W.L. Diver
39. Electrophoretic Studies on Agricultural Pests (1989)[‡]
    Edited by H.D. Loxdale and J. den Hollander
40. Evolution, Systematics, and Fossil History of the Hamamelidae (2 volumes) (1989)[‡]
    Edited by P.R. Crane and S. Blackmore
41. Scanning Electron Microscopy in Taxonomy and Functional Morphology (1990)[‡]
    Edited by D. Claugher
42. Major Evolutionary Radiations (1990)[‡]
    Edited by P.D. Taylor and G.P. Larwood
43. Tropical Lichens: Their Systematics, Conservation and Ecology (1991)[‡]
    Edited by G.J. Galloway
44. Pollen and Spores: Patterns and Diversifiction (1991)[‡]
    Edited by S. Blackmore and S.H. Barnes
45. The Biology of Free-Living Heterotrophic Flagellates (1991)[‡]
    Edited by D.J. Patterson and J. Larsen
46. Plant–Animal Interactions in the Marine Benthos (1992)[‡]
    Edited by D.M. John, S.J. Hawkins and J.H. Price
47. The Ammonoidea: Environment, Ecology and Evolutionary Change (1993)[‡]
    Edited by M.R. House
48. Designs for a Global Plant Species Information System (1993)[‡]
    Edited by F.A. Bisby, G.F. Russell and R.J. Pankhurst
49. Plant Galls: Organisms, Interactions, Populations (1994)[‡]
    Edited by M.A.J. Williams
50. Systematics and Conservation Evaluation (1994)[‡]
    Edited by P.L. Forey, C.J. Humphries and R.I. Vane-Wright
51. The Haptophyte Algae (1994)[‡]
    Edited by J.C. Green and B.S.C. Leadbeater
52. Models in Phylogeny Reconstruction (1994)[‡]
    Edited by R. Scotland, D.I. Siebert and D.M. Williams
53. The Ecology of Agricultural Pests: Biochemical Approaches (1996)**
    Edited by W.O.C. Symondson and J.E. Liddell
54. Species: the Units of Diversity (1997)**
    Edited by M.F. Claridge, H.A. Dawah and M.R. Wilson
55. Arthropod Relationships (1998)**
    Edited by R.A. Fortey and R.H. Thomas

56. Evolutionary Relationships among Protozoa (1998)**
    Edited by G.H. Coombs, K. Vickerman, M.A. Sleigh and A. Warren
57. Molecular Systematics and Plant Evolution (1999)‡‡
    Edited by P.M. Hollingsworth, R.M. Bateman and R.J. Gornall
58. Homology and Systematics (2000)‡‡
    Edited by R. Scotland and R.T. Pennington
59. The Flagellates: Unity, Diversity and Evolution (2000)‡‡
    Edited by B.S.C. Leadbeater and J.C. Green
60. Interrelationships of the Platyhelminthes (2001)‡‡
    Edited by D.T.J. Littlewood and R.A. Bray
61. Major Events in Early Vertebrate Evolution (2001)‡‡
    Edited by P.E. Ahlberg
62. The Changing Wildlife of Great Britain and Ireland (2001)‡‡
    Edited by D.L. Hawksworth
63. Brachiopods Past and Present (2001)‡‡
    Edited by H. Brunton, L.R.M. Cocks and S.L. Long
64. Morphology, Shape and Phylogeny (2002)‡‡
    Edited by N. MacLeod and P.L. Forey
65. Developmental Genetics and Plant Evolution (2002)‡‡
    Edited by Q.C.B. Cronk, R.M. Bateman and J.A. Hawkins
66. Telling the Evolutionary Time: Molecular Clocks and the Fossil Record (2003)‡‡
    Edited by P.C.J. Donoghue and M.P. Smith
67. Milestones in Systematics (2004)‡‡
    Edited by D.M. Williams and P.L. Forey
68. Organelles, Genomes and Eukaryote Phylogeny (2004)‡‡
    Edited by R.P. Hirt and D.S. Horner
69. Neotropical Savannas and Seasonally Dry Forests: Plant Diversity, Biogeography and Conservation (2006)‡‡
    Edited by R.T. Pennington, G.P. Lewis and J.A. Rattan
70. Biogeography in a Changing World (2006)‡‡
    Edited by M.C. Ebach and R.S. Tangney
71. Pleurocarpous Mosses: Systematics & Evolution (2006)‡‡
    Edited by A.E. Newton and R.S. Tangney
72. Reconstructing the Tree of Life: Taxonomy and Systematics of Species Rich Taxa (2006)‡‡
    Edited by T.R. Hodkinson and J.A.N. Parnell
73. Biodiversity Databases: Techniques, Politics, and Applications (2007)‡‡
    Edited by G.B. Curry and C.J. Humphries
74. Automated Taxon Identification in Systematics: Theory, Approaches and Applications (2007)‡‡
    Edited by N. MacLeod

75. Unravelling the algae: the past, present, and future of algal systematics (2008)[‡‡]
    Edited by J. Brodie and J. Lewis
76. The New Taxonomy (2008)[‡‡]
    Edited by Q.D. Wheeler
77. Palaeogeography and Palaeobiogeography: Biodiversity in Space and Time (in press)[‡‡]
    Edited by P. Upchurch, A. McGowan and C. Slater

---

[a] Published by Clarendon Press for the Systematics Association.
[*] Published by Academic Press for the Systematics Association.
[‡] Published by Oxford University Press for the Systematics Association.
[**] Published by Chapman & Hall for the Systematics Association.
[‡‡] Published by CRC Press for the Systematics Association.